AF361019

Toxins: Mr Hyde or Dr Jekyll?

Toxins: Mr Hyde or Dr Jekyll?

Editors

Daniel Ladant
Gilles Prévost
Michel R. Popoff
Evelyne Benoit

MDPI • Basel • Beijing • Wuhan • Barcelona • Belgrade • Manchester • Tokyo • Cluj • Tianjin

Editors

Daniel Ladant
Unité Biochimie des
Interactions
Macromoléculaires,
Institut Pasteur
Paris, France

Gilles Prévost
Unité UR-7290 Virulence
Bactérienne Précoce ,
Institut de Bactériologie
Strasbourg, France

Michel R. Popoff
Unité Bactéries Anaérobies et
Toxines, Institut Pasteur
Paris, France

Evelyne Benoit
SIMoS, EMR 9004
CNRS/CEA, CEA de Saclay,
Université Paris-Saclay
Gif-sur-Yvette, France

Editorial Office
MDPI
St. Alban-Anlage 66
4052 Basel, Switzerland

This is a reprint of articles from the Special Issue published online in the open access journal *Toxins* (ISSN 2072-6651) (available at: https://www.mdpi.com/journal/toxins/special_issues/toxins_Hyde).

For citation purposes, cite each article independently as indicated on the article page online and as indicated below:

LastName, A.A.; LastName, B.B.; LastName, C.C. Article Title. *Journal Name* **Year**, *Volume Number*, Page Range.

ISBN 978-3-0365-6904-8 (Hbk)
ISBN 978-3-0365-6905-5 (PDF)

Contents

About the Editors

Daniel Ladant

Dr. Daniel Ladant is Director of Research at French CNRS (National Center for Scientific Research) and the head of the "Biochemistry of Macromolecular Interactions" unit at the Institut Pasteur, Paris, France. He obtained a Ph.D. in Microbiology in 1989 and a Habilitation à Diriger des Recherches (HDR) in 1999 from the Université Paris Diderot, Paris, France. His research focuses on the study of the molecular mechanisms that underlie protein–protein and protein–membrane interactions, using as a model system of a bacterial toxin, the adenylate cyclase, CyaA, produced by *Bordetella pertussis*, the causative agent of whooping cough. He has characterized the structure, function, biogenesis, and mechanisms of action of CyaA. He has also exploited this basic knowledge to develop various applications in vaccinology (engineering CyaA as an antigen-delivery vehicle) and biotechnology (designing a CyaA-based two-hybrid technique, BACTH). He has published more than 100 articles in peer-reviewed journals, written over 40 review articles or book chapters, and co-authored 14 patents.

Gilles Prévost

Dr. Gilles Prévost is an assistant professor–hospital practitioner (C.ex), head of the Unit Research 7290, and responsible for the master's programme "Biologie–Santé" at the University of Strasbourg, France. He obtained a Ph.D. in Cellular and Molecular Biology in 1988 and a Habilitation à Diriger des Recherches in 1996. His main research work deals with the molecular epidemiology of bacteria and bacterial simplification, and he currently focuses on the development of antibiotics with new structural and beneficial aspects of bacterial toxins. Of note, he has participated in work to identify and characterize the large family of staphylococcal bicomponent leucotoxins. He has primarily concentrated on the molecular and cellular biology of Panton–Valentine leucocidin, a virulence factor with potentially severe consequences. He has published more than 100 PubMed-referenced articles, has an *h-index* 46, has written more than 35 reviews and book chapters, and has co-authored 3 patents.

Michel R. Popoff

Dr. Michel R. Popoff, DVM 1973, Ph.D. 1985, h-index 63. Assistant Professor at the Institut Pasteur, Paris, France. Head of the research unit "Anaerobic bacteria and toxins" from 2002 to 2019 and head of the French National Reference Center "Anaerobic bacteria and botulism" from 1998 to 2017. His laboratory is focused on *Clostridium* toxins through genetic and biological activity analysis and has investigated the regulation of toxin synthesis in *Clostridium botulinum* and *Clostridium tetani*. In recent years, he has analyzed the molecular mechanism of the actin depolymerizing *C. sordellii* lethal toxin and clostridial binary toxins, the pore-forming *C. perfringens* epsilon toxin, and the passage of botulinum neurotoxins through the intestinal barrier. A member of the French Veterinary Academia and co-editor of the 3rd and 4th editions of the *Sourcebook of Bacterial Protein Toxins*, Academic Press, (2006 and 2015) and several books. Editor-in-chief of the section Bacterial Toxins of the journal *Toxins*.

Evelyne Benoit

Dr. Evelyne Benoit completed her Ph.D. in Neuroscience at the age of 26 and her habilitation to lead research in 2010 at Paris-XI-Orsay University (France). She is a senior researcher at the CNRS and is responsible for "Ion transfers and pathophysiology of sensory and neuromuscular conduction" studies. To gain a better understanding of these processes, she studies the mechanism of action of emerging neurotoxins and uses some of these bioactive agents because of their specificity of action. She has published 117 papers in reputed journals, has written 41 book chapters, and has co-authored 2 patents. From 2016 to 2021, she was the President of the French Society of Toxinology (http://sfet.asso.fr/international/).

Preface to "Toxins: Mr Hyde or Dr Jekyll?"

Toxins are biologically active substances produced by most kinds of living organisms, bacteria, fungi, plants, and animals. They present a vast diversity of molecular structures and target a wide variety of receptors involved in a range of physiological processes. As toxins are selected during evolution to acquire/improve their disabling/lethal effects, they display finely tuned functional properties often associated with high affinities and selectivity. Moreover, toxins are valuable tools to unravel cellular processes due to their extreme specificity for cell surface and/or intracellular targets. Therefore, toxins are very attractive compounds because of their Janus-like character; while they mostly act as deadly poisons like monstrous Mr. Hyde, they can also be tamed into good remedies like admirable Dr. Jekyll. As such, they have been primarily investigated not only for the light they can throw on fundamental physiological processes but also for their potential therapeutic applications. This book, emerging from the 27th Annual Meeting of the French Society of Toxinology (SFET, http://sfet.asso.fr/international), will be of great interest for those in the scientific community who want to know more about the fascinating world of toxins. We would like to thank all the individuals who contributed to the meeting and to this book.

Daniel Ladant, Gilles Prévost, Michel R. Popoff, and Evelyne Benoit
Editors

Editorial

Editorial of the Special Issue "Toxins: Mr Hyde or Dr Jekyll?"

Daniel Ladant [1], Gilles Prévost [2], Michel R. Popoff [3] and Evelyne Benoit [4,*]

[1] Institut Pasteur, Unité Biochimie des Interactions Macromoléculaires, 25-28 Rue du Docteur Roux, F-75015 Paris, France
[2] Institut de Bactériologie, Unité UR-7290 Virulence Bactérienne Précoce, ITI InnoVec, 3 Rue Koeberlé, F-67000 Strasbourg, France
[3] Institut Pasteur, Unité Toxines Bactériennes, 25-28 Rue du Docteur Roux, F-75015 Paris, France
[4] CEA, Institut des Sciences du Vivant Frédéric Joliot, Département Médicaments et Technologies pour la Santé (DMTS), Service d'Ingénierie Moléculaire pour la Santé (SIMoS), Université Paris-Saclay, EMR 9004 CNRS/CEA, F-91191 Gif-sur-Yvette, France
* Correspondence: evelyne.benoit@cea.fr; Tel.: +33-1-6908-5685

Citation: Ladant, D.; Prévost, G.; Popoff, M.R.; Benoit, E. Editorial of the Special Issue "Toxins: Mr Hyde or Dr Jekyll". *Toxins* **2023**, *15*, 142. https://doi.org/10.3390/toxins15020142

Received: 20 January 2023
Accepted: 7 February 2023
Published: 10 February 2023

The 27th Annual Meeting of the French Society of Toxinology (SFET, http://sfet.asso.fr/international (accessed since 1 September 2022)) was held on 9–10 December 2021 as a virtual meeting (e-RT27). The central theme selected for this meeting, "Toxins: Mr Hyde or Dr Jekyll?", gave rise to three thematic sessions: the first on plant toxins, algal toxins and mycotoxins; the second on animal toxins; and the third on bacterial toxins. All sessions were aimed at emphasizing the latest findings on this topic. Apart from this central theme, a "miscellaneous" session was dedicated to recent results obtained in toxinology. Ten speakers from seven countries (Australia, Brazil, Burkina Faso, France, Germany, The Netherlands and the United States of America) were invited as international experts to present their work, and other researchers and students presented theirs through 15 shorter lectures and 20 posters. Of the ca. 80 participants who registered, 38% were foreigners, a value highlighting the international attractiveness of the SFET meeting.

The Special Issue "Toxins: Mr Hyde or Dr Jekyll?" includes ten articles.

All the abstracts of the ten invited speaker's lectures, as well as those of the 12 shorter lectures and 20 posters, are available in the report from the 27th Meeting of Toxinology [1]. Owing to a donation from MDPI's *Toxins*, two prizes of 300EUR each were awarded to the best oral communication and best poster, both selected from a vote by all the invited speakers to the meeting and the SFET Board of Directors. The two winners of the best oral communication award (Barbara Ribeiro: Functional impact of BeKm-1, a high-affinity hERG blocker, on cardiomyocytes derived from human-induced pluripotent stem cells) and from the best poster award (Anne-Cécile Van Baelen: Characterization of immune animal toxins for the functional study of angiotensin receptors) are mentioned in the meeting report [1].

Clostridial neurotoxins, botulinum neurotoxins (BoNTs) and tetanus neurotoxins (TeNTs) are produced by sporulating and anaerobic bacteria from the environment. The synthesis of clostridial neurotoxins is a highly regulated process by complex regulatory networks including alternative sigma factors, two-component systems, non-coding RNA, quorum-sensing system, and the interplay with general metabolism. Environmental and nutritional factors control the regulation of clostridial neurotoxin synthesis but are still poorly understood. Notably, amino acid-to-peptide metabolism transition seems to be an important regulatory factor. Comparative regulation of BoNT and TeNT synthesis is reviewed in the article of Popoff and Brüggemann [2].

Various bacterial pathogens produce toxins that target cyclic nucleotide monophosphate (CNMPs). Certain toxins exhibit nucleotidyl cyclase activities that are activated by eukaryotic factors (adenylate cyclase (AC) of *Bordetella pertussis* CyaA, or *Bacillus anthracis* edema factor EF). Davi et al. propose a robust and efficient in vitro assay for the detection of AC activity based on the spectrometry detection of cyclic AMP after chromatographic separation on aluminum oxide. This assay is very sensitive (fmol level) and can be used in complex media [3].

Bacterial toxins and virulence factors are involved in potent pathogenicity mechanisms. One of the host targets is the immune system. However, hosts have generated sensors to detect the pathogenic potential of microbes. Indeed, the innate immunity senses microbial-associated molecular patterns (MAMPs), notably virulence factors, to distinguish between pathogens and commensals. The innate immunity can persist in host and it is called the innate immune memory of trained immunity. The toxins and virulence factors that target Rho-GTPases activate the NLRP3 inflammasome. The release of interleukins, such as IL-1β and IL-18, are associated with the induction of the adaptive immune memory. The innate immunity and trained immunity triggered by virulence factors that modify Rho-GTPases are discussed in the article by Torre and Boyer [4].

A particular group of bacterial toxins concern the toxin superantigens. These toxins recognize specific receptors on both antigen-presenting cells and T lymphocytes leading to the activation of a large proportion of T lymphocytes (up to 20%). Toxin superantigens are involved in several mild-to-severe diseases in humans and animals. Toxin superantigens are mainly produced by *Staphylococcus aureus*. They constitute a very heterogeneous family ranging from 20 to 89% in similarity. The review by Truant et al. concerns the repertoire of *Staphylococcus* superantigens including their functional and pathological properties [5].

Clostridium botulinum C and D produce a toxin or enzyme, called C3, that specifically inactivates the eukaryotic protein Rho by ADP-ribosylation. C3 enters only certain cell types such as monocyte-derived cells including macrophages, osteoclasts, and dendritic cells. Thus, inactivated C3 at the enzymatic site can be used as a transporter for the specific delivery of cargo into macrophages and dendritic cells. Fellermann et al. investigated the selectivity of C3 delivery by using eGFP coupled to non-toxic C3. Thereby, C3 is a promising molecule to specifically deliver small compounds into macrophages and dendritic cells [6].

The killer strains of *Debaryomyces hansenii* and *Wickerhamomyces anomalus* species secrete antimicrobial proteins called killer toxins active against selected fungal phytopathogens. Czarnecka et al. investigated the role of plasma membrane pleiotropic drug-resistant (PDR) transporters (Pdr5p and Snq2p) in the mechanism of defense against killer toxins. They showed that the Snq2p efflux pump influences the immunity to the *W. anomalus* BS91 killer toxin and that the activity of the killer toxins of *D. hansenii* AII4b, KI2a, MI1a and CBS767 strains is regulated by other transporters than those influencing the *W. anomalus* killer toxin activity. This is the first study that reports the involvement of PDR transporters in the cell membrane of susceptible microorganisms in resistance to killer yeasts' toxins [7].

Gambierol is a polycyclic ether toxin produced by the dinoflagellate *Gambierdiscus toxicus* that is a potent neurotoxin by blocking voltage-gated potassium channels. Benoit et al. have analyzed the effect of gambierol on single rat fetal (F19–F20) adrenomedullary-cultured chromaffin cells [8]. Using whole-cell voltage-clamp recording, gambierol was shown to block only a fraction of the total outward K^+ current by slowing down their activation kinetics. It was found that gambierol did not affect calcium-activated K^+ (K_{Ca}) or ATP-sensitive K^+ (K_{ATP}) channels. These results highlight the important modulatory role played by K_{Ca} channels in the control of exocytosis from fetal (F19–F20) adrenomedullary chromaffin cells.

Acid-sensing ion channels (ASICs) are voltage-independent H^+-gated cation channels largely expressed in the nervous system of rodents and humans. At least six isoforms (ASIC1a, 1b, 2a, 2b, 3 and 4) associate into homo- or heterotrimers to form functional channels with highly pH-dependent gating properties. Verkets et al. reviewed the pharmacological profiles of animal peptide toxins targeting ASICs, including PcTx1 from tarantula and related spider toxins, APETx2 and APETx-like peptides from sea anemone, and mambalgin from snake, as well as the dimeric protein snake toxin MitTx that have all been instrumental to understanding the structure and the pH-dependent gating of rodent- and human-cloned ASICs and to study the physiological and pathological roles of native

ASICs in vitro and in vivo [9]. ASICs are expressed all along the pain pathways and the pharmacological data clearly support a role for these channels in pain.

α-Bungarotoxin is a large, 74 amino acid toxin containing five disulfide bridges, initially identified in the venom of *Bungarus multicinctus* snake. Similar to most large toxins, the chemical synthesis of α-bungarotoxin is challenging, explaining why all previous reports use purified or recombinant α-bungarotoxin. However, only chemical synthesis allows easy insertion of non-natural amino acids or new chemical functionalities. Herein, Brun et al. described a procedure for the chemical synthesis of a fluorescent-tagged α-bungarotoxin [10]. An azide-modified Cy5 fluorophore was coupled to the toxin. Using automated patch-clamp recordings, the fluorescent synthetic α-bungarotoxin blocked acetylcholine-mediated currents in response to muscle nicotinic receptor activation in TE671 cells. Thus, synthetic fluorescent-tagged α-bungarotoxin retains excellent properties for binding onto muscle nicotinic receptors.

Author Contributions: D.L., G.P., M.R.P. and E.B. equally contributed to the conceptualization, the writing—original draft preparation, and the writing—review and editing. All authors have read and agreed to the published version of the manuscript.

Conflicts of Interest: The authors declare no conflict of interest.

References

1.	Ladant, D.; Marchot, P.; Diochot, S.; Prévost, G.; Popoff, M.R.; Benoit, E. Report from the 27th (Virtual) Meeting on Toxinology, "Toxins: Mr Hyde or Dr Jekyll?", Organized by the French Society of Toxinology, 9–10 December 2021. *Toxins* **2022**, *14*, 110. [CrossRef] [PubMed]
2.	Popoff, M.R.; Brüggemann, H. Regulatory Networks Controlling Neurotoxin Synthesis in *Clostridium botulinum* and *Clostridium tetani*. *Toxins* **2022**, *14*, 364. [CrossRef] [PubMed]
3.	Davi, M.; Sadi, M.; Pitard, I.; Chenal, A.; Ladant, D. A Robust and Sensitive Spectrophotometric Assay for the Enzymatic Activity of Bacterial Adenylate Cyclase Toxins. *Toxins* **2022**, *14*, 691. [CrossRef] [PubMed]
4.	Torre, C.; Boyer, L. Effector-Triggered Trained Immunity: An Innate Immune Memory to Microbial Virulence Factors? *Toxins* **2022**, *14*, 798. [CrossRef] [PubMed]
5.	Noli Truant, S.; Redolfi, D.M.; Sarratea, M.B.; Malchiodi, E.L.; Fernández, M.M. Superantigens, a Paradox of the Immune Response. *Toxins* **2022**, *14*, 800.
6.	Fellermann, M.; Stemmer, M.; Noschka, R.; Wondany, F.; Fischer, S.; Michaelis, J.; Stenger, S.; Barth, H. *Clostridium botulinum* C3 Toxin for Selective Delivery of Cargo into Dendritic Cells and Macrophages. *Toxins* **2022**, *14*, 711. [CrossRef] [PubMed]
7.	Czarnecka, M.; Połomska, X.; Restuccia, C.; Żarowska, B. The Role of Plasma Membrane Pleiotropic Drug Resistance Transporters in the Killer Activity of *Debaryomyces hansenii* and *Wickerhamomyces anomalus* Toxins. *Toxins* **2022**, *14*, 180. [CrossRef]
8.	Benoit, E.; Schlumberger, S.; Molgó, J.; Sasaki, M.; Fuwa, H.; Bournaud, R. Gambierol Blocks a K(+) Current Fraction without Affecting Catecholamine Release in Rat Fetal Adrenomedullary Cultured Chromaffin Cells. *Toxins* **2022**, *14*, 254. [CrossRef] [PubMed]
9.	Verkest, C.; Salinas, M.; Diochot, S.; Deval, E.; Lingueglia, E.; Baron, A. Mechanisms of Action of the Peptide Toxins Targeting Human and Rodent Acid-Sensing Ion Channels and Relevance to Their In Vivo Analgesic Effects. *Toxins* **2022**, *14*, 709. [CrossRef]
10.	Brun, O.; Zoukimian, C.; Oliveira-Mendes, B.; Montnach, J.; Lauzier, B.; Ronjat, M.; Béroud, R.; Lesage, F.; Boturyn, D.; De Waard, M. Chemical Synthesis of a Functional Fluorescent-Tagged α-Bungarotoxin. *Toxins* **2022**, *14*, 79. [CrossRef] [PubMed]

Conference Report

Report from the 27th (Virtual) Meeting on Toxinology, "Toxins: Mr Hyde or Dr Jekyll?", Organized by the French Society of Toxinology, 9–10 December 2021

Daniel Ladant [1], Pascale Marchot [2], Sylvie Diochot [3], Gilles Prévost [4], Michel R. Popoff [5] and Evelyne Benoit [6,*]

[1] Institut Pasteur, Unité Biochimie des Interactions Macromoléculaires, 25-28 rue du Docteur Roux, 75015 Paris, France; daniel.ladant@pasteur.fr

[2] Laboratoire Architecture et Fonction des Macromolécules Biologiques, CNRS/Aix-Marseille Université, Faculté des Sciences—Campus Luminy, 13288 Marseille, France; pascale.marchot@univ-amu.fr

[3] Institut de Pharmacologie Moléculaire et Cellulaire, Université Côte d'Azur, CNRS, Sophia Antipolis, 06560 Valbonne, France; diochot@ipmc.cnrs.fr

[4] Institut de Bactériologie, Unité UR-7290 Virulence Bactérienne Précoce, ITI InnoVec, 3 rue Koeberlé, 67000 Strasbourg, France; prevost@unistra.fr

[5] Institut Pasteur, Unité Toxines Bactériennes, 25-28 rue du Docteur Roux, 75015 Paris, France; popoff2m@gmail.com

[6] CEA, Institut des Sciences du Vivant Frédéric Joliot, Département Médicaments et Technologies pour la Santé (DMTS), Service d'Ingénierie Moléculaire pour la Santé (SIMoS), Université Paris-Saclay, EMR 9004 CNRS/CEA, 91191 Gif-sur-Yvette, France

* Correspondence: evelyne.benoit@cea.fr; Tel.: +33-1-6908-5685

Abstract: The French Society of Toxinology (SFET) organized its 27th annual meeting on 9–10 December 2021 as a virtual meeting (e-RT27). The central theme of this meeting was "Toxins: Mr Hyde or Dr Jekyll?", emphasizing the latest findings on plant, fungal, algal, animal and bacterial toxins during 10 lectures, 15 oral communications (shorter lectures) and 20 posters shared by ca. 80 participants. The abstracts of lectures and posters, as well as the winners of the best oral communication and poster awards, are presented in this report.

Keywords: animal toxin; bacterial toxin; marine toxin; medical application; plant toxin; toxin function/activity; toxin receptor/target; toxin structure

Citation: Ladant, D.; Marchot, P.; Diochot, S.; Prévost, G.; Popoff, M.R.; Benoit, E. Report from the 27th (Virtual) Meeting on Toxinology, "Toxins: Mr Hyde or Dr Jekyll?", Organized by the French Society of Toxinology, 9–10 December 2021. *Toxins* **2022**, *14*, 110. https://doi.org/10.3390/toxins14020110

Received: 11 January 2022
Accepted: 25 January 2022
Published: 1 February 2022

Publisher's Note: MDPI stays neutral with regard to jurisdictional claims in published maps and institutional affiliations.

1. Acknowledgments

We warmly acknowledge the contribution of all those people who work daily at ensuring the national and international shinning of the French Society of Toxinology (SFET) and those who made the 27th Meeting on Toxinology a success. We also address special thanks to our sponsors who, this year again, supported our meeting (Figure 1).

Thanks for their financial support :

Thanks for its moral support :

Figure 1. Sponsor logos.

2. Preface

Toxins are biologically active substances produced by most kinds of living organisms, bacteria, fungi, plants and animals. They present a vast diversity of molecular structures and target a wide variety of receptors involved in a range of physiological processes. As toxins were selected during evolution to acquire/improve their disabling/lethal effect, they display finely tuned functional properties often associated with high affinities and selectivity. Moreover, toxins are valuable tools to unravel cellular processes due to their extreme specificity for cell surface and/or intracellular targets. Therefore, toxins are very attractive compounds because of their Janus-like character, being both poisons and remedies, and as such, they have been primarily investigated not only for the light they can throw on fundamental physiological processes but also for their potential therapeutic applications.

This 27th Meeting on Toxinology of the SFET was held on 9–10 December 2021 as a virtual meeting (e-RT27). The central theme selected for this meeting, "Toxins: Mr Hyde or Dr Jekyll?", gave rise to three thematic sessions: the first one on plant toxins, algal toxins and mycotoxins; the second one on animal toxins; and the third one on bacterial toxins, all sessions were aimed at emphasizing the latest findings on this topic. Apart from this central theme, a "miscellaneous" session was dedicated to recent results obtained in Toxinology. Ten speakers from seven countries (Australia, Brazil, Burkina Faso, France, Germany, the Netherlands and the United States of America) were invited as international experts to present their work, and other researchers and students presented theirs through 15 shorter lectures and 20 posters. Of the ca. 80 participants who registered, 38% were foreigners, a value highlighting the international attractiveness of the SFET meetings.

Owing to a donation from MDPI-Toxins, two prizes of EUR 300 each were awarded to the best oral communication and the best poster (Figure 2), both selected from a vote by all the participants to the meeting.

Figure 2. Winners of the "Best Oral Communication" and "Best Poster" Awards at the 27th Meeting of the French Society of Toxinology (SFET).

Last but not least, we warmly thank the Editors of MDPI—Toxins for supporting the publication of a Special Issue also entitled "Toxins: Mr Hyde or Dr Jekyll?", and gathering this meeting report, along with peer-reviewed original articles and reviews. We believe that this Special Issue will be attractive to all, including those of our colleagues who could not attend the e-RT27 meeting, and that it will represent a comprehensive source of information for researchers and students in Toxinology.

3. Scientific and Organizing Committees (SFET Board of Directors)

Julien Barbier, CEA de Saclay, Gif-sur-Yvette, France
Evelyne Benoit, CEA de Saclay, Gif-sur-Yvette, France
Alexandre Chenal, Institut Pasteur, Paris, France
Michel De Waard, L'institut du thorax, Nantes, France
Sylvie Diochot, Institut de Pharmacologie Moléculaire et Cellulaire, Valbonne, France
Sébastien Dutertre, Institut des Biomolécules Max Mousseron, Montpellier, France
Daniel Ladant, Institut Pasteur, Paris, France
Christian Legros, Université d'Angers, Angers, France
Pascale Marchot, CNRS/Aix-Marseille Université, Marseille, France
Gilles Prévost, Institut de Bactériologie, Université de Strasbourg, France
Michel R. Popoff, Institut Pasteur, Paris, France
Loïc Quinton, Université de Liège, Liège, Belgium

4. Invited Lectures (When More Than One Author, the Underlined Name Is That of the Presenter)

4.1. The Plant with the Scorpion Sting: Novel Pain-Causing Toxins from the Australian Giant Stinging Tree

Irina Vetter [1,4,*], Edward K. Gilding [1], Sina Jami [1], Jennifer R. Deuis [1], Mathilde R. Israel [1,2], Peta J. Harvey [1], Aaron G. Poth [1], Fabian B.H. Rehm [1], Jennifer L. Stow [1], Samuel D. Robinson [1], Kuok Yap [1], Darren L. Brown [1], Brett R. Hamilton [3], David Andersson [2], David J. Craik [1], Thomas Durek [1]

[1] Institute for Molecular Bioscience, The University of Queensland, Brisbane, QLD 4072, Australia.
[2] Wolfson Centre for Age-Related Diseases, Institute of Psychiatry, Psychology & Neuroscience, King's College London, London, SE5 8AF, United Kingdom.
[3] Centre for Advanced Imaging, The University of Queensland, Brisbane, QLD 4072, Australia.
[4] School of Pharmacy, The University of Queensland, Brisbane, QLD 4102, Australia.
* Correspondence: i.vetter@uq.edu.au

Abstract: Stinging Trees from Australasia produce remarkably persistent and painful stings upon contact of their stiff epidermal hairs, called trichomes, with mammalian skin. *Dendrocnide*-induced acute pain typically lasts for several hours, and intermittent painful flares can persist for days and weeks. Our recent work shows that the venoms of Australian *Dendrocnide* species contain heretofore-unknown pain-inducing peptides that potently activate sensory neurons and delay the inactivation of voltage-gated sodium channels. These neurotoxins localize specifically to the stinging hairs and are mini-proteins of 4 kDa whose 3D structure is stabilized in an inhibitory cystine knot motif, a characteristic shared with neurotoxins found in spider and cone snail venoms. Our results provide an intriguing example of inter-kingdom convergent evolution of animal and plant venoms with shared modes of delivery, molecular structure and pharmacology.

Keywords: pain; peptide toxin; sodium channel; stinging nettle

4.2. Evidences That Pinnatoxin-G Crosses Intestinal, Hemato-Encephalic and Maternofetal Barriers to Reach Central and Peripheral Nicotinic Acetylcholine Receptors

Denis Servent [1,*], **Carole Malgorn** [1], **Sophie Gil** [2], **Christelle Simasotchi** [2], **Anne-Sophie Hérard** [3], **Thierry Delzescaux** [3], **Robert Thai** [1], **Peggy Barbe** [1], **Mathilde Keck** [1], **Fabrice Beau** [1], **Armen Zakarian** [4], **Vincent Dive** [1], **Jordi Molgó** [1]

[1] Université Paris-Saclay, CEA, Département Médicaments et Technologies pour la Santé (DMTS), Service d'Ingénierie Moléculaire pour la Santé (SIMoS), ERL CNRS 9004, F-91191 Gif-sur-Yvette, France.
[2] Université de Paris, UMR S1139, Faculté de Pharmacie de Paris, France.
[3] Université Paris-Saclay, UMR 9199, CNRS, CEA, MIRCen, Fontenay-aux-Roses, France.
[4] University of California Santa Barbara, Department of Chemistry and Biochemistry, California 93106, USA.
* Correspondence: denis.servent@cea.fr

Abstract: The warming of ocean temperatures and the increase in nutrients has driven an intensification of harmful algal bloom events. During these active dinoflagellate proliferations, phycotoxins may accumulate in shellfish tissues and can be transferred into fish, marine mammals and ultimately to humans. Cyclic imines, produced by various species of marine dinoflagellates, constitute a widely distributed group of phycotoxins, including pinnatoxins. Pinnatoxin-G (PnTx-G) produced by the cosmopolitan dinoflagellate *Vulcanodinium rugosum* is considered the precursor of other pinnatoxins, and it is a fast-acting toxin in the mouse bioassay. PnTx-G is regularly detected in European coastal environments and in contaminated shellfish samples and may represent a human health risk. In this work, exploiting the ability of PnTx-G to be chemically synthesized and radiolabeled, we studied in vivo the toxicokinetics of [^{3}H]-PnTx-G and its capacity to interact with neuronal and muscle-type nicotinic acetylcholine receptors (nAChRs). The biodistribution of [^{3}H]-PnTx-G, using high-resolution digital radio-imaging after oral or intravenous administration to rats, revealed the presence of the radiolabeled toxin in various peripheral organs, as well as in specific regions of the central nervous system, highlighting its property to cross both the intestinal and blood–brain barriers. In addition, we demonstrate that PnTx-G crosses the materno–fetal barrier, first in rats by detecting the radiolabeled toxin in embryos after its injection to pregnant rats and secondly in humans, using a perfused ex vivo cotyledon model and mass spectrometry analysis. Furthermore, to assess the nAChR subtypes labeled by [^{3}H]-PnTx-G, we performed competition experiments on brain and embryos sections in the presence of selective nAChR antagonists, revealing the major role of muscle-type and α7 subtype in the peripheral and central labeling, respectively. In conclusion, this work shows that PnTx-G efficiently crosses the intestinal, blood–brain and placental barriers to interact with central and peripheral nAChRs, supporting its in vivo effects. The mechanism used by pinnatoxins to cross these physiological barriers and the possible involvement of a receptor-mediated process are still under investigation.

Keywords: biodistribution; nicotinic acetylcholine receptor; pinnatoxin

4.3. A Blue Mountains Funnel-Web Spider Toxin Expressed under Control of a Hemolymph-Specific Promoter Increases Fungal Lethality against Insecticide-Resistant Malaria-Vector Mosquitoes

Étienne Bilgo [1,*], **Brian Lovett** [2], **Raymond St. Leger** [2], **Abdoulaye Diabate** [1]

[1] Institut de Recherche en Sciences de la Santé/Centre Muraz, Bobo-Dioulasso, Burkina Faso.
[2] Department of Entomology, University of Maryland, College Park, MD 20742, USA.
* Correspondence: bilgo02@yahoo.fr

Abstract: The continued success of malaria control efforts requires the development, study and implementation of new technologies that circumvent insecticide resistance. We previously demonstrated that fungal pathogens could provide an effective delivery system for mosquitocidal or malariacidal biomolecules. In this study, we firstly compared genes from arthropod predators encoding insect-specific sodium, potassium and calcium channel blockers for their ability to improve the efficacy of Metarhizium against wild-caught, insecticide-resistant anophelines. Toxins expressed under the control of a hemolymph-specific promoter increased fungal lethality to mosquitoes at spore dosages as low as one conidium per mosquito. One of the most potent, the EPA-approved Hybrid (Ca^{2+}/K^+ channel blocker), was studied for pre-lethal effects. These included reduced blood-feeding behavior, with almost 100% of insects infected with ca. six spores unable to transmit malaria within five days post-infection, surpassing the World Health Organization threshold for successful vector control agents. Furthermore, recombinant strains co-expressing Hybrid toxin and AaIT (Na^+ channel blocker) produced synergistic effects, requiring 45% fewer spores to kill half of the mosquitoes in five days as single toxin strains. Secondly, through a semifield trial in a MosquitoSphere (a contained, near-natural environment) in Soumousso, a region of Burkina Faso where malaria is endemic, we confirmed the proof of our concept in the field. The expression of Hybrid toxin increased fungal lethality and the likelihood that insecticide-resistant mosquitoes would be eliminated from this site. In summary, our results identified a repertoire of toxins with different modes of action that improve the utility of entomopathogens as a technology that is compatible with existing insecticide-based control methods.

Keywords: biotechnology; malaria; mosquito; mycology; toxin

4.4. Bee Venom and Its Component Apamin Have Anti-Parkinsonian Properties in Animal Models of Parkinson's Disease

Marianne Amalric [1,*], **Nathalie Turle-Lorenzo** [1], **Christophe Melon** [2], **Nicolas Maurice** [2]

[1] Laboratoire de Neurosciences Cognitives (LNC) UMR 7291, CNRS, Aix-Marseille Univ., Marseille, France.
[2] Institut de Biologie du Développement de Marseille (IBDM) UMR 7288, NeuroMarseille, Aix-Marseille Univ., CNRS, Marseille, France.
* Correspondence: marianne.amalric@univ-amu.fr

Abstract: Parkinson's disease (PD) is an age-related neurological disorder that affects more than 6.3 million people worldwide. PD is characterized by progressive loss of dopaminergic (DA) neurons in the substantia nigra (SN), which leads to a wide range of debilitating motor, cognitive and neuropsychiatric symptoms. L-DOPA therapy is effective on motor symptoms and motor fluctuations but leads over the years to dyskinesia. Calcium-activated potassium channels (SK channels) have recently emerged as alternative therapeutic targets because they regulate the neuronal firing of midbrain dopamine neurons. SK channels (in particular SK_2/SK_3 subtypes) are highly expressed in the basal ganglia (BG). These brain structures are dysregulated by DA neuronal degeneration. Potassium channels are privileged targets because they regulate neuronal excitability in order to maintain balance in the input-output circuits of the BG. Our study aimed to assess the behavioral effects of apamin, a selective SK_2/SK_3 channel blocker peptide extracted from bee venom, and compare them with those produced by bee venom, in pharmacological and 6-hydroxydopamine (6-OHDA) lesional models of PD in rats. Both bee venom and apamin reversed haloperidol-induced catalepsy, a model of akinetic symptoms, while a co-treatment with the SK opener CYPPA prevented this antiakinetic effect. After extensive unilateral 6-OHDA nigrostriatal lesions, acute and subchronic administration of bee venom and apamin reduced forelimb asymmetry in the cylinder test and apomorphine-induced rotations revealing an antiparkinsonian action on motor symptoms. In another rat model of partial bilateral DA nigrostriatal lesions, apamin also reduced deficits revealed on anxiety, social interaction and visuospatial memory. The neural substrates of these effects were investigated by in vivo electrophysiological recordings of neuronal activity of the BG output structure substantia nigra pars reticulata (SNr). Bee venom restored the balance between the inhibitory and excitatory influence exerted by the trans-striatal direct and indirect pathways that were disrupted by the pharmacological blockade of DA receptors. These results suggest that bee venom and apamin restore the functional properties of the basal ganglia circuitry in PD conditions and emphasize the crucial role of potassium channels (Ca^2-dependent) in these anti-parkinsonian effects. Supported by CNRS, AMU, Fondation de France, Association France Parkinson.

Keywords: bee venom; dopamine; Parkinson's disease

4.5. Slithering Stem Cells—Understanding Snake Venom Production Using Organoids

Jens Puschhof [1,2,*], Yorick Post [1], Joep Beumer [1], Harald M. Kerkkamp [3], Julien Slagboom [4], Buys De Barbanson [1,2], Nienke R. Wevers [5], Xandor Spijkers [5,6], Thomas Olivier [5], Taline D. Kazandijan [7], Stuart Ainsworth [7], Carmen Lopez Iglesias [8], Willine van de Wetering [1,8], Maria C. Heinz [2,9], Ravian L. Van Ineveld [2,10], Regina G.D.M. Van Kleef [11], Harry Begthel [1], Jeroen Korving [1], Yotam E. Bar-Ephraim [1,2], Walter Getreuer [12], Anne C. Rios [2,10], Remco H.S. Westerink [11], Hugo J.G. Snippert [2,9], Alexander Van Oudenaarden [1,2], Peter J. Peters [8], Freek J. Vonk [3], Jeroen Kool [4], Michael K. Richardson [3], Nicholas R. Casewell [7], Hans Clevers [1,2,10]

[1] Hubrecht Institute, Royal Netherlands Academy of Arts and Sciences (KNAW) and UMC Utrecht, 3584 CT Utrecht, The Netherlands.
[2] Oncode Institute, Hubrecht Institute, 3584 CT Utrecht, The Netherlands.
[3] Institute of Biology Leiden, Department of Animal Science and Health, 2333 BE Leiden, The Netherlands.
[4] Division of BioAnalytical Chemistry, Department of Chemistry and Pharmaceutical Sciences, Vrije Universiteit Amsterdam, 1081 LA Amsterdam, The Netherlands.
[5] Mimetas BV, Organ-on-a-chip Company, 2333 CH Leiden, The Netherlands.
[6] Department of Translational Neuroscience, University Medical Center, 3584 CG, Utrecht, The Netherlands.
[7] Centre for Snakebite Research & Interventions, Parasitology Department, Liverpool School of Tropical Medicine, Liverpool, L3 5QA, UK.
[8] The Maastricht Multimodal Molecular Imaging institute, Maastricht University, 6229 ER Maastricht, The Netherlands.
[9] Molecular Cancer Research, Center for Molecular Medicine, University Medical Center Utrecht, Utrecht University, 3584 CX Utrecht, The Netherlands.
[10] The Princess Maxima Center for Pediatric Oncology, 3584 CS Utrecht, The Netherlands.
[11] Neurotoxicology Research Group, Division of Toxicology, Institute for Risk Assessment Sciences (IRAS), Utrecht University, 3584 CL Utrecht, The Netherlands.
[12] Serpo, 2288 ED Rijswijk, The Netherlands.
* Correspondence: jens.puschhof@dkfz-heidelberg.de

Abstract: Recent advances in organoid technology have proven this system to be a valuable tool in understanding human organ development and pathologies. These adult stem cell-derived cultures closely recapitulate the structural and functional properties of their organ of origin. Here, we expand the organoid technology toolbox by describing a protocol to culture non-mammalian organoids derived from a snake venom gland. The complexity of venom production, composition and function remain largely unknown for many species. Organoids derived from an *Aspidelaps lubricus* venom gland can be long-term expanded and histologically resemble the gland. Expression of typical venom-related transcripts (three-finger toxins and Kunitz-type protease inhibitors) can be detected in proliferating organoids with RNA sequencing. Single-cell RNA sequencing reveals distinct venom-expressing cell types, as well as proliferating cells with features of mammalian stem cells. By using mass spectrometry, we identified peptides in the culture medium supernatant that match the composition of the crude venom of the same species. Venom gland organoids furthermore consist of specialized secretory cells visible by transmission electron microscopy. The system enables the investigation of venom production and function on a cellular level in controlled conditions and without the need for experimental animals. This study describes the adaption of organoid technology to a non-mammalian species, providing a model to understand the complexity of the snake venom gland.

Keywords: in vitro; organoid; snake; venom gland

4.6. Prospection of Animal Toxins in Drug Discovery—Challenges and Perspectives

Karla de Castro Figueiredo Bordon *

University of São Paulo, School of Pharmaceutical Sciences of Ribeirão Preto, Ribeirão Preto, São Paulo, Brazil.

* Correspondence: karlabordon@yahoo.com.br

Abstract: Venomous animals may cause severe medical complications and untimely death, but their venoms are also sources of molecules acting on the nervous, cardiovascular, immune and other systems. Animal toxins are often used as pharmacological tools for the validation of therapeutic targets, but they are also used as cosmeceuticals and diagnostic tools in the design of new therapeutic agents and to improve drug libraries. Some drugs used in the therapy of many disorders, such as diabetes and cardiovascular diseases, were developed based on the molecular structures of animal toxins. Captopril was the first animal toxin-based drug approved for human use. Scorpion neurotoxins are known to be responsible for the pathological manifestations of scorpionism. They are a threat to human health but may serve as leads for the development of new therapies against current and emerging diseases. The scorpion toxin chlorotoxin is undergoing clinical phase trials as a fluorescent molecular probe that paints tumors, while the CPP-Ts peptide is a potential intranuclear delivery tool targeting cancer cells. The bioprospection of neurotoxins that block KV1.3 potassium channels may lead to the development of drugs to treat autoimmune diseases since these channels are found in macrophages, platelets and T cells. The overexpression of KV1.3 in lymphocytes is related to diseases such as atherosclerosis, some types of cancers, obesity and autoimmune diseases such as Crohn's disease, dermatitis, psoriasis and rheumatoid arthritis, among others. Related to snake toxins, a heterologous fibrin sealant, comprising a cryoprecipitate rich in fibrinogen extracted from the blood of buffaloes in association with a serine protease from rattlesnake, is a biodegradable biological product that reduces surgical time, promotes faster postoperative recovery, is highly adhesive and can also be used as an adjuvant in suture procedures and as a carrier for drug delivery. The use of biotechnological tools, such as the heterologous expression of peptides and proteins, enables the production of biologically active molecules in sufficient quantity for the evaluation and development of new medicines. Recently, a serine protease was recombinantly expressed with functional and structural integrity and showed fibrinogenolytic activity and inhibition against the hEAG1 channel, highly expressed in tumor cells, in a mechanism independent of its catalytic activity. Non-PEGylated and PEGylated forms of this enzyme share similar kinetic and functional characteristics, and the latter showed no evidence of immunogenicity or cytotoxicity, even at high concentrations. Given the examples, approaches to improve the druggability of animal toxins are fruitful fields for future research.

Keywords: biological dressing; heterologous expression; immunosuppression; PEGylation; poison; toxin

4.7. Redirecting the Target of Muscarinic Toxin

Shoji Maeda [1,7,*], Jun Xu [1], Francois Marie N. Kadji [2], Mary J. Clark [3], Jiawei Zhao [4], Naotaka Tsutsumi [5], Junken Aoki [2], Roger K. Sunahara [3], Asuka Inoue [2], K. Christopher Garcia [1,5,6], Brian K. Kobilka [1]

[1] Stanford University School of Medicine, Stanford, CA 94305, USA;
[2] Tohoku University, Sendai, Japan;
[3] University of California San Diego School of Medicine, La Jolla, CA 92093, USA;

[4] Tsinghua University, Beijing, China;
[5] Stanford University School of Medicine, Stanford, CA 94305, USA;
[6] Howard Hughes Medical Institute; [7] University of Michigan, MI 48109, USA.
* Correspondence: shojim@umich.edu

Abstract: Muscarinic acetylcholine receptors are prototypical class-A GPCRs that are distributed throughout the human body and play critical roles in the maintenance of fundamental human physiology by responding to the neurotransmitter acetylcholine. Five muscarinic receptor subtypes (M1R–M5R) have been identified in humans with distinct G protein coupling preferences, distribution profile and physiological roles. With their conserved orthosteric ligand-binding site and the presence of an allosteric ligand-binding site at the extracellular region, muscarinic receptors were extensively studied as a model system for the subtype-selective targeting at a GPCR through small molecule allosteric modulators. Animals have evolved venoms to hunt prey and/or run from predators. These venoms have been a rich source to isolate natural products that modulate numerous protein functions, including GPCRs. Peptide and small protein toxins are emerging modalities because of their superior selectivity and stability. Muscarinic toxin belongs to the three-finger toxin family and is one of the best-characterized peptide toxins produced by venomous snakes. A member of muscarinic toxin, MT7, shows extremely high selectivity to M1R over other muscarinic receptors and elicits an inhibitory effect. We solved the crystal structure of the M1R–MT7 complex and revealed the exquisite design of this toxin to exclusively fit into M1R by selectively engaging at the allosteric site with residues unique to M1R. By using the structural information and the surface display platform, we succeeded in converting the subtype selectivity of MT7 to another muscarinic receptor, M2R. Furthermore, we obtained conformational selective three-finger proteins exclusively targeting the active conformation of muscarinic receptors. Our data indicate a promise of repurposing this natural toxin scaffold to a broader range of target systems.

Keywords: muscarinic acetylcholine receptor; protein engineering; three-finger toxin

4.8. Synthetic AB-Toxins for Targeted Pharmacological Modulation of Cancer and Immune Cells

Holger Barth *

University of Ulm Medical Center, Institute of Pharmacology and Toxicology, Albert-Einstein-Allee 11, 89081 Ulm, Germany.
* Correspondence: holger.barth@uni-ulm.de

Abstract: Bacterial AB-toxins are highly toxic proteins because of their unique modular structure: these toxins bind to mammalian cells by a specific receptor binding (B) subunit and deliver their enzymatically active (A) subunit into the host cell cytosol via their intrinsic translocation (T) subunit. In the cytosol, the A subunit modifies its specific cellular substrate, which alters cellular structures and/or functions and is the reason for the clinical symptoms of the particular toxin-associated disease. Because of their intracellular, highly specific and extremely potent mode of action, some AB-toxins serve as valuable molecular tools in pharmacology and cell biology for the targeted modulation of cell functions and are used as drugs. In past years, we and others used the non-toxic B and T subunits of various toxins to deliver pharmacologically active proteins (e.g., enzymes) and peptides into the cytosol of cells and developed cell type-selective transport systems to modulate functions of immune and cancer cells. However, for some cell types, there are no selective AB toxins available by nature, and therefore, we aimed to develop "synthetic" AB-toxins. In one approach, supramolecular transporter systems were generated, where avidin with four biotin-binding sites serves as a central binding platform for three biotinylated binding peptides that selectively bind to receptors on target cells and trigger internalization of the transporter molecule, and biotinylated cargo molecules that act as "A subunit" in the cytosol. Proof-of-concept was provided by using clostridial C3 Rho-inhibitor as cargo enzyme and the peptide somatostatin as a ligand to selectively target human lung cancer cells in vitro and in the HET/CAM model for a human xenograft lung tumor that overexpresses somatostatin receptor. This strategy also served for the generation of a novel cell type-selective Rho-inhibitor for human neutrophils. This universal "LEGO-like" modular approach represents a promising platform for the generation of cell-type selective synthetic AB-toxins for the targeted pharmacological modulation of cell functions.

Keywords: bacterial AB-toxin; cellular uptake; targeted drug delivery; receptor binding; membrane translocation; avidin-biotin technology; (semi)synthetic AB-type molecule

4.9. Phenotypic Screening for Anti-Toxin Molecules Selects Inhibitors of Intracellular Trafficking with Broad-Spectrum Anti-Infectious Properties

Daniel Gillet *

Université Paris-Saclay, CEA, INRAE, Département Médicaments et Technologies pour la Santé (DMTS), SIMoS, 91191 Gif-sur-Yvette, France.
* Correspondence: daniel.gillet@cea.fr

Abstract: In the aftermath of the anthrax letter attacks in the fall of 2001, the French authorities mandated the French health products sanitary safety agency (AFSSAPS) to determine the needs for medical countermeasures (MCM) against biothreat agents. A working group under the lead of Prof. Dominique Dormont established the cruel lack of medications against bacterial and plant toxins prone to be used as bioweapons, among which the plant toxin ricin. Then, the growing concerns about antimicrobial resistance led to suggest that antitoxin drugs should be developed to treat common public health bacterial infections as an alternative to antibiotics. In these respects, we developed two phenotypic screenings of small chemical molecules in 2005 and 2015; the first was against ricin toxin, and the second (in collaboration with Emmanuel Lemichez, Institut Pasteur) was against the cytotoxic necrotizing factor CNF1, produced by uropathogenic and extra-intestinal strains of *Escherichia coli*. The first screening was designed to select molecules capable of rescuing cells from ricin-induced toxicity. The second screening was designed to select molecules inhibiting the degradation of CNF1-induced activated Rac1. Overall, the anti-ricin screening identified three original inhibitors of intracellular trafficking. Retro-1 and Retro-2 act on the retrograde trafficking pathway (collaboration with Ludger Johannes, Institut Curie). In particular, Retro-2 interferes with the ERES protein Sec16A and the circulation of syntaxin-5 along the anterograde and retrograde routes. ABMA acts on the endolysosomal pathway by the accumulation of late endosomes and alteration of the autophagic flux. The CNF1 screening identified C910 as an original inhibitor of early endosome sorting function. Most interestingly, all four inhibitors displayed a broad spectrum of inhibition against a series of bacterial toxins, but also viruses, bacteria and parasites exploiting these pathways for intoxication or infection of cells. There are at least eight examples published or under the submission of in vivo protection of mice by these compounds against intoxications and/or infections. Compound optimization by medicinal chemistry was performed. The challenges of turning these molecules into drugs are discussed. Nevertheless, 16 years of research by us and others showed that intracellular trafficking pathways might be interesting druggable targets to fight intracellular toxins and pathogens.

Keywords: bacterial toxin; broad-spectrum; intracellular trafficking; ricin toxin; small chemical inhibitor

4.10. NLRP3 Inflammasome Sensing of RhoGTPase-Activating Toxins during Bacteremia

Laurent Boyer *

C3M, INSERM U1065, Université Cote d'Azur, Nice, France.
* Correspondence: boyerl@unice.fr

Abstract: The detection of the activities of pathogen-encoded virulence factors by the innate immune system has emerged as a new paradigm of pathogen recognition. Much remains to be determined regarding the molecular components contributing to this defense mechanism in mammals and its importance during infection. Our team showed the central role of the IL-1β signaling axis in controlling the Escherichia coli burden in the blood in response to the sensing of the RhoGTPase-activating toxin CNF1. Using the CNF1 toxin, we provided evidence of the role of the NLRP3 inflammasome in sensing the activity of bacterial toxins and virulence factors that activates host RhoGTPases. We demonstrated that this activation relies on monitoring of the toxin activity on the RhoGTPase Rac2. We also showed that the NLRP3 inflammasome is activated by a signaling cascade involving the P21-activated kinases (Pak)-1/2. The Pak1-mediated phosphorylation of threonine-659 of NLRP3 was necessary for NLRP3-Nek7 interaction, inflammasome activation and IL-1ß cytokine maturation. Furthermore, inhibition of the Pak1-NLRP3 axis diminished the bacterial clearance of CNF1-expressing E. coli during bacteremia. Altogether, our results established Pak1/2 as critical regulators of the NLRP3 inflammasome and revealed the role of the Pak1-NLRP3 signaling axis in vivo during bacteremia.

Keywords: bacterial toxin; inflammasome; NLPR3 RhoGTPase

5. Oral Presentations (When More Than One Author, the Underlined Name Is That of the Presenter)

5.1. Screening an In-House Library of Isoquinoline Alkaloids with Fluorescent Probes for Discovering New Ligands of Voltage-Gated Na^+ and Ca^{2+} Channels

<u>Claire</u> **Legendre** [1,*], **Jacinthe Frangieh** [1,2], **Léa Réthoré** [1], **Quentin Coquerel** [1], **Daniel Henrion** [1], **César Mattei** [1], **Ziad Fajloun** [2,4], **Anne-Marie Le Ray** [3], **Christian Legros** [1]

[1] Laboratory Mitochondrial and Cardiovascular Pathophysiology (MITOVASC), CNRS UMR 6015—INSERM U1083, CHU Angers, Université d'Angers, Angers, France.
[2] Laboratory of Applied Biotechnology (LBA3B), Azm Center for Research in Biotechnology and its Applications, EDST, Lebanese University, 1300 Tripoli, Lebanon.
[3] SONAS Laboratory, SFR QUASAV, Université d'Angers, Angers, France.
[4] Department of Biology, American University of Beirut, 11-0236 Beirut, Lebanon.
* Correspondence: claire.legendre@univ-angers.fr

Abstract: The isoquinoline alkaloids (IA) is a large chemical group of natural compounds with various structures and acting on multiple pharmacological targets of therapeutical interest, including ion channels, such as voltage-gated Na+ (Nav) and Ca2+ (Cav) channels. These ion channels are established molecular targets for the development of therapeutical agents in cardiovascular diseases and particularly arrhythmia. Here, we screened 62 IA from an in-house vegetal chemical library, using a model of excitable cells, the rat pituitary GH3b6 cells, which endogenously express both Nav and Cav channels, to find new blockers for these channels. Moreover, we demonstrated that Nav activation by selective neurotoxins induced an intracellular Ca^{2+} concentration $[Ca^{2+}]_i$ elevation mediated by Cav channels, highlighting crosstalk between Nav and Cav channels that we used for pharmacological studies. We first tested these IA for their abilities to inhibit batrachotoxin (BTX)-induced depolarization using fluorescent voltage-sensor probes (VSP) and identified two oxoaporphines, namely liriodenine and oxostephanine, which abolished BTX-induced VSP signal with IC_{50} values in the micromolar range. While the blocking activity of liriodenine was already reported, the activity of oxostephanine on Nav channels has never been described yet. Interestingly, oxostephanine differs from liriodenine only by a methoxy group on the benzyl part of the skeleton. This subclass of IA might constitute a new group of ligands of Nav channels. We confirmed the blocking effect of both molecules in a Na+ influx assay using the Na^+ fluorescent probe ANG-2. Since liriodenine is also known to block Cav channels, we hypothesized that oxostephanine probably targets Cav channels. Thus, we investigated the effects of both IA on L-type Cav channels (LTCC) expressed in GH3b6 cells. In order to activate LTCC, we used a chemical depolarization with KCl or the agonist Bay K8644 and monitored $[Ca^{2+}]_i$ change with the fura-2 probe. Our results showed that liriodenine and oxostephanine induced a concentration-dependent inhibition of KCl- and Bay K8644-evoked Ca^{2+} responses, with similar IC_{50} values in the micromolar range. In addition, this interaction of liriodenine and oxostephanine on LTCC was also highlighted by their ability to inhibit veratridine (VTD)- and BTX-induced $[Ca^{2+}]_i$ elevation. In conclusion, our data showed that liriodenine and oxostephanine, two oxoaporphine alkaloids, inhibit Nav and Cav channels with similar potency. The oxoaporphine skeleton might bring an interesting pharmacophore for structure-function relationship studies for designing more selective ligands toward Cav and Nav channels and for developing new antiarrhythmic therapeutical leads.

Keywords: Cav channel; isoquinoline alkaloid; Nav channel

5.2. The Venom of the Lebanese Viper, Montivipera bornmuelleri, Contains Vasoactive Compounds

Jacinthe Frangieh [1,2,*], **Joohee Park** [1], **Ziad Fajloun** [2], **Loïc Quinton** [3], **Riyad Sadek** [4], **Daniel Henrion** [1], **César Mattei** [1], **Christian Legros** [1]

[1] Laboratory Mitochondrial and Cardiovascular Pathophysiology—MITOVASC, CNRS UMR 6015, INSERM U1083, CHU Angers, Université d'Angers, Angers, France.
[2] Laboratory of Applied Biotechnology (LBA3B), Azm Center for Research in Biotechnology and its Applications, EDST, Lebanese University, 1300 Tripoli, Lebanon.
[3] Laboratory of Mass Spectrometry, Department of Chemistry, MolSys Research Unit, University of Liège, 4000 Liège, Belgium.
[4] Department of Biology, American University of Beirut, 11-0236 Beirut, Lebanon.
* Correspondence: jacynthefrangieh@gmail.com

Abstract: Snake venoms are rich mixtures of polypeptides that target, among other physiological networks, the cardiovascular system. Snake toxins such as phospholipase A2, natriuretic peptides, and bradykinin-potentiating peptides exert various effects on the cardiovascular system (hypotension, vasorelaxation, etc.). Some of these toxins were developed as drugs with antihypertensive properties, such as Captopril™. We previously showed that the venom of the viper *Montivipera bornmuelleri* induces relaxation of rat aortic rings. Here, we aim to identify and characterize the vasoactive compounds of this venom. We fractionated the venom by HPLC. The fractions were assayed on the endothelial MS1 cell line and CHO cells lines expressing muscarinic receptors M1, M3 or M5. We screened for compounds able to inhibit acetylcholine-induced intracellular Ca^{2+} rise, using the fluorescence probe FURA-2. In addition, wire myography on isolated mesenteric arteries was used. Proteomic analysis was performed to characterize selected fractions. The crude venom exhibited a vasorelaxant effect on mesenteric arteries. Among 23 fractions of *Mb* venom, one fraction, namely P14, was selected for its ability to reduce acetylcholine-induced Ca^{2+} rise in MS1 cells. The proteomic analysis allowed us to identify a novel peptidyl toxin (P14) as a homolog of the vascular endothelial growth factor (VEGF). However, P14 was unable to antagonize acetylcholine-induced Ca^{2+} rise in cells expressing muscarinic receptors, suggesting that this peptide interferes with other target(s) of the Ca^{2+} signaling pathway in endothelial cells. Further studies will be carried out to characterize the molecular targets of P14 and its effects on vascular function.

Keywords: calcium signaling; *Montivipera bornmuelleri* venom; vascular function

5.3. Lipids and Cholesterol Mediate the Cytotoxicity of Spider Peptides

Javier Moral-Sanz [1], **Sergey Kurdyukov** [2], **Ana Vela-Sebastián** [1], **Zoltan Dekan** [3], **Thomas Kremsmayr** [4], **Markus Muttenthaler** [3,4], **Paul F. Alewood** [3], **Gregory G. Neely** [2], **Évelyne Deplazes** [3], **Maria P. Ikonomopoulou** [1,3,*]

[1] Madrid Institute for Advanced Studies in Food, Madrid, E28049, Spain.
[2] University of Sydney, Camperdown, NSW 2006, Australia.
[3] The University of Queensland, St. Lucia, QLD 4072, Australia.
[4] University of Vienna, Vienna 1090, Austria.
* Correspondence: maria.ikonomopoulou@imdea.org

Abstract: Spider gomesin peptides target melanoma cells of BRAF-mutation with minimum effect on non-transformed neonatal foreskin fibroblasts. We hypothesized that the selectivity of gomesin peptides towards melanoma of BRAF mutation could be influenced by interactions with lipids or cholesterol present in cell membranes. In order to elucidate upon this, we employed a multidisciplinary approach, including CRISPR/Cas9 genome-wide screening, to identify key players underlying the antiproliferative mechanisms of gomesins. We also used a combination of fluorescence spectroscopy to measure membrane dipole changes, electrical impedance spectroscopy (EIS) with tethered bilayer membranes (tBLMs) to quantify alterations in membrane conductance and membrane thickness, and cell viability assays. Gomesin and variants showed weak binding to POPC membranes alone. However, the presence of POPS and cholesterol significantly improved the binding of the peptides and lessened membrane disruption. In addition, the cytotoxicity of gomesin was blunted by increasing concentrations of cholesterol in both melanoma cells and fibroblasts. Conversely, cholesterol depletion potentiated the cytotoxicity of peptides in fibroblasts to almost the levels originally observed in melanoma cells. In conclusion, we postulate a specific role of cholesterol in the selective cytotoxic profile of gomesin in melanoma of BRAF mutation.

Keywords: cytotoxicity; melanoma; spider peptide

5.4. ExoU and ExoY, Two Toxins from Pseudomonas aeruginosa with Different Mechanisms of Action and Biological Effects

__Vincent Deruelle__ [1,*], **Dorothée Raoux-Barbot** [1], **Undine Mechold** [1], **Philippe Huber** [2]

[1] Unité de Biochimie des Interactions Macromoléculaires, Institut Pasteur, Département de Biologie Structurale et Chimie, CNRS UMR 3528, Paris, France.
[2] Center for Immunology of Viral, Auto-immune, Hematological and Bacterial diseases (IMVA-HB/IDMIT), Université Paris-Saclay, INSERM, CEA, Fontenay-aux-Roses, France.
* Correspondence: vincent.deruelle@pasteur.fr

Abstract: *Pseudomonas aeruginosa* is a ubiquitous and opportunistic Gram-negative bacterium. It is a leading cause of nosocomial infections and is responsible for acute and chronic infections thanks to a broad panel of virulence factors. Among them, the Type 3 Secretion System (T3SS) plays an essential role in the pathogenicity of the bacterium. This system pierces the cell membrane and allows the delivery of toxins into the host cytoplasm. Four T3SS effectors have been identified—ExoS, ExoU, ExoT and ExoY. Most strains inject two or three of these effectors simultaneously, with ExoU and ExoS being mutually exclusive. Each T3SS effector is inactive when injected and requires a host cell factor to be activated. Strains injecting ExoU toxin induce the most severe pathologies because the damage caused by the toxin is rapid and irreversible, leaving no time for proper clinical management. Its mechanism of action is well described. Inside host cells, ExoU is a phospholipase that interacts with a specific lipid at the cytosolic side of the plasma membrane, allowing its activation and leading to cell necrosis. However, several aspects of ExoU activation and its trafficking in human cells upon injection remain elusive. Thanks to genetic approaches and analyses by microscopy, we identified a favored anterograde transport, led by the co-chaperone DNAJC5, which targets ExoU to the plasma membrane. Inactivation of DNAJC5 gene disrupts ExoU-dependent toxicity in vitro and in vivo. Unlike ExoU, the role of ExoY in the pathogenicity of *P. aeruginosa* is still controversial. This toxin is a nucleotidyl cyclase that interacts with the eukaryotic filamentous actin (F-actin) to be activated and thereby to massively convert nucleotide triphosphates into their cyclic form. It was reported that ExoS toxicity to the epithelial cells is enhanced in the absence of ExoY and that in many ExoU+ strains, ExoY is not active. Thus, ExoY seems to limit the action of other toxins. We will discuss our advances in exploring the role of ExoY by studying its effects in infected cells.

Keywords: CRISPR-Cas9; DNAJC5; ExoU; ExoY; virulence

5.5. Moving Forward Tetanus Therapy: Two Exceptionally Potent humAbs Are Effective for Prophylaxis and Treatment in Mice

Marco Pirazzini [1,*], **Alessandro Grinzato** [1], **Davide Corti** [2], **Sonia Barbieri** [2], **Oneda Leka** [1], **Francesca Vallese** [1], **Marika Tonellato** [1], **Chiara Silacci-Fregni** [3], **Luca Piccoli** [3], **Eaazhisai Kandiah** [4], **Giampietro Schiavo** [5,6], **Giuseppe Zanotti** [1], **Antonio Lanzavecchia** [3,7], **Cesare Montecucco** [1,8]

[1] Department of Biomedical Sciences, University of Padova, Via Ugo Bassi 58/B, Padova, 35131, Italy.
[2] Humabs BioMed SA, 6500 Bellinzona, Switzerland.
[3] Institute for Research in Biomedicine, Università della Svizzera Italiana, 6500 Bellinzona, Switzerland.
[4] European Synchrotron Radiation Facility, 71 avenue des Martyrs, F-38000 Grenoble, France.
[5] Department of Neuromuscular Diseases, Queen Square Institute of Neurology, University College London, London, WC1N 3BG, UK.
[6] UK Dementia Research Institute, University College London, London, WC1E 6BT UK.
[7] Fondazione Istituto Nazionale Genetica Molecolare c/o Fondazione IRCCS Cà Granda Ospedale Maggiore Policlinico di Milano, Via Francesco Sforza 35, 20122 Milano, Italy.
[8] Institute of Neuroscience, National Research Council, Via Ugo Bassi 58/B, Padova, 35131, Italy.
* Correspondence: marco.pirazzini@unipd.it

Abstract: Tetanus neurotoxin (TeNT) is the causative agent of tetanus, a life-threatening disease of vertebrates, including humans, characterized by neurogenic muscle rigidity and spasticity. Although tetanus can be prevented by a very effective vaccine, a worldwide clinical practice in the emergency rooms is the administration of anti-TeNT immunoglobulins (TIG), which are used both for prophylaxis, to avoid tetanus development in wounded patients, and for therapy, to treat patients already carrying tetanus symptoms. TIG is produced from the blood of hyperimmune individuals, either humans or horses (in developing countries). As such, it exposes patients to several possible side effects, including infections by still-unknown pathogens as well as dangerous anaphylactic reactions. Human monoclonal antibodies (humAbs), which are emerging as superior therapeutics against several diseases, could overcome the drawbacks of TIG. Here, we screened the immortalized memory B cells pooled from the blood of immunized human donors and isolated two humAbs, dubbed TT104 and TT110, which display an unprecedented neutralization ability against TeNT. We determined the epitopes recognized by TT104 and TT110 via cryo-EM and defined how they interfere with the mechanism of neuron intoxication. These analyses pinpointed two novel mechanistic aspects of TeNT activity in neurons and unraveled at the same time the molecular bases of TT104 and TT110 exceptional neutralization ability. Crucially, the combination of TT104 and TT110 display a prophylactic activity in mice when injected long before TeNT, and the two Fab derivatives (TT104-Fab and TT110-Fab) neutralize TeNT in post-exposure experiments. Of note, in both these two paradigms of experimental tetanus, the humAbs and the Fabs show an activity fully comparable to TIG. Therefore, TT104 and TT110 humAbs and their Fab derivatives meet all requirements for being considered for prophylaxis and therapy of human tetanus and are ready for clinical trials.

Keywords: monoclonal antibody; spastic paralysis; tetanus neurotoxin; tetanus prophylaxis

5.6. CyaA Toxin Cell Invasion Involves Membrane-Active and Calmodulin-Binding Properties of Its Translocation Domain

Alexis Voegele [1], Mirko Sadi [1], Darragh P. O'Brien [1], Pauline Gehan [2], Dorothée Raoux-Barbot [1], Maryline Davi [1], Sylviane Hoos [1], Sébastien Brulé [1], Bertrand Raynal [1], Patrick Weber [1], Ariel Mechaly [1], Ahmed Haouz [1], Nicolas Rodriguez [2], Patrice Vachette [3], Dominique Durand [3], Sébastien Brier [1], Daniel Ladant [1], <u>Alexandre Chenal</u> [1,*]

[1] Institut Pasteur, CNRS UMR 3528, 75015 Paris, France.
[2] Sorbonne Université, École Normale Supérieure, PSL University, CNRS, Laboratoire des biomolécules, LBM, 75005 Paris, France.
[3] Université Paris-Saclay, CEA, CNRS, Institute for Integrative Biology of the Cell (I2BC), 91198, Gif-sur-Yvette, France.
* Correspondence: chenal@pasteur.fr

Abstract: The molecular mechanisms and forces involved in the translocation of bacterial toxins into host cells are still a matter of intense research. *Bordetella pertussis*, the causative agent of whooping cough, produces an adenylate cyclase (CyaA) toxin that plays an essential role in the early stages of respiratory tract colonization. CyaA displays a unique intoxication pathway of human cells via a direct translocation of its catalytic domain (AC) across the plasma membrane. Once in the cytosol, AC impairs the physiology of immune cells, leading to cell death. We showed that the P454 peptide (CyaA residues 454–484) is able to translocate across membranes and interact with calmodulin. The key residues involved in membrane-active and calmodulin-binding properties were identified. The mutational analysis demonstrates that these residues play a crucial role in CyaA translocation into target cells. We propose that after CyaA binding to target cells, the P454 segment destabilizes the plasma membrane, translocates across the lipid bilayer and binds calmodulin (Figure 3). Trapping of CyaA by the calmodulin:P454 interaction in the cytosol may assist the entry of AC by converting the stochastic motion of the polypeptide chain through the membrane into an efficient vectorial chain translocation into host cells.

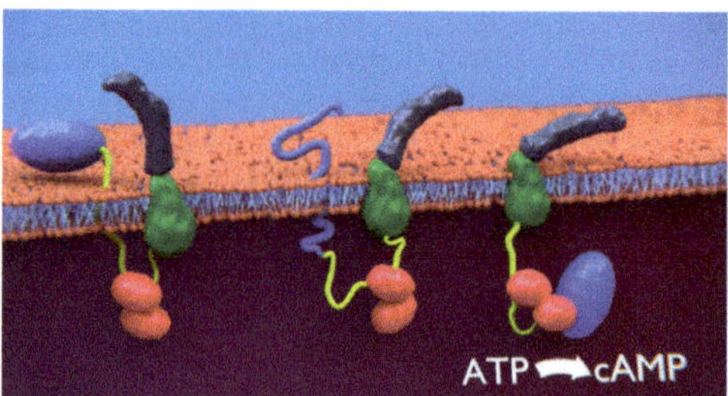

Figure 3. The membrane-active segment (yellow) of the CyaA toxin translocates across the plasma membrane and binds calmodulin (red), which assists the entry and refolding of the catalytic domain (blue) into host cells while the hydrophobic and acylation domains (green) interact with the membrane, and the C-terminal Repeat-in-Toxin domain (grey) remains in the extra-cellular milieu. The cAMP production ultimately leads to cell death.

Keywords: CyaA; adenylate cyclase toxin; membrane-active peptide; membrane translocation; entropic pulling; calmodulin-binding peptide

References

1. Voegele, A., Sadi, M., O'Brien, D. P., Gehan, P., Raoux-Barbot, D., Davi, M., Hoos, S., Brûlé, S., Raynal, B., Weber, P., Mechaly, A., et.al. A High-Affinity Calmodulin-Binding Site in the CyaA Toxin Translocation Domain is Essential for Invasion of Eukaryotic Cells. *Adv. Sci.* **2021**, 8, 2003630.

2. O'Brien, D. P., Cannella, S. E., Voegele, A., Raoux-Barbot, D., Davi, M., Douche, T., Matondo, M., Brier, S., Ladant, D., Chenal, A. Post-translational acylation controls the folding and functions of the CyaA RTX toxin. *FASEB J.* **2019**, 33, fj201802442RR.

3. O'Brien, D. P., Durand, D., Voegele, A., Hourdel, V., Davi, M., Chamot-Rooke, J., Vachette, P., Brier, S., Ladant, D., Chenal, A. Calmodulin fishing with a structurally disordered bait triggers CyaA catalysis. *PLoS Biol.* **2017**, 15, e2004486.

4. Voegele, A., Subrini, O., Sapay, N., Ladant, D., Chenal, A. Membrane-Active Properties of an Amphitropic Peptide from the CyaA Toxin Translocation Region. *Toxins* **2017**, 9, 369.

5.7. Is the Regulation of Toxinogenesis the Same in Clostridium botulinum and Clostridium tetani?

Diana Chapeton-Montes [1], **Holger Brüggemann** [2], <u>**Michel R. Popoff**</u> [1,*]

[1] Bacterial Toxins, Institut Pasteur, Paris, France.
[2] Department of Biomedicine, Aarhus University, Aarhus, Denmark.
* Correspondence: popoff2m@gmail.com

Abstract: Botulinum neurotoxins (BoNTs) and tetanus neurotoxin (TeNT) share similar structures and molecular modes of action, albeit BoNTs target motor-neuron endings leading to a flaccid paralysis (botulism) and TeNT interacts with central inhibitory interneurons causing spastic paralysis (tetanus). BoNTs and TeNT are produced by anaerobic and sporulating bacteria of the *Clostridium* genus, *C. botulinum* and *C. tetani*, respectively. The synthesis of BoNTs and TeNT is a highly regulated process. Among the diverse *C. botulinum* types, toxinogenesis was mainly investigated in *C. botulinum* A. In both *C. botulinum* A and *C. tetani*, a gene coding for an alternative sigma factor lies upstream of the neurotoxin gene and is required for the expression of the toxin gene. *C. botulinum* and *C. tetani* contain numerous two-component systems (TCSs) (39 in *C. botulinum* A and 30 in *C. tetani*), which are major regulatory systems in response to environmental conditions in prokaryotes. Albeit most TCS genes are homologous in both species, *C. botulinum* A and *C. tetani* use distinct sets of TCSs in the regulation of toxin synthesis. Only one TCS, which is homologous in *C. botulinum* A and *C. tetani*, has a similar function of negative regulation of toxin synthesis in both species. In contrast to *C. botulinum* A, inorganic phosphate and carbonate are environmental factors controlling toxin synthesis in *C. tetani*. In addition, a non-coding RNA downstream the *tent* gene negatively regulates TeNT synthesis. No homologous RNA sequence was found in *C. botulinum* A. Thereby, *C. botulinum* and *C. tetani* are two environmental bacteria that use distinct regulatory pathways to synthesize potent neurotoxins.

Keywords: *Clostridium botulinum*; *Clostridium tetani*; botulinum neurotoxin; tetanus neurotoxin; toxin synthesis regulation

References

1. Connan, C.; Brueggemann, H.; Mazuet, C.; Raffestin, S.; Cayet, N.; Popoff, M.R. Two-component systems are involved in the regulation of Botulinum neurotoxin synthesis in *Clostridium botulinum* type A strain hall. *PLoS ONE* **2012**, 7, e41848.

2. Chapeton-Montes, D.; Plourde, L.; Deneve, C.; Garnier, D.; Barbirato, F.; Colombié, V.; Demay, S.; Haustant, G.; Gorgette, O.; Schmitt, C.; et al. Tetanus toxin synthesis is under the control of a complex network of regulatory genes in *Clostridium tetani*. *Toxins* **2020**, *12*, 328.

3. Brüggemann, H.; Chapeton-Montes, D.; Plourde, L.; Popoff, M.R. Identification of a non-coding RNA and its putative involvement in the regulation of tetanus toxin synthesis in *Clostridium tetani*. *Sci. Rep.* **2021**, *11*, 4157.

5.8. In Vivo Spatiotemporal Control of Voltage-Gated Ion Channels by Engineered Photoactivatable Peptidic Toxins

Jérôme Montnach [1,2], **Laila Ananda Blömer** [2,3], **Ludivine Lopez** [1,2,4], **Luiza Filipis** [2,3], **Hervé Meudal** [5], **Aude Lafoux** [6], **Sébastien Nicolas** [1,2], **Duong Chu** [7], **Cécile Caumes** [4], **Rémy Béroud** [4], **Chris Jopling** [8], **Frank Bosmans** [9], **Corinne Huchet** [6], **Céline Landon** [5], **Marco Canepari** [2,3], <u>**Michel De Waard**</u> [1,2,4,*]

[1] L'institut du thorax, INSERM, CNRS, UNIV NANTES, F-44007 Nantes, France.
[2] Laboratory of Excellence Ion Channels, Science & Therapeutics, F-06560 Valbonne, France.
[3] Laboratoire Interdisciplinaire de Physique, Université Grenoble Alpes, CNRS UMR 5588, 38402 Saint Martin-d'Hères cedex, France.
[4] Smartox Biotechnology, 6 rue des Platanes, F-38120 Saint-Egrève, France.
[5] Center for Molecular Biophysics, CNRS, rue Charles Sadron, CS 80054, Orléans 45071, France.
[6] Therassay Platform, IRS2-Université de Nantes, Nantes, France.
[7] Queen's University Faculty of Medicine, Kingston, ON, Canada.
[8] Institut de Génomique Fonctionnelle, 141 rue de la Cardonille, 34094 Montpellier, France.
[9] Department of Basic and Applied Medical Sciences, Ghent University, Ghent, Belgium.
* Correspondence: michel.dewaard@univ-nantes.fr

Abstract: Photoactivatable drugs targeting ligand-gated ion channels opened up new opportunities for light-guided therapeutic interventions. Photoactivable toxins targeting ion channels have the potential to control excitable cell activities with low invasiveness and high spatiotemporal precision. We developed caged HwTxIV-Nvoc, a UV light-cleavable and photoactivatable peptide that targets voltage-gated sodium (Na_V) channels. We first validated physico-chemical parameters of photolysis, and by using a high-throughput patch-clamp system (SyncroPatch364, Nanion), we validated the activity and photosensitivity of caged HwTxIV-Nvoc in vitro in HEK293 cells. We further pursued our investigations in ex vivo brain slices and in vivo on mice neuromuscular junctions and zebrafish models. We found that caged HwTxIV-Nvoc enables precise spatiotemporal control of neuronal Na_V channel function under all conditions tested. We developed multiple photoactivatable toxins and therefore demonstrated the broad applicability of the toxin-photoactivation technology as a tool for experiments but also as a relevant clinical approach in disease management.

Keywords: ion channel; photopharmacology; toxin

5.9. Functional Impact of BeKm-1, a High-Affinity hERG Blocker, on Cardiomyocytes Derived from Human-Induced Pluripotent Stem Cells

Stephan De Waard [1,2,#], **Jérôme Montnach** [1,#], <u>**Barbara Ribeiro**</u> [1], **Sébastien Nicolas** [1], **Virginie Forest** [1], **Flavien Charpentier** [1], **Matteo Elia Mangoni** [2,3], **Nathalie Gaborit** [1], **Michel Ronjat** [1,2], **Gildas Loussouarn** [1], **Patricia Lemarchand** [1], **Michel De Waard** [1,2,4,*]

[1] L'institut du thorax, INSERM, CNRS, Université de Nantes, F-44007 Nantes, France.
[2] LabEx Ion Channels, Science & Therapeutics, F-06560 Valbonne, France.
[3] Institut de Génomique Fonctionnelle, CNRS, INSERM, Université de Montpellier, F34094 Montpellier, France.
[4] Smartox Biotechnology, 6 rue des Platanes, F-38120 Saint-Egrève, France.
* Correspondence: michel.dewaard@univ-nantes.fr

Abstract: IKr current, a major component of cardiac repolarization, is mediated by human Ether-à-go-go-Related Gene (hERG, Kv11.1) potassium channels. The blockage of these channels by pharmacological compounds is associated with drug-induced long QT syndrome (LQTS), which is a life-threatening disorder characterized by ventricular arrhythmias and defects in cardiac repolarization that can be illustrated using cardiomyocytes derived from human-induced pluripotent stem cells (hiPS-CMs). This study was meant to assess the modification in hiPS-CMs excitability and contractile properties by BeKm-1, a natural scorpion venom peptide that selectively interacts with the extracellular face of hERG, by opposition to reference compounds that act onto the intracellular face. By using an automated patch-clamp system, we compared the affinity of BeKm-1 for hERG channels with some reference compounds. We fully assessed its effects on the electrophysiological, calcium handling and beating properties of hiPS-CMs. By delaying cardiomyocyte repolarization, the peptide induces early afterdepolarizations and reduces spontaneous action potentials, calcium transients and contraction frequencies, therefore recapitulating several of the critical phenotype features associated with arrhythmic risk in drug-induced LQTS. BeKm-1 exemplifies an interesting reference compound in the integrated hiPS-CMs cell model for all drugs that may block the hERG channel from the outer face. Being an easily modifiable peptide, it will serve as an ideal molecular platform for the design of new hERG modulators displaying additional functionalities.

Keywords: BeKm-1; hERG; hiPS-cardiomyocyte; LQTS

5.10. An Agonist of the CXCR4 Receptor Is Therapeutic for the Neuroparalysis Induced by Bungarus Snake Envenoming

Marco <u>Stazi</u> [1,*], Federico Fabris [1], Kae Yi Tan [2], Aram Megighian [1], Alessandro Rubini [1], Andrea Mattarei [3], Samuele Negro [1], Giorgia D'Este [1], Florigio Lista [4], Ornella Rossetto [1], Choo Hock Tan [5], Cesare Montecucco [1,6]

[1] Department of Biomedical Sciences, University of Padova, Via Ugo Bassi 58/B, Padova 35131, Italy.
[2] Department of Molecular Medicine, Faculty of Medicine, University of Malaya, Kuala Lumpur 50603, Malaysia.
[3] Department of Pharmaceutical and Pharmacological Sciences, University of Padova, Padova 35131, Italy.
[4] Center of Medical and Veterinary Research of the Ministry of Defense, Policlinico Militare, Via Santo Stefano Rotondo 4, Rome 00184, Italy.
[5] Department of Pharmacology, Faculty of Medicine, University of Malaya, Kuala Lumpur 50603, Malaysia.
[6] CNR Institute of Neuroscience, Padova 35131, Italy.
* Correspondence: marco.stazi@studenti.unipd.it

Abstract: Snake envenoming is a major but neglected human tropical disease. Among venomous snakes, kraits (snakes of the *Bungarus* genus) are medically important venomous species that cause reversible peripheral neuroparalysis. Hospitalization and use of antivenoms derived from an animal immunized with *Bungarus* venoms are the primary therapies that prevent death from early-onset respiratory paralysis. There is a general consensus that additional and non-expensive treatments, which can be delivered even long after the snakebite, are needed. Traumatic or toxic degeneration of peripheral motor neurons, with ensuing neuroparalysis, is characterized by the activation of a pro-regenerative intercellular signaling program. A major player is the intercellular signaling axis consisting of the chemokine CXCL12a, produced by perisynaptic Schwann cells and acting on the CXCR4 receptor expressed on the damaged neuronal axons. The CXCR4 agonist NUCC-390 was recently found to promote axonal growth. Here, we tested its efficacy on the neuroparalysis induced by the venoms of three major krait species, i.e., *Bungarus caeruleus*, *B. multicinctus* and *B. candidus* that are prevalent in Asia. These venoms cause a complete degeneration of motor axon terminals. Functional recovery of the neuromuscular junction was assessed by electrophysiological recordings and by imaging. We report that NUCC-390 administration to venom-injected mice greatly accelerates the recovery from paralysis. These data candidate NUCC-390 to be tested as novel therapeutics to reduce death by respiratory deficits and to improve the recovery of normal neuromuscular physiology, thus reducing the human and hospital costs of envenoming. NUCC-390 can be administered any time after the snakebite and has great potential of becoming a non-expensive addition to the currently available antivenom treatments whose efficacy is limited to a short period after snakebite. This drug is expected to decrease the long and expensive post-snakebite mechanical ventilation phase.

Keywords: *Bungarus* venom; CXCR4 agonist; neuromuscular junction

5.11. Neutralization of Crotamine by Polyclonal Antibodies against Two Rattlesnake Venoms and a Novel Recombinant Fusion Protein

Roberto Ponce-López [1], Alejandro Olvera-Rodríguez [1,*], Miguel Borja-Jiménez [2], Edgar Neri-Castro [1], Leticia Olvera-Rodríguez [1], Alejandro Alagón [1]

[1] Instituto de Biotecnología, Universidad Nacional Autónoma de México, Cuernavaca, Morelos, México.
[2] Facultad de Ciencias Biológicas, Universidad Juárez del Estado de Durango, Gómez Palacio, Durango, México.
* Correspondence: alejandro.olvera@ibt.unam.mx

Abstract: Current pit-viper antivenoms for human use in Mexico have shown a low level of protection against one neurotoxin found in some rattlesnakes (*Crotalus* spp.), crotamine. This toxin has a low molecular weight (~5 kDa) and is well known for its spastic paralysis symptom provoked in mice with no evidence of neutralization. Recently, it was reported that crotamine is the major toxin found in some rattlesnake venoms such as *C. molossus nigrescens* (~50%) and *C. oreganus helleri* (62%). On the other hand, sphingomyelinase D (SMD) is a highly immunogenic enzyme (MW: ~30 kDa) found in *Loxosceles* spp. spider venom. In this study, we aimed to neutralize the crotamine-induced main symptom of paralysis by employing, as immunogens, two rattlesnake venoms and a novel recombinant fusion protein made from crotamine and SMD, used as a carrier. **Methods**—Crotamine cDNA was synthetized from venom gland mRNA of a *C. m. nigrescens* individual from Mexico, while one plasmid containing *L. reclusa* SMD was available in the lab. We cloned the sphingomyelinase D and crotamine in tandem into the expression vector pQE30 to transform competent Origami *Escherichia coli* cells for the production of the protein. By using the recombinant protein and the whole venoms of *C. o. helleri* and *C. m. nigrescens*, we performed separate immunization protocols in rabbits to measure the antibody recognition to crotamine (ELISA and Western blot) and the neutralization capacity in mice against the main toxin-induced paralysis symptom. **Results**—The fusion protein was obtained at ~10 mg/L of bacterial culture with the expected 37.5 kDa molecular mass as analyzed by SDS-PAGE and Western-blot. The recombinant protein and the two whole crotamine-enriched venoms generated antibodies with cross-reactivity against crotamine from up to seven species. The three experimental antivenoms were able to neutralize the paralysis symptom provoked by crotamine in the mice model. **Discussion/Conclusion**—We showed, for the first time, the neutralization of the crotamine-induced spastic paralysis symptom by three experimental antivenoms. The recombinant SMD-Crotamine fusion protein as well as crotamine-enriched venoms could be good candidate immunogens for the improvement of Mexican antivenoms.

Keywords: antivenom; crotamine; rattlesnake; sphingomyelinase-D

Acknowledgments: Fordecyt 303045.

5.12. Anti-Cancer Effect of the Moroccan cobra (Naja haje) Venom and Its Fractions against Hepatocellular Carcinoma in 3D Cell Culture

Ayoub Lafnoune [1,2,*], Su-Yeon Lee [3], Jin-Yeong Heo [4], Salma Chakir [1,5], Khadija Daoudi [1,2], Bouchra Darkaoui [1,2], Aziz Hmyene [5], Rachida Cadi [2], Khadija Mounaji [2], David Shum [4], Haeng-Ran Seo [3], Naoual Oukkache [1]

[1] Laboratoire des Venins et Toxines, Département de Recherche, Institut Pasteur du Maroc, 1 place Louis Pasteur, Casablanca 20360, Morocco.

[2] Laboratoire Physiopathologie, Génétique Moléculaire & Biotechnologie, Faculté des Sciences Ain-Chock, Hassan II University of Casablanca, B.P 5366 Maarif, Casablanca 20000, Morocco.
[3] Cancer Biology Research Laboratory, Institut Pasteur Korea, 16, Daewangpangyo-ro 712 beon-gil, Bundang-gu, Seongnam-si, Gyeonggi-do, 13488 Rep. of Korea.
[4] Screening Discovery Platform, Institut Pasteur Korea, 16, Daewangpangyo-ro 712 beon-gil, Bundang-gu, Seongnam-si, Gyeonggi-do, 13488 Rep. of Korea.
[5] Laboratoire de Biochimie Environnement et Agroalimentaire, Faculté des Sciences et Techniques de Mohammedia, Mohammedia, Morocco.
* Correspondence: ayoublafnoune@gmail.com

Abstract: Hepatocellular carcinoma (HCC) is the most common primary liver cancer in adults, the fifth most common malignancy worldwide and the third leading cause of cancer-related death. An alternative to the surgical treatments and drugs, such as sorafenib, commonly used in medicine, is necessary to overcome this public health problem. In this study, we determine the anticancer effect on HCC of Moroccan cobra *Naja haje* venom and its fraction obtained by gel filtration chromatography against Huh7.5 cancer cell line. Cells were grown together with WI38 human fibroblast cells, LX2 human hepatic stellate cell line, and human endothelial cells (HUVEC) in MCTS (multi-cellular tumor spheroids) models. The hepatotoxicity of venom and its fractions were also evaluated using the normal hepatocytes cell line (Fa2N-4 cells). Our results showed that an anti HCC activity of *N. haje* venom and, more specifically, the F7 fraction of gel filtration chromatography exhibited the greatest anti-hepatocellular carcinoma effect by decreasing the size of MCTS. This effect is associated with low toxicity against normal hepatocytes. These results strongly suggest that the F7 fraction of *N. haje* venom obtained by gel filtration chromatography possesses the ability to inhibit cancer cell proliferation. More research is needed to identify the specific molecule(s) responsible for the anticancer effect and investigate their mechanism of action.

Keywords: anticancer molecule; hepatocellular carcinoma; multicellular tumor spheroid; *Naja haje*; venom

6. Poster Presentations (When More than One Author, the Underlined Name Is That of the Presenter)

6.1. Circadian Variations in Inflammatory Response and Oxidative Stress during Experimental Scorpion Envenomation

Fares Daachi, <u>Sonia Adi-Bessalem</u> *, Amal Megdad-Lamraoui, Fatima Laraba-Djebari

University of Science and Technology Houari Boumediene (USTHB), Faculty of Biological Sciences, Laboratory of Cellular and Molecular Biology, BP 32 El-Alia, Bab Ezzouar, Algiers, Algeria.
* Correspondence: soniabessalem@hotmail.com

Abstract: The mammalian circadian clock orchestrates diverse physiological processes by synchronizing with the nervous and cardiovascular systems, immune response, and metabolic homeostasis. Little is known about the circadian rhythm in the immune response in vivo settings during envenomation—few studies have demonstrated that a circadian pattern might exist. The aim of this study is to investigate whether circadian rhythm affects the inflammation and oxidative stress in envenomed model. In this study, the systemic inflammatory response is induced by the administration of a sublethal dose of *Androctonus australis hector* (Aah) venom to mice during light (1 HALO) and dark (18 HALO) periods. The hypothalamic–pituitary–adrenal (HPA) axis activity was evaluated by measurement of adrenocorticotropic (ACTH) and corticosterone plasma hormones levels as well as by analysis of CD68 immunohistochemistry staining. The inflammation was assessed by the measurement of the serum level of the proinflammatory cytokines, IL-17 and IL-6, evaluation of neutrophil infiltration and oxidative/nitrosative stress markers (Nitrites, MDA, H_2O_2, CAT and GSH). Glucose level and activities of AST and ALT as well as histopathological analysis of hepatic tissue, were also performed in the two groups of animals (1 HALO and 18 HALO). Obtained results showed a high increase of serum ACTH and corticosterone levels as well as positive immunostaining of CD68 cells during the dark period indicating activation of the HPA axis associated with local inflammation. The results also showed day-night variations, with significantly high levels of nitrite, hydrogen peroxide, myeloperoxidase and lipid peroxidation during the daytime compared with the nighttime. Significant variations in catalase activity and GSH levels are also observed with the highest evening values. Furthermore, an increase in IL-6 levels was observed during the active phase, while no differences were observed in the IL-17 levels between day and night times. These results indicate also that Aah venom induced a time-dependent increase of metabolic parameters during the dark phase and severe hepatocellular injury. In conclusion, the daily variation of the HPA axis activation and inflammatory response appears to be closely related to the circadian moment of envenomation. Future studies should investigate the molecular mechanisms resulting in this circadian rhythmicity.

Keywords: circadian variation; inflammation; oxidative stress; scorpion envenomation

6.2. Synthesis and Characterization of μ-Conotoxins as Molecular Probes for the Detection of Voltage-Gated Sodium Channel Pore Blockers

Rómulo Aráoz [1,2,*], Giovanny Covaleda-Cortés [1], Denis Servent [1,2]

[1] Université Paris Saclay, CEA, INRAE, Département Médicaments et Technologies pour la Santé (DMTS), SIMoS, 91191 Gif-sur-Yvette, France.
[2] CNRS, ERL9004, 91191 Gif-sur-Yvette, France.
* Correspondence: romulo.araoz@cea.fr

Abstract: Voltage-gated sodium channels (Na_V) play critical functional roles by controlling, in excitable and non-excitable cells, the action potential initiation/propagation, cell motility, and proliferation. In humans, there are nine subtypes of Na_V channels—1.1 to 1.9—with Na_V1.4 being the muscle subtype. Na_V channel dysfunctions are associated with neurological, cardiovascular, muscular and psychiatric disorders. Marine phytoplankton and freshwater cyanobacteria are able to synthesize a series of neurotoxins targeting Na_V channels, among them, saxitoxin, a potent site-1 pore blocker classified as a chemical bioweapon because of its high lethality for humans. Receptor binding assays are suitable analytical techniques for ligand screening owing to their extremely high resolution and sensitivity, their fast analysis and easy automation, which facilitates the development of high-throughput protocols. There is, however, a lack of toxin-tracers for Na_V toxins detection. The marine cone snail of the genus *Conus* produces a series of conotoxins for prey hunting, among them, μ-conotoxin peptides, with an exquisite affinity for Na_V channels. Wild-type and biotinylated synthesis of five μ-conotoxins was performed using a Protein Technologies Prelude synthesizer. Biotinylation was performed on-column after Fmoc deprotection of the N-terminal residue of the μ-conotoxins. After refolding, the resulting conopeptides were purified by reverse-phase HPLC. The affinity of each μ-conotoxin for the Na_V1.4 channel was determined. Two biotinylated μ-conopeptides with high affinity for Na_V channels were identified as potential toxin tracers for the detection of site-1 Na_V pore blockers by ligand binding assay.

Keywords: cyanotoxin; ligand-binding assay; μ-conopeptide; phycoyoxin; sodium channel

Acknowledgments: This research was supported by the NRBC-E Program project MULTITOX (Fiche N° H35 to RA). The authors acknowledge the INTERREG Atlantic Area (ALERTOX-NET EAPA_317/2016 project to DS), and the LABEX LERMIT (DETECTNEUROTOX project, CDE 2017–001173—RD 91 to RA)

6.3. Multi-Scale Evaluation of Spider Toxins as Potential Anti-Nociceptive Agents: Example of Cyriotoxin-1a

Evelyne Benoit [1,*], Michel De Waard [2], Rémy Béroud [3], Michel Partiseti [4], Denis Servent [1]

[1] Service d'Ingénierie Moléculaire pour la Santé, ERL CNRS/CEA 9004, Gif-sur-Yvette, France.
[2] L'institut du thorax, INSERM UMR 1087/CNRS UMR 6291, Nantes, France.
[3] Smartox Biotechnology, Saint-Egrève, France.
[4] Sanofi R & D, Integrated Drug Discovery—High Content Biology, Vitry-sur-Seine, France.
* Correspondence: evelyne.benoit@cea.fr

Abstract: Our expertise is the identification and structural and functional characterization of original natural toxins targeting receptors and ion channels, which may have applications in human health. In this context, and in collaboration with *L'institut du thorax*, Smartox Biotechnology and Sanofi R & D, we are interested in the multi-scale evaluation (from the cell in vitro to the organism in vivo) of spider toxins as potential anti-nociceptive agents, as illustrated by the example of Cyriotoxin-1a (CyrTx-1a). $Na_V1.7$ channel subtype is highly expressed in the dorsal root ganglia (DRG) of the sensory nervous system and plays a central role in the pain signaling process. We investigated a library prepared from original venoms of 117 different animals to identify new selective inhibitors of this target. We used high-throughput screening of the venom library, using automated patch-clamp experiments on human voltage-gated sodium channel subtypes, and then in vitro and in vivo electrophysiological experiments to characterize the active peptides that were purified, sequenced and chemically synthesized. Analgesic effects were evaluated in mice in vivo. We identified and further characterized CyrTx-1a, a novel peptide isolated from *Cyriopagopus schioedtei* spider venom. This 33 amino acids toxin belongs to the inhibitor cystine knot structural family and inhibits $hNa_V1.1$–1.3 and 1.6–1.7 in the low nanomolar range, compared to the micromolar range for $hNa_V1.4$–1.5 and 1.8. CyrTx-1a was 920 times more efficient at inhibiting tetrodotoxin (TTX)-sensitive than TTX-resistant sodium currents recorded from adult mouse DRG neurons in vitro and approximately 170 times less efficient than huwentoxin-IV at altering mouse skeletal neuromuscular excitability in vivo. CyrTx-1a exhibited an analgesic effect in mice by significantly increasing reaction time in the hot-plate assay.

Keywords: anti-nociceptive toxin; mouse dorsal root ganglia neuron; mouse tactile and hot pain sensitivity; pain; voltage-gated sodium channel subtype

6.4. Imaging of Muscular Nicotinic Receptor Distribution Using a Synthetic Fluorescent Analogue of α-Bungarotoxin

<u>Oliver</u> **Brun** [1,5,6,*], **Claude Zoukimian** [3], **Barbara Ribeiro De Oliveira Mendes** [1], **Jérôme Montnach** [1], **Michel Ronjat** [1,2], **Rémy Béroud** [3], **Didier Boturyn** [4], **Frédéric Lesage** [5,7], **Michel De Waard** [1,2,3]

[1] L'institut du thorax, INSERM, CNRS, UNIV NANTES, Nantes, France.
[2] LabEx "Ion Channels, Science & Therapeutics", Valbonne, France.
[3] Smartox Biotechnology, 6 rue des Platanes, F-38120 Saint-Egrève, France.
[4] Department of Molecular Chemistry, Univ. Grenoble Alpes, CNRS, 570 rue de la chimie, CS 40700, Grenoble 38000, France.
[5] Research Center, Montreal Heart Institute, Montreal, QC, Canada.
[6] Faculty of Medicine, Université de Montréal, Montreal, QC, Canada.
[7] Department of Electrical Engineering, École Polytechnique de Montréal, Montréal, QC, Canada.
* Correspondence: oliver.brun@univ-nantes.fr

Abstract: Characterizing the distribution of drug receptors through optical imaging can be challenging, as the availability of suitable fluorescent probes is limited. Over history, natural toxins have benefited science through an ever-growing source of biologically active peptides, many of which show great potential as probes to study the distribution of their associated receptors. Here, we aim at developing a synthetic, fluorescent analog of α-bungarotoxin (α-BgTx), a highly potent, selective and irreversible inhibitor of the muscle-type nicotinic receptors (mnAChR), through the addition of a Cy5 tag. We first performed functional validation of this Cy5-α-BgTx analog by high-throughput patch-clamp (SyncroPatch364, Nanion), looking at the inhibition of acetylcholine-mediated currents on TE671 cells. It, therefore, appeared that the addition of the Cy5 tag still generates a peptide with nanomolar affinity. In order to further ensure that natural binding properties of the peptide were preserved, we synthesized a peptide analog of the α-BgTx binding site on the mnAChR (BSpep). Preincubation of Cy5-α-BgTx with BSpep resulted in the capture of the α-BgTx and complete blockage of its inhibitory effect. The fluorescent toxin was then used to study the distribution of mnAChR on fixed TE671 cells, allowing us to visualize both its intra and extracellular distribution. Thus, we were able to synthesize a functional, synthetic analog of α-BgTx that can be used to localize mnAChRs, highlighting the potential of toxin-derived biopeptide engineering as a tool for receptor imaging.

Keywords: α-bungarotoxin; peptide engineering; receptor imaging

6.5. Proteomic Analysis of the Predatory Venom of Conus striatus Reveals Population-Specific Glycosylation Patterns in Kappa A-Conotoxins

Claudia Murraciole-Bich, <u>Sébastien Dutertre</u> *
Institut des Biomolécules Max Mousseron (IBMM), Université de Montpellier, CNRS, ENSCM, Montpellier, France.
* Correspondence: sebastien.dutertre@umontpellier.fr

Abstract: Animal venoms are a rich source of pharmacological compounds with ecological and evolutionary significance, as well as with therapeutic and biotechnological potentials. Among the most promising venomous animals, cone snails produce potent neurotoxic venom to facilitate prey capture and defend against aggressors. The main objective of this study was to investigate and decipher the composition of the predatory venom injected by *Conus striatus*, one of the largest and most widely distributed piscivorous species. Standard LC-MS analysis provided an overview of native venom compounds, whereas LC-MS/MS followed by bioinformatic analysis identified many conotoxin sequences. Kappa A-conotoxins, which contain three disulfide bridges and complex glycosylation, are by far the dominant compounds in predatory venom. Remarkably, different and population-specific glycosylation patterns were detected in specimens from the Pacific and Indian oceans, providing a new level of intraspecific variations for the first time.

Keywords: conotoxin; *Conus striatus*; glycosylation; KappA-conotoxin

6.6. Why Is Androctonus australis Venom from Far more Studied Than A. mauritanicus One? Does the Answer Reside in the Venom Composition?

<u>**Lou Freuville**</u> **[1,*], Fernanda Gobbi Amorim [1], Alain Brans [2], Rudy Fourmy [3], Aude Violette [3], Loïc Quinton [1]**

[1] Laboratory of Mass Spectrometry, MolSys Research Unit, University of Liège, Liège, Belgium.
[2] Centre for Protein Engineering, University of Liège, B 4000 Liège, Belgium.
[3] Alphabiotoxine Laboratory sprl, Barberie 15, Montroeul-au-bois 7911 Belgium.
* Correspondence: lfreuville@uliege.be

Abstract: Scorpions are characterized by the great complexity and diversity of their venom, which has evolved over 400 million years into a well-elaborated library of powerful toxins mostly targeting, with high specificity, ion channels. Moreover, protein or non-protein compounds found in such venoms display antibacterial, antiviral or again anticancer activity, converting them from toxic molecules to therapeutic and benefic ones. Even if scorpions constitute a target of choice for many researchers, the high number of identified species (>2000) imposes that some of them remain little-known. It is even still the case for the Buthidae family, which constitutes the most human medically relevant species, due to the high concentration of potent ion-channel toxins in their venoms. As an example, despite their phylogenetical proximity, *Androctonus australis* is the heart of 365 publications (Pubmed), whereas *A. mauritanicus* is only considered in five of them. In this work, we aim at comparing the venom profiles of these scorpions at the protein level. A first macroscopic overview of the protein content of each venom was evaluated by 1D SDS-PAGE and showed no notable difference. During the second time, the crude venoms of the two scorpions were analyzed with bottom-up proteomics. Toxins were classically reduced to remove the disulfide bonds, alkylated to avoid their reformation and finally digested with trypsin. The LC-MS/MS spectra were realized with M-Class UPLC coupled to a Q-Exactive spectrometer. Data analysis was performed with the bioinformatics dedicated software "Peaks X+", with the use of an extracted database from Uniprot built with the keyword "Buthidae". As preliminary results, *A. mauritanicus* presents 163 Top proteins against 142 for *A. australis* with this database. They share 100 peptides and 63 proteins in common. A table with the comparison of the classification of proteins in each species is presented below (Table 1). In the light of the protein classification, some groups of non-toxins are unexplored in the scorpion venoms, such as the Kunitz-type family and Cysteine-rich venom proteins. Therefore, the proteomic approaches can still open perspectives for the study of these groups, which is relevant to highlight them for future explorations. Comparing a well-studied scorpion species with another less studied but phylogenetically close is interesting to highlight the evolution, diversity and complexity of venom.

Table 1. Comparison of the classification of proteins between *A. australis* and *A. mauritanicus*.

	A. australis (%)	*A. mauritanicus* (%)
Toxins	54.9	72.4
Non-toxins	16.9	12.3
Cellular components	26.1	13.5
Proteins not assigned	2.1	1.8

Keywords: *Androctonus mauritanicus*; LC-MS/MS; scorpion venom

6.7. New Proteomic Approach for Inventorying the Snake Venom Arsenal

Fernanda Gobbi Amorim [1,*], **Damien Redureau** [1], **Nicholas Casewell** [2], **Loïc Quinton** [1]

[1] Laboratory of Mass Spectrometry, MolSys Research Unit, University of Liège, Liège, Belgium.
[2] Centre for Snakebite Research and Interventions, Liverpool School of Tropical Medicine, Pembroke Place, Liverpool, UK.
* Correspondence: fgamorim@uliege.be

Abstract: Multi-Enzymatic Limited Digestion (MELD) is a new methodology that applies synergic and time-limited digestion of multiple enzymes, representing a versatile yet straightforward approach for a new generation of proteomics methodology (MORSA et al., 2019). However, until now, this approach has not been applied to venomic studies, which may be useful to define the venom composition. The generation of a higher number of peptides per protein during the MELD digestion increases the quality of protein sequencing and, consequently, identification. Herein we applied the MELD strategy for the venomics of two snake species: *Echis ocellatus* (EoV) and *Dendroaspis polylepis* (DpV). Ten micrograms of each venom were reduced/alkylated followed by two different digestion protocols: (1) trypsin and (2) MELD (Trypsin, GluC and Chymotrypsin). The digested materials were analyzed in a Q-Exactive™ Plus Mass Spectrometer with protein identification performed by Peaks Studio X+ using the Uniprot and transcriptomic databases. MELD showed more peptides/proteins identified for both venoms compared to the trypsin protocol. In EoV, 82.8% were identified as toxins and 17.2% as non-toxins, compared to the trypsin protocol, which resulted in 69.2% toxins, 24% non-toxins and 6.9% for cellular components (CC). In DpV, MELD showed coverage of 26.2% for toxins, 39.9% for non-toxins and 33.9% for CC, while for trypsin, we obtained 23.3% for toxins, 37.4% for non-toxins and 39.3%

for CC. MELD was able to identify new components in both venoms. Fifty-one percent of the EoV were metalloproteinases (MP), while DpV showed a high content of nerve growth factor (22%), which was not identified using the transcriptome database. The highest number of mass spectra was obtained for a metalloproteinase (tr | Q2UXQ4) for *E. ocellatus*, in which the MELD approach led to four times more mass spectra. For *D. polylepis*, dendrotoxin I (P00979) showed the highest number of mass spectra, and the trypsin-based approach yields two times more mass spectra. As the number of mass spectra gives a rough evaluation of the toxin concentration, the two cited toxins are among the most produced in the venoms. MELD presented a different coverage according to the presence of high molecular mass content in the venom arsenal. This strategy can be applied to identify new groups of venom components. It represents an innovative strategy for venomics, opening new perspectives for sequencing and inventorying the venom arsenal.

Keywords: proteomics; snake; toxin; venom

6.8. Nano-Rattlers against Cancer: Snake Venom-Loaded Nanoparticles against Tumoral Cell Lines

Jorge Jimenez-Canale [1,*], Nayelli-Guadalupe Teran-Saavedra [1], Enrique-Fernando Velazquez-Contreras [1], Martin-Samuel Hernandez-Zazueta [2], Hector-Manuel Sarabia-Sainz [3], Daniel Fernandez-Quiroz [4], Alexel-Jesus Burgara-Estrella [5], Jose-Andrei Sarabia-Sainz [5]

[1] Departamento de Investigacion en Polimeros y Materiales, Universidad de Sonora, blvd. Luis Encinas s/n, CP 83000, Hermosillo, Sonora, Mexico.
[2] Departamento de Investigacion y Posgrado en Alimentos, Universidad de Sonora, blvd. Luis Encinas s/n, CP 83000, Hermosillo, Sonora, Mexico.
[3] Departamento de Ciencias del Deporte y de la Actividad Fisica, Universidad de Sonora, blvd. Luis Encinas s/n, CP 83000, Hermosillo, Sonora, Mexico.
[4] Departamento de Ingenieria Quimica y Metalurgia, Universidad de Sonora, blvd. Luis Encinas s/n, CP 83000, Hermosillo, Sonora, Mexico.
[5] Departamento Investigacion en Fisica, Universidad de Sonora, blvd. Luis Encinas s/n, CP 83000, Hermosillo, Sonora, Mexico.
* Correspondence: jorgejimzc@gmail.com

Abstract: Research in drug delivery systems has led to the development and improvement of materials that affect the pharmaceutical effect of bioactive components, broadening the options of treatment for several diseases, including cancer. Chitosan (Cs) has been firmly established as a biocompatible and biodegradable low-toxic polymer able to form complexes with bioactive agents, making them promising drug delivery vehicles. Additionally, some snake venom toxins such as phospholipases A2 (PLA2s), serine proteinases (SVSPs) and metalloproteinases (SVMPs) were reported to present cytotoxic activity in different tumor cell lines, making them an auspicious option to be used as cancer pharmaceuticals. In the present study, we identified the major proteins in the venom of a northern black-tailed rattlesnake (*Crotalus molossus molossus*) and cytotoxic activity against T-47D breast carcinoma cells were evaluated. Afterward, the venom was loaded into a Cs NPs system through the ionotropic gelation process with tripolyphosphate (TPP), obtaining particles of 415.9 ± 21 nm and zeta potential of $+28.3 \pm 1.1$ mV. The venom-loaded Cs-ALG complex was able to deliver the venom into the breast carcinoma cells through endocytosis, inhibiting their viability and inducing morphological changes. Although more studies are required, we suggest the potential use of *C. m. molossus* venom toxins entrapped within polymer nanoparticles for the future development and research of tumor pharmaceuticals.

Keywords: chitosan; nanomedicine; nanoparticle; snake venom

6.9. Adenosine Amplifies Granulopoiesis and Systemic Inflammatory Response after Scorpion Envenomation

Asma Kaddache *, Louiza Bechohra, Fatima Laraba-Djebari, Djelila Hammoudi-Triki

Department of Cellular and Molecular Biology, University of Science and Technology Houari Boumediene (USTHB), BP 32 El-Alia, Bab Ezzouar, Algiers, Algeria.
* Correspondence: akaddache@usthb.dz

Abstract: Scorpion envenomation (SE) triggers granulopoiesis, resulting in heightened production of neutrophils that, in turn, exacerbate tissue damage. However, the mechanisms that sustain their production and recruitment to the injured tissue are unclear. Experimental evidence suggests that alarmines, inter alia adenosine, can modulate proliferation and differentiation of hematopoietic progenitors as well as mature neutrophils mobilization. Indeed, adenosine accumulates in the extracellular space in response to cell damage and can modulate immune cells through its transmembrane receptors A1, A2A, A2B, and A3. However, despite increasing data incriminating adenosine in many models of inflammation, little is known about its effect in the inflammatory response caused by the venom of *Androctonus australis hector* (Aah). **Aim**—In this study, we focused on deciphering molecular signals that promote neutrophil production and recruitment after SE by targeting adenosine as a potential candidate. **Methods**—Using a mouse model of envenomation and a pharmacological strategy, we prevented adenosine signaling by 1,3,7-triméthylxanthine, a nonselective inhibitor of adenosine A1 and A2A receptors. We first characterized the effect of this treatment on granulopoiesis in the bone marrow and then evaluated the expanse of inflammation to various tissues, mainly of the cardiorespiratory system. To this end, medullar MPO activity was assessed as a marker of progenitor cells proliferation and differentiation. In parallel, differential count of leucocytes and histopathological analysis of heart and lungs were carried out. **Results**—Inhibition of the action of adenosine helped to attenuate venom-induced granulopoiesis and reduced the accumulation of neutrophils in the blood. The damages caused by neutrophils in the lungs and heart were improved to a large extent. **Conclusion**—These results highlight the involvement of adenosine in the inflammatory response induced after SE via its A1 and A2A receptors and could help identify a new target to modulate inflammation and prevent cardiorespiratory failure in envenomed patients.

Keywords: adenosine; granulopoiesis; scorpion envenomation

6.10. Comparative Study of Two Thrombin-Like Molecules from Vipera lebetina Venom

Amel <u>Kadi-Saci</u> *, Fatima Laraba-Djebari *

University of Science and Technology Houari Boumediene (USTHB), Faculty of Biological Sciences, Laboratory of Cellular and Molecular Biology, BP 32 El-Alia, Bab Ezzouar, Algiers, Algeria. * Correspondence: amel_saci@yahoo.fr; flaraba@hotmail.com; flaraba@usthb.dz

Abstract: Snake venoms contain a wide range of components that are able to initiate or inhibit several steps of coagulation or platelet aggregation. Two thrombin-like enzymes, VLCV and VLCII (45 and 60 kDa, respectively) with fibrinogenolytic and pro-aggregating activities, are purified from *Vipera lebetina* venom. These molecules of interest are involved in platelet adhesion to fibrinogen by activating signaling pathways via αIIbβ3 integrins and thus binding to fibrinogen. Results also showed that both molecules present a high pro-aggregating effect with a similar mechanism of thrombin. The incubation of heparin with VLCV or VLCII has no inhibition on induced platelet aggregation by these molecules, suggesting that both enzymes have similar functional characteristics. The use of PMSF also has no effect on the activity of both molecules, indicating that the induced platelet aggregation by VLCV or VLCII involves an independent mechanism of catalytic activity. Results also indicate that both molecules VLCII and VLCV present a mechanism of action involving the thrombin and/or ADP pathways via PAR1/PAR4/P2Y12 receptors, as well as the αIIbβ3 integrin, leading to adhesion and platelets activation. Both thrombin-like VLCII and VLCV could be used as potential targets for the development of new drugs.

Keywords: biomolecule; coagulation; pro-aggregating; snake venom; thrombin-like

6.11. A comparative Study of the Histological Changes Induced by the Venoms of the Moroccan Snakes Cerastes cerastes and Naja haje

Soukaina Khourcha [1,2,*], Salma Chakir [1,2], Ayoub Lafnoune [1], Bouchra Darkaoui [1], Khadija Daoudi [1], Fatima Chgoury [1], Abdelaziz Hmyene [2], Naoual Oukkache [1]

[1] Laboratory of Venoms and Toxins, Institut Pasteur of Morocco, 1 place Louis Pasteur, Casablanca 20360, Morocco.
[2] Laboratory of Biochemistry, Environment and Food Technology, Faculty of Sciences and Technologies, BP 146, Mohammedia 20650, Morocco.
* Correspondence: khourcha.soukaina9@gmail.com

Abstract: In Morocco, ophidian envenomations are perpetrated by eight species of snakes belonging to the families of vipers and elapids, including *Cerastes cerastes* (Cc) and *Naja haje* (Nh), respectively. The present work offers a comparative study between the two venoms from a toxicological and physiopathological standpoint. We first determined the toxicity of the Cc and Nh venoms by an LD50 test and then carried out an anatomopathological study on Swiss mice in order to detect the signs of envenomation by each venom. The organs of the envenomed mice were then removed for a histological study to determine the main systemic alterations induced by the venoms. The results of our comparative study showed large disparities between the two venoms in terms of toxicity, revealing that the Nh venom is more toxic than the Cc venom. This disparity is also evident in the MHD test and the histopathological study that shows that the Nh venom exerts a cytotoxic activity on the brain, heart, lungs, liver and kidneys, while that of Cc leads to the formation of hemorrhagic foci and lesions on the liver, kidneys and heart.

Keywords: characterization; histology; physiopathology; snake; toxicity; venom

6.12. Modulation of Induced Neuro–Immuno–Inflammatory Response by K⁺ Neurotoxin: Involvement of Serotonin and Histamine Receptors

Amina Ladjel-Mendil [1,*], Marie-France Martin-Eauclaire [2], Fatima Laraba-Djebari [1]

[1] University of Science and Technology Houari Boumediene (USTHB), Faculty of Biological Sciences, Laboratory of Cellular and Molecular Biology, BP 32, El–Alia Bab Ezzouar, 16111 Algiers, Algeria.
[2] Aix-Marseille University, CRN2M CNRS UMR 7290, IFR Jean-Roche, Faculté de Médecine Secteur Nord, boulevard Pierre Dramard, 13916 Marseille cedex 20, France.
* Correspondence: mendilamina@yahoo.fr

Abstract: The neuro–immuno–inflammatory response triggered by neurotoxins is a key event in the pathogenesis of scorpion envenomation. In the present study, this response was evaluated in cardiac, pulmonary and brain tissues of intoxicated mice with kaliotoxin; a neurotoxin derived from *Androctonus australis hector* scorpion venom. The involvement of serotoninergic and histaminergic pathways in the systemic inflammatory response following kaliotoxin administration was also investigated. Obtained results revealed that the injected kaliotoxin by intracerebroventricular route induces an important immune-inflammatory response in brain, cardiac and pulmonary tissues. This response is mainly characterized by local features such as edema formation, inflammatory cell infiltration and imbalanced redox status. These effects are correlated with severe tissue alterations and a concomitant increase in metabolic enzymes in sera. Pretreatment of mice with antagonists of serotonin (5HT) and histamine (H1) receptors markedly attenuated these alterations in all the studied tissues. Serotonin and histamine-specific receptors seem to be the main pharmacological targets involved in the neural and systemic inflammatory processes. These findings could help to understand better the role of serotonin and histamine in scorpion venom-induced inflammatory response and pave the way to new therapies targeting 5HT and H1 receptors in order to attenuate the induced neuro–immuno–inflammatory disturbances that may be encountered in severe grades of scorpion envenoming.

Keywords: 5HT and H1 receptors; immunoinflammatory response; kaliotoxin; oxidative stress; tissue injury

6.13. High-Throughput Screening of Spider Venoms for Identification of Active Compound in Voltage-Gated Sodium Channel

Ludivine Lopez [1,2,*], Jérôme Montnach [1], Sébastien Nicolas [1], Lucie Jaquillard [2], Rémy Béroud [2], Michel De Waard [1]

[1] L'institut du thorax, INSERM UMR 1087/CNRS UMR 6291, LabEx "Ion Channels, Science & Therapeutics", F-44007 Nantes, France.
[2] Smartox Biotechnology, 6 rue des Platanes, F38120 Saint Egrève, France.
* Correspondence: ludivine.lopez@univ-nantes.fr

Abstract: Dysfunctions of voltage-gated sodium channels (Nav) have been associated with many pathologies such as cardiac diseases, neuropathic pain and epilepsy. In order to study the role of these channels in diseases or to restore their function, selective molecules targeting ion channels are needed. Only a few molecules selective to a single sodium channel isoform have been discovered. This makes these channel types important targets in drug discovery. Animal venoms, and especially spider venoms, contain various peptides that target ion channels. By screening spider venoms on Nav isoforms, we aimed to identify new peptidic toxins targeting specifically one of them. We performed a screening of about thirty spider venoms on Nav1.3, Nav1.4, Nav1.5 and Nav1.6 isoforms using the automated patch-clamp (APC) technique (SyncroPatch364, Nanion). All venoms were preliminarily fractionated in 64 fractions and tested on each Nav. Fractions of interest are those that reduce sodium peak current (by at least 30%), slow-down inactivation or increase late sodium current. False-positive fractions were excluded based on the detection of material in HPLC or mass spectrometry. The primary screening identified 220 fractions active on at least one isoform. Some of these fractions were then selected for purification and tested again on the automatic patch-clamp (74 for Nav1.3, 42 for Nav1.4, 28 for Nav1.5 and 23 for Nav1.6). The most interesting peptides were sequenced, synthesized and characterized. This study suggests that among a large number of toxins present in venoms, close to 50% of them target sodium channels with specificity for each sodium channel isoform.

Keywords: 5HT screening; automated patch-clamp; spider toxin

6.14. Gambierol Action on K⁺ Currents and Catecholamine Release in Cultured Single Chromaffin Cells from Fetal Rat Adrenal Medulla

Jordi Molgó [1,2,*], Roland Bournaud [2], Sébastien Schlumberger [2], Makoto Sasaki [3], Haruhiko Fuwa [4], Denis Servent [1], Evelyne Benoit [1,2]

[1] Université Paris-Saclay, CEA, INRAE, Institut des sciences du vivant Frédéric Joliot, Département Médicaments et Technologies pour la Santé (DMTS), Service d'Ingénierie Moléculaire pour la Santé (SIMoS), ERL CNRS 9004, F-91191 Gif-sur-Yvette, France.
[2] CNRS, Laboratoire de Neurobiologie Cellulaire et Moléculaire-UPR 9040, F-91198 Gif-sur-Yvette, France.
[3] Tohoku University, Graduate School of Life Sciences, Sendai, Japan.
[4] Chuo University, Faculty of Science and Engineering, Department of Applied Chemistry, 1-13-27 Kasuga, Bunkyo, Tokyo 112-8551, Japan.
* Correspondence: jordi.molgo@cea.fr

Abstract: Gambierol, characterized by a transfused octacyclic polyether core, was first isolated and chemically typified from cultured *Gambierdiscus toxicus* dinoflagellates collected in French Polynesia. Subsequently, distinct groups in Japan and the USA, using various strategies, achieved their total chemical synthesis. This allowed detailed studies on its mode of action. Gambierol inhibits voltage-gated K⁺ (K$_V$) channels in various excitable and non-excitable cells, as well as in motor nerve terminals of the skeletal neuromuscular junction. In the present study, we investigated first the effects of nanomolar concentrations of gambierol on K^+ current of cultured chromaffin cells from fetal rat adrenal medulla using perforated patch-clamp current recordings. Our results show that gambierol only blocked a small component of the total K^+ current and affected neither calcium-activated K^+ (K_{Ca}) nor ATP-sensitive K^+ (K_{ATP}) channels, as revealed using apamin and iberiotoxin (selective K_{Ca} channel blockers) and glibenclamid (K_{ATP} channel blocker). After inhibiting K_{ATP} and K_{Ca} channel activation, the gambierol concentration blocking 50% of the K^+ current component (IC_{50}) was 7.6 ± 1.1 nM. The repolarizing phase of all-or-none elicited action potentials, recorded under current-clamp conditions (triggered by 1-ms depolarizing pulses), was sensitive to the action of gambierol but insensitive to the action of apamin and iberiotoxin, indicating that K_{Ca} channels do not participate in the modulation of action potential duration triggered by short depolarizing pulses. The use of simultaneous patch-clamp and single-cell amperometry allowed controlling the membrane potential and detecting exocytosis events (with a carbon electrode polarized to $+650$ mV to allow the oxidation of released catecholamines). Such recordings revealed that gambierol did not modify the membrane potential following 14-s depolarizing current steps and did not significantly increase the number of exocytotic catecholamine release events with respect to controls. The addition of K_{Ca} channel blockers (in the continuous presence of gambierol) enhanced the membrane depolarization by about 15 mV (during the 14 s current step), and at the same time, increased significantly the number of exocytotic events related to catecholamine secretion. Such enhanced depolarization induced by the K_{Ca} channel blockers probably brings the membrane potential above the activation threshold of high-voltage activated Ca_V channels, triggering both Ca^{2+} influx and subsequent catecholamine secretion. These results emphasize the diversity of K_V channels in chromaffin cells from fetal rat adrenal medulla and highlight the modulatory role played by K_{Ca} channels in the control of exocytosis in the absence of splanchnic innervation.

Keywords: catecholamine release; cultured chromaffin cell; fetal rat adrenal medulla; gambierol; potassium current

6.15. Sprouting and Convergent and Stable Polyinnervation Characterize Human Orbicularis oculi Muscles Treated for Blepharospasm with Repeated Botulinum Type A Neurotoxin Injections

Brigitte Girard [1], Aurélie Couesnon [2], Émmanuelle Girard [3], Jordi Molgó [2,4,*

[1] Service d'Ophtalmologie—Hôpital Tenon, 4 rue de la Chine 75020 Paris, Sorbonne Université, UPMC, France.
[2] Institut des Neurosciences Paris-Saclay, UMR 9197, CNRS/Université Paris-Sud, 91198 Gif-sur-Yvette cedex, France.
[3] Institut NeuroMyoGene, CNRS UMR 5310, INSERM U1217, Université Lyon1, 8 avenue Rockefeller, 69008 Lyon, France.
[4] Université Paris-Saclay, CEA, INRAE, Institut des sciences du vivant Frédéric Joliot, Département Médicaments et Technologies pour la Santé, Service d'Ingénierie Moléculaire pour la Santé, ERL CNRS 9004, F-91191 Gif-sur-Yvette, France.
* Correspondence: jordi.molgo@cea.fr

Abstract: Currently, a number of pathological conditions, including movement disorders, are treated with botulinum type-A neurotoxin (BoNT/A). This neurotoxin is known to block quantal acetylcholine (ACh) release from motor nerve terminals of the skeletal neuromuscular junction (NMJ), causing temporary muscle paralysis. The treatment with BoNT/A has shown its clinical efficacy in the management of benign blepharospasm, a form of focal dystonia. Due to the slow functional recovery of neuromuscular transmission after BoNT/A-treatment, patients with blepharospasm receive recurrent (every 2–3 months) intramuscular toxin injections in the affected muscles. Regardless of this limitation, the toxin securely relieves patients from their dystonia symptoms and considerably enriches their quality of life. Nevertheless, in less than 4% of the blepharospasm patients treated at Tenon Hospital, BoNT/A-treatment was no longer effective in relieving the patient's symptoms, and it was compulsory to perform the upper myectomy of the *Orbicularis* *oculi* muscle. In this study, we used surgical waste muscle specimens from 14 patients treated with repeated injections of either Dysport®, abobotulinumtoxinA Ipsen or Xeomin® incobotulinumtoxinA, Merz. These muscle specimens were compared to others, obtained in normal subjects (naïve of BoNT/A) during blepharoplasty. The morphological study was performed blinded to the BoNT/A sample treatment. Neuromuscular specimens analyzed by confocal laser scanning microscopy, using fluorescent staining and immune-labeling of presynaptic proteins, revealed that the pattern of innervation, the muscle nicotinic ACh receptors (nAChR) and the NMJs exhibited marked differences in BoNT/A-treated muscles (regardless of the toxin clinically used), with respect to controls. The control muscles were constantly innervated by a single motor axon (mono-innervated), and NMJs were relatively simple with the typical nAChR array. In contrast, most of the BoNT/A-treated muscles exhibited profuse and stable poly-neuronal innervation, with two unambiguous features: one in which multiple axons innervated a single muscle fiber, the other in which distinct motor axons converged to a unique endplate (convergent innervation). The ability to increase the proportion of poly-innervated muscle fibers may be related to either the stimulation of nerve sprouting (due to muscle inactivity caused by BoNT/A) or the absence of synapse elimination. During development, the structures composing the NMJ undergo rapid formation and elimination. In rodents, unessential synapses are eliminated. We previously reported synapse elimination in mature rodent NMJs following a single BoNT/A injection. The new findings reported here raise a number of questions about the origin and factors contributing to the plasticity changes observed and the expected detrimental functioning of NMJs and muscle fibers.

Keywords: botulinum neurotoxin type A; innervation pattern; sprouting

6.16. Synthetic Analogues of Huwentoxin-IV Spider Peptide with Altered Human Na$_V$1.7/Na$_V$1.6 Selectivity Ratios

Ludivine Lopez [1], **Jérôme** <u>**Montnach**</u> [1,*], **Barbara Oliveira-Mendes** [1], **Baptiste Thomas** [2], **Cécile Caumes** [2], **Sébastien Nicolas** [1], **Denis Servent** [3], **Charles Cohen** [4], **Rémy Béroud** [2], **Evelyne Benoit** [3], **Michel De Waard** [1,2,5]

[1] L'institut du thorax, INSERM, CNRS, UNIV NANTES, Nantes, France.
[2] Smartox Biotechnology, Saint-Egrève, France.
[3] Université Paris Saclay, CEA, Institut des sciences du vivant Frédéric Joliot, Département Médicaments et Technologies pour la Santé (DMTS), Service d'Ingénierie Moléculaire pour la Santé (SIMoS), ERL CNRS/CEA 9004, Gif-sur-Yvette, France.
[4] Xenon Pharmaceuticals, Burnaby, British Columbia, Canad. [5] LabEx "Ion Channels, Science and Therapeutics", Valbonne, France.
* Correspondence: jerome.montnach@univ-nantes.fr

Abstract: Huwentoxin-IV (HwTx-IV), a peptide discovered in the venom of the Chinese bird spider *Cyriopagopus schmidti*, has been reported to be a potent antinociceptive compound due to its action on the genetically validated Na$_V$1.7 pain target. Using this peptide for antinociceptive applications in vivo suffers from one major drawback, namely its negative impact on the neuromuscular system. Although it was studied only recently, this effect appears to be due to an interaction between the peptide and the Na$_V$1.6 channel subtype located at the presynaptic level. The aim of this work was to investigate how HwTx-IV could be modified in order to alter the original human (h) Na$_V$1.7/Na$_V$1.6 selectivity ratio of 23. Nineteen HwTx-IV analogs were chemically synthesized and tested for their blocking effects on the Na$^+$ currents flowing through these two channel subtypes stably expressed in cell lines. The dose–response curves for these analogs were generated, thanks to the use of an automated patch-clamp system. Several key amino acid positions were targeted owing to the information provided by earlier structure–activity relationship (SAR) studies. Among the analogs tested, the potency of HwTx-IV E^{4}K was significantly improved for hNa$_V$1.6, leading to a decreased hNa$_V$1.7/hNa$_V$1.6 selectivity ratio (close to 1). Similar decreased selectivity ratios, but with increased potency for both subtypes, were observed for HwTx-IV analogs that combine a substitution at position 4 with a modification of amino acid 1 or 26 (HwTx-IV E^{1}G/E^{4}G and HwTx-IV E^{4}K/R^{26}Q). In contrast, increased selectivity ratios (>46) were obtained if the E^{4}K mutation was combined to an additional double substitution (R^{26}A/Y^{33}W) or simply by further substituting the C-terminal amidation of the peptide by a carboxylated motif, linked to a marked loss of potency on hNa$_V$1.6 in this latter case. These results demonstrate that it is possible to significantly modulate the selectivity ratio for these two channel subtypes in order to improve the potency of a given analog for hNa$_V$1.6 and/or hNa$_V$1.7 subtypes. In addition, selective analogs for hNa$_V$1.7, possessing better safety profiles, were produced to limit neuromuscular impairments.

Keywords: huwentoxin-IV; selectivity; sodium channel

6.17. Characterization of a Highly Neutralizing Single Monoclonal Antibody to Botulinum Neurotoxin Type A

Christine Rasetti-Escargueil [2,*], Sébastien Brier [1], Anne Wijkhuisen [3], Stéphanie Simon [3], Maud Marechal [2], Émmanuel Lemichez [2], Michel R. Popoff [2]

[1] Biological NMR Technological Plateform, Institut Pasteur, CNRS UMR 3528, 75015 Paris, France.
[2] Institut Pasteur, Toxines Bactériennes, CNRS ERL6002, 75015 Paris, France.
[3] Université Paris-Saclay, CEA, INRAE, Département Médicaments et Technologies pour la Santé, 91191 Gif-sur-Yvette, France.
* Correspondence: christine.rasetti-escargueil@pasteur.fr

Abstract: Monoclonal antibodies (mAbs) represent an alternative and safer way to treat botulism, a fatal flaccid paralysis due to botulinum neurotoxins (BoNTs). In addition, mAbs offer the advantage of being produced in a reproducible manner. We previously identified a unique and potent mouse mAb (TA12) targeting BoNT/A1 with high affinity and neutralizing activity. **Methods**—We characterized the molecular basis of TA12 neutralization by combining Hydrogen/Deuterium eXchange Mass Spectrometry (HDX-MS) with site-directed mutagenesis and neutralization studies. **Results**—The TA12 recognized a conformational epitope located at the interface between the HCN and HCC subdomains of the BoNT/A1 receptor-binding domain (HC). This interface shares common structural features with the ciA-C2 VHH epitope and lies on the face opposite recognized by ciA-C2 and the CR1/CR2 neutralizing mAbs. The single substitution of N1006 was sufficient to affect TA12 binding to HC, confirming the position of the epitope. We further uncovered that the TA12 epitope overlaps with the BoNT/A1 binding site for both the neuronal receptor synaptic vesicle glycoprotein 2 isoform C (SV2C) and the GT1b ganglioside. In vivo neutralization studies confirmed that TA12 is highly neutralizing the most potent subtype BoNT/A1 with medium neutralizing activity against BoNT/A2 and A3 subtypes and low neutralizing activity against BoNT/A5. **Conclusion**—Hence, TA12 potently blocks the entry of BoNT/A1 into neurons by interfering simultaneously with the binding of SV2C and, to a lower extent, GT1b. Our study reveals the unique neutralization mechanism of TA12 and emphasizes the benefit of single mAbs to treat botulism type A.

Keywords: botulinum neurotoxin; epitope; monoclonal antibody

6.18. Steatoda nobilis: A Comparative Study including Full-Body MALDI Imaging and Application to the Venomics Field

Damien Redureau [1], Mathieu Tiquet [1], John P. Dunbar [2], Fernanda Gobbi Amorim [1], Michel M. Dugon [2], Loïc Quinton [1,*]

[1] Mass Spectrometry Laboratory, MolSys RU, University of Liège, 4000 Liège, Belgium.
[2] Venom Systems & Proteomics Lab, School of Natural Sciences, Ryan Institute, National University of Ireland Galway, H91 TK33 Galway, Ireland.
* Correspondence: loic.quinton@uliege.be

Abstract: The Noble false widow spider *Steatoda nobilis* is a member of the Theridiidae family, akin to the "true" Black widows of the genus Latrodectus. *S. nobilis* is rapidly expanding its geographic range throughout Europe and parts of the Americas, particularly in and around human dwellings. *S. nobilis* has also been shown to be of medical significance in the UK and in Ireland, where a growing number of severe cases of envenomation have occurred over the past five years [Dunbar, 2021]. The comparison of venom composition of male and female *S. nobilis*, using a proteo-transcriptomic approach, demonstrates that male venom contained a lower quantity and diversity of latrotoxins. In order to illustrate the comparison of both male and female *S. nobilis* profiles, we additionally present the first whole-body imaging of a spider using MALDI mass spectrometry. This proof-of-concept allows one to compare the anatomy of females and males based on molecular markers (specific m/z distribution on spider slices). This is the first time this approach has been performed on a spider and could allow identifying regionalization on toxin production in venom glands when applied to other venomous animal groups. In the next year, this technology could also drive the identification of the spatial activity of toxin in tissue slice after envenomation by comparing the distribution of toxin in the tissue and the spatial distribution of substrate of some toxin/enzyme or again markers of inflammatory response and cellular death.

Keywords: MALDI imaging; *Steatoda nobilis*; venomics

6.19. Characterization of Innovative Animal Toxins for the Functional Study of Angiotensin Receptors

Anne-Cécile Van Baelen [1,*], **Denis Servent** [1], **Nicolas Gilles** [1], **Philippe Robin** [1]

Université Paris-Saclay, CEA, Département Médicaments et Technologies pour la Santé (DMTS), SIMoS, F-91191 Gif-sur-Yvette, France.

* Correspondence: anne-cecile.vanbaelen@cea.fr

Abstract: An overactivation of the renin–angiotensin system (RAS) with an increase in angiotensin II level is the main cause of arterial hypertension. AT1 receptor antagonists are the goal standard for the treatment of this pathology. Nevertheless, other RAS-related GPCRs (AT2, Mas and MrgD) are also involved in the regulation of blood pressure but are under-exploited due to the lack of knowledge of their physiological properties. Furthermore, the development of biased ligand activating only certain intracellular signaling pathways is in full swing. The discovery of new ligands with original functions would be of great interest to better understand these receptors. Toxins from venomous animals are reticulated peptides that display antagonist, agonist or allosteric functions on GPCRs, and because of their high target specificity and selectivity, they constitute a large reservoir of unexploited promising ligands. The screening of an 800 toxin library was performed on the AT2 receptor by the mean of a radioligand binding assay, and it highlighted three original toxins which originate from different organisms with different structural profiles. These toxins display a micromolar affinity on AT2 and AT1 receptors, and so far, pharmacological trials are more advanced on this receptor. A-CTX-cMila is a small cyclic toxin of seven residues from the cone snail *Conus miliaris*, demonstrating a competitive antagonism on AT1 receptor with micromolar affinity, blocking both G-protein and β-arrestin pathways. A-CTX-MilVa, from the same cone snail, is composed of 20 residues and seems to be an AT1 agonist activating Gq protein pathway. Finally, A-TRTX-Por1a from the spider *Poecilotheria ornata*, is a three disulfide bridges peptide of 29 residues, which also appears to display an agonist activity on the AT1 receptor. These preliminary results describe the first animal toxins active on the RAS and highlight animal venoms as a prolific source of angiotensin receptor binders. A-CTX-cMila, A-CTX-MilVa and A-TRTX-Por1a, with their interesting structural and pharmacological profiles, may represent promising tools for the study of angiotensin receptors. More investigations are needed to characterize these toxins fully, especially on AT2. This work highlights a new family of GPCRs targeted by toxins.

Keywords: angiotensin; GPCR; toxin

6.20. Mast Cells Modulate the Immune Response and Redox Status of Intestinal Tissue in Induced Venom Pathogenesis

Nehla Zerarka-Chabane *, Fatima Laraba-Djebari, Djelila Hammoudi-Triki

University of Science and Technology Houari Boumediene (USTHB), Faculty of Biological Sciences, Laboratory of Cellular and Molecular Biology, Bab-Ezzouar, 16111 Algiers, Algeria.
* Correspondence: chabane.nehla88@gmail.com

Abstract: The inflammatory response caused by scorpion venoms components is a key event in the pathogenesis of scorpion envenomation. The binding of neurotoxins on their targets leads to a massive release of inflammatory mediators, such as cytokines, eicosanoids and bioactive amines from activated immune cells. Moreover, mast cells undergo maturation and polarization in gut mucosal surfaces and can alter intestinal permeability, an important factor in many inflammatory mucosal disorders. The c-Kit ligand, stem cell factor (SCF), is a major regulator of mast cell migration, development, growth, survival and activation. This study was conducted to investigate the effect of mast cells in modulating intestinal inflammation induced by scorpion venom. NMRI mice were injected by a sublethal dose of *Androctonus australis hector* (Aah) venom (0.5 mg/kg, subcutaneously) and sacrificed 24 h after envenomation; blood and organs were then collected for further analysis. The assessment of inflammatory response and the evaluation of redox status biomarkers were performed in the small intestine and colonic homogenates. Hematoxylin–eosin staining was used for general histological assessment, and toluidine blue staining was performed for identification and quantification of mast cells. The results revealed that Aah venom induced a significant increase in inflammatory cell infiltration, nitrite oxide levels and malondialdehyde (MDA) concomitantly with a significant decrease in catalase and glutathione S-transferase activities companied by intestinal tissue alteration. Furthermore, the percentage of mast cells was increased in the peritoneal cavity. Pharmacological inhibition of tyrosine kinase receptors alleviated these alterations. These findings could help to better understand the mechanisms involved in scorpion envenoming syndrome and develop potential drugs targeting mast cells for the management of inflammatory disorders of the gut.

Keywords: mast cell; redox status; scorpion venom

Author Contributions: All authors equally contributed to the conceptualization, the writing—original draft preparation and the writing—review and editing. All authors have read and agreed to the published version of the manuscript.

Funding: Not applicable.

Institutional Review Board Statement: Not applicable.

Informed Consent Statement: Not applicable.

Data Availability Statement: Not applicable.

Review

Regulatory Networks Controlling Neurotoxin Synthesis in *Clostridium botulinum* and *Clostridium tetani*

Michel R. Popoff [1,*] **and Holger Brüggemann** [2]

[1]   Bacterial Toxins, Institut Pasteur, 75724 Paris, France
[2]   Department of Biomedicine, Aarhus University, 8000 Aarhus, Denmark; brueggemann@biomed.au.dk
*   Correspondence: popoff2m@gmail.com

Citation: Popoff, M.R.; Brüggemann, H. Regulatory Networks Controlling Neurotoxin Synthesis in *Clostridium botulinum* and *Clostridium tetani*. *Toxins* **2022**, *14*, 364. https://doi.org/10.3390/toxins14060364

Received: 14 April 2022
Accepted: 21 May 2022
Published: 24 May 2022

Publisher's Note: MDPI stays neutral with regard to jurisdictional claims in published maps and institutional affiliations.

Abstract: *Clostridium botulinum* and *Clostridium tetani* are Gram-positive, spore-forming, and anaerobic bacteria that produce the most potent neurotoxins, botulinum toxin (BoNT) and tetanus toxin (TeNT), responsible for flaccid and spastic paralysis, respectively. The main habitat of these toxigenic bacteria is the environment (soil, sediments, cadavers, decayed plants, intestinal content of healthy carrier animals). *C. botulinum* can grow and produce BoNT in food, leading to food-borne botulism, and in some circumstances, *C. botulinum* can colonize the intestinal tract and induce infant botulism or adult intestinal toxemia botulism. More rarely, *C. botulinum* colonizes wounds, whereas tetanus is always a result of wound contamination by *C. tetani*. The synthesis of neurotoxins is strictly regulated by complex regulatory networks. The highest levels of neurotoxins are produced at the end of the exponential growth and in the early stationary growth phase. Both microorganisms, except *C. botulinum* E, share an alternative sigma factor, BotR and TetR, respectively, the genes of which are located upstream of the neurotoxin genes. These factors are essential for neurotoxin gene expression. *C. botulinum* and *C. tetani* share also a two-component system (TCS) that negatively regulates neurotoxin synthesis, but each microorganism uses additional distinct sets of TCSs. Neurotoxin synthesis is interlocked with the general metabolism, and CodY, a master regulator of metabolism in Gram-positive bacteria, is involved in both clostridial species. The environmental and nutritional factors controlling neurotoxin synthesis are still poorly understood. The transition from amino acid to peptide metabolism seems to be an important factor. Moreover, a small non-coding RNA in *C. tetani*, and quorum-sensing systems in *C. botulinum* and possibly in *C. tetani*, also control toxin synthesis. However, both species use also distinct regulatory pathways; this reflects the adaptation of *C. botulinum* and *C. tetani* to different ecological niches.

Keywords: *Clostridium tetani*; *Clostridium botulinum*; botulinum neurotoxin; tetanus neurotoxin; toxin gene regulation; two-component system; small RNA

Key Contribution: *Clostridium botulinum* and *Clostridium tetani* produce potent neurotoxins under the control of complex regulatory networks. This review summarizes and compares the regulation of toxin synthesis in both microorganisms.

1. Introduction

Clostridium botulinum and *Clostridium tetani* are Gram-positive, spore-forming, and anaerobic rod-shaped bacteria that produce the most potent toxins among bacterial, animal, and plant toxins. *C. tetani* synthesizes the tetanus neurotoxin (TeNT), which is responsible for an often fatal spastic paralysis in humans and animals, whereas *C. botulinum* produces botulinum neurotoxins (BoNTs), which induce severe flaccid paralysis in vertebrates. Although TeNT and BoNTs lead to opposite clinical symptoms, they use a similar molecular mechanism of action. Indeed, TeNT and BoNTs deliver into target neurons their intracellularly active domain (light chain, Lc), which exerts a specific protease activity towards one of the three SNARE (soluble N-ethylmaleimide-sensitive factor attachment protein receptor) proteins: synaptobrevin or VAMP (vesicle-associated membrane protein), SNAP25 (synaptosomal-associated protein 25), and syntaxin. This leads to the inhibition of exocytosis of synaptic vesicles containing neurotransmitters. However, through their heavy chains (Hc), which recognize distinct cell surface receptors, TeNT and BoNTs undergo distinct trafficking in the host and target different neuronal cell types. Thereby, BoNTs target the motoneuron endings of the peripheral nervous system and block the release of acetylcholine, leading to a flaccid paralysis (botulism). In contrast, TeNT enters the central nervous system via retrograde transport through motoneurons and specifically inhibits the release of neurotransmitters (glycine, gamma aminobutyric acid (GABA)) in inhibitory interneurons, thus disrupting the negative control exerted by inhibitory interneurons over motoneurons, resulting in spastic paralysis (tetanus) [1,2].

2. Diversity of Clostridial Neurotoxins and Neurotoxigenic Bacterial Strains

BoNTs constitute a family of neurotoxins that share a similar structure and mode of action. They are classically divided into seven toxinotypes (A, B, C, D, E, F, and G) based on their neutralization by corresponding specific polyclonal antibodies. At the genetic level, two additional types, H or F/A, and X, have been described [3,4]. Each BoNT type is subdivided into subtypes based on amino acid sequence variations (0.9–36.2%, in most cases >2.6%). More than 40 subtypes have been identified [5]. Most BoNTs are produced by *C. botulinum* strains that are classified into physiologically and genetically distinct groups: group I including proteolytic strain types A, B, F; group II including non-proteolytic strain types B, E, F; group III corresponding to *C. botulinum* types C and D; and group IV, referred to as *Clostridium argentinense* type G. BoNTs are also synthesized by some other clostridial species, such as BoNT/E by atypical *Clostridium butyricum* strains and BoNT/F by atypical *Clostridium baratii* strains [5]. Moreover, a *Paraclostridium bifermentans* strain produces a BoNT-like neurotoxin (PMP1), which is insect-specific, whereas BoNTs are only active in vertebrates [6]. Sequences related to *bont* genes have been found in the genomes of a few other non-clostridial bacteria, such as *bont*/Wo or *bont*/I from *Weisenella oryzae*, a bacterium isolated from fermented rice, *bont*/J (*ebont*/F or *bont*/En) in an *Enterococcus faecalis* strain isolated from a cow, and Cp1 in *Chryseobacterium piperi* isolated from sediment [7–12]. However, the production of BoNT by non-clostridial strains has not been reported and their activity remains to be characterized.

In contrast to the heterogeneity of BoNTs- and BoNT-producing clostridia, TeNT is highly conserved [13,14]. Only a few amino acid variations were observed, notably a four-amino-acid insertion in four strains, in a study that investigated 37 *C. tetani* strains [13]. TeNT-producing *C. tetani* strains show a high level of genomic conservation. The population of *C. tetani* is divided into two major closely related clades and most strains belong to clade 1 [13].

3. Genetic Organization of Clostridial Neurotoxin Genes

BoNT-producing clostridia synthesize non-toxic proteins (associated-non-toxic proteins, ANTPs), which assemble with BoNTs through non-covalent bounds to form large-sized botulinum complexes (also referred to as progenitor toxins). A main characteristic of botulinum complexes is that they are stable at acidic pH and dissociate at alkaline pH (≥7) into free BoNT and ANTPs [15]. The genes encoding BoNTs and ANTPs are clustered in a DNA fragment called the botulinum locus. The genetic organization of the *bont* locus was initially determined in *C. botulinum* type C, where it is localized on a phage [16]. It consists of two operons transcribed in opposite directions. One operon contains *bont* and, located immediately upstream, *ntnh*, which encodes for the non-toxic non-hemagglutinin (NTNH) protein. NTNH shares a similar size and structure with BoNT, but NTNH lacks the catalytic site of BoNT. The interlocked association of NTNH with BoNT confers high resistance to acidic and protease degradation, while each protein separately is sensitive to proteolysis [17]. The *ntnh-bont* operon is highly conserved in all BoNT-producing clostridia, suggesting that *ntnh* and *bont* derive from a common ancestor by gene duplication [18–20].

The second operon is more divergent; it contains either hemagglutinin (HA) genes, including *ha70*, *ha17*, and *ha33*, or *orfX* genes (*orfX1*, *orfX2*, *orfX3*). The botulinum loci with *orfX-bont* genes encompass an additional gene in the first operon, *p47*, located upstream of *ntnh* [19,21]. A gene, *botR*, encoding an alternative sigma factor is localized in the botulinum loci upstream of the two operons in *C. botulinum* C and D and between the two operons in the other *C. botulinum* types, but it is lacking in *C. botulinum* E (Figure 1) [22,23]. HA complexes have been found to facilitate the passage of BoNT through the intestinal barrier by disrupting E-cadherin intercellular junctions between intestinal epithelial cells [24–26]. Up to now, no function has been attributed to OrfX and P47 proteins.

In *C. tetani*, the *tent* gene is localized on large plasmids. In comparison to the chromosome, these plasmids exhibit a higher degree of diversity. A homologous gene to *botR*, called *tetR*, lies immediately upstream of *tent*. No genes related to *C. botulinum* ANTP genes were identified in *C. tetani* genomes (Figure 1) [13,27,28].

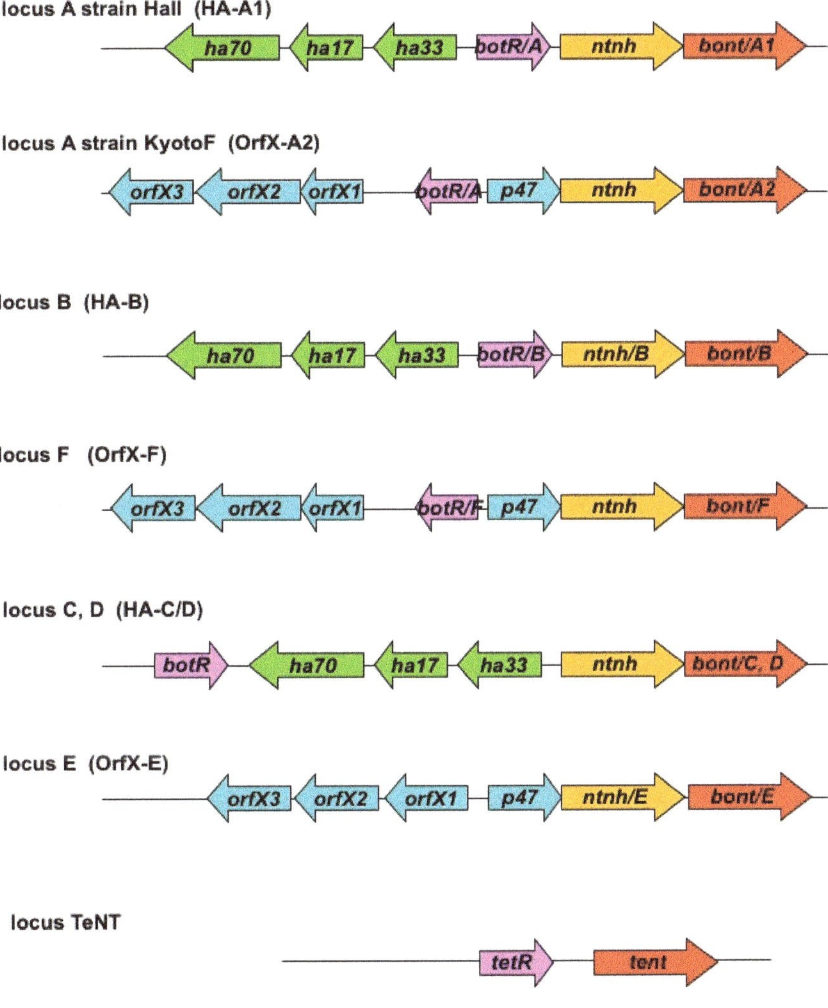

Figure 1. Genetic organization of neurotoxin genes in representative *Clostridium botulinum* and *Clostridium tetani* strains. red, neurotoxin genes; orange, *ntnh* genes; purple, regulatory genes; green, *ha* genes; cyan, *orfX* and *p47* genes.

4. Alternative Sigma Factors

BotR and TetR are the first factors that have been identified to be involved in the regulation of neurotoxin synthesis in *C. botulinum* and *C. tetani*. Overexpression of *bot/R* and *tetR* in *C. botulinum* and *C. tetani*, respectively, enhances neurotoxin synthesis (approximately 10-fold based on overexpression with a multicopy vector; toxin levels were quantified using toxin titration by mouse lethal activity), and inversely, their partial repression reduces toxin production (approximately 10-fold, based on experiments using an antisense mRNA strategy) [22,27]. Indeed, BotR and TetR are positive regulators of neurotoxin synthesis, and BotR also regulates the HA production in *C. botulinum* A [22]. Interestingly, overexpression of *botR* from *C. botulinum* type C in *C. tetani* stimulates TeNT synthesis, indicating a common mechanism of action of BotR and TetR [27]. BotR/A from *C. botulinum* A binds to conserved motifs in the two promoters of both operons, *ntnh-bont/A* and *ha*, in an RNA polymerase core enzyme-dependent manner. These motifs are also conserved in the promoter of *tent* and facilitate the binding of the TetR-RNA polymerase core enzyme complex. Thereby, BotR/A and TetR drive the transcription of the corresponding neurotoxin genes, and in addition, BotR/A controls that of the *ha* genes in *C. botulinum* [29]. Homologs of BotR and TetR have been characterized in *Clostridioides difficile* and *Clostridium perfringens*, where they positively control toxin synthesis. TcdR from *C. difficile* regulates the production of Toxin A (TcdA) and Toxin B (TcdB), and UviA controls the synthesis of a *C. perfringens* bacteriocin. BotR, TetR, TcdR, and UviA, which are interchangeable, as tested by in vitro transcription, are assigned to a distinct subgroup (group 5) of alternative sigma factors based on their targeted DNA motifs [30,31].

BotR and TetR are concomitantly expressed with their corresponding neurotoxin genes, showing a maximum level of expression at the end of the exponential growth phase and beginning of the stationary phase. The transcription levels of *botR* or *tetR* are approximately 100-fold less than those of *ntnh-bont*, *has*, or *tent*, respectively, as it is usually observed between regulatory and target genes [32,33]. Levels of neurotoxins secreted into the culture supernatant follow a similar pattern, including a progressive accumulation during the exponential growth, reaching a maximum level at the beginning of the stationary phase and maintaining a stable toxin level. Since *botR* and *tetR* are located in close proximity to neurotoxin and ANTP genes, they primarily have an impact on these target genes. However, *botR* and *tetR* might also regulate other distantly located genes, as is observed with most of the regulatory genes that have widely pleiotropic effects. Indeed, in *C. botulinum* A, *botR* seems to control the expression of at least 21 genes, 15 and 6 being up- and downregulated, respectively [34]. In contrast to primary sigma factors that are required for controlling house-keeping genes, alternative sigma factors are involved in controlling growth phase transitions, such as from the exponential growth to stationary phase, in response to environmental factors, including nutritional factors and stress conditions such as oxidative stress or heat shock. Their regulatory activity also comprises morphological differentiation, flagellar biosynthesis, and sporulation. In most cases, environmental bacteria that are exposed to a wide range of external factors employ more sigma factors than obligate pathogens or commensals, which are adapted to a specific host compartment and have a more restricted environment. Many obligate pathogens have lost sigma factors after host adaptation, such as *Mycobacterium leprae*, a strictly obligate pathogen that contains four sigma factors, compared to 13 sigma factors in *Mycobacterium tuberculosis*, which can live in the environment [35,36]. Indeed, *C. botulinum* A and *C. tetani* possess 18 and 25 sigma factors, respectively [37], compared to seven in *Escherichia* coli K12, six in *Shigella flexneri*, and three in *Helicobacter pylori*, the two latter species being strict human pathogens [38]. Group 4 alternative sigma factors, also called sigma factors of the extracytoplasmic function (ECF) family, are mainly involved in sensing and responding to extracellular signals and regulate cell envelope functions (transport, secretion, bacterial cell wall stress response) [31,36,39]. Group 5 alternative sigma factors, including BotR and TetR, are distantly related to the other sigma factors based on amino sequence similarity [30,31,36]. They control toxin synthesis and possibly other bacterial functions in response to yet unknown environmental stimuli or conditions.

In group III *C. botulinum* strains, the *bont* locus is located on a phage, and, in contrast to group I and II *C. botulinum* strains, where *botR* lies between the two *has/orfX* and *ntnh-bont* operons, *botR/C* or *D* is upstream of both *has* and *ntnh-bont* operons. The binding motifs recognized by BotR/A in the promoters of the *has* and *ntnh-bont* operons in *C. botulinum* A are conserved in the corresponding promoter regions of *C. botulinum* C and D [29]. BotR/C and D play likely a similar regulatory role of *has* and *ntnh-bont* operons as in *C. botulinum* A, but possibly do not respond to the same environmental stimuli, since group III *C. botulinum* strains have different physiological properties—for example, a higher optimal growth temperature; they are more often associated with animals than group I and II *C. botulinum* [40–42].

Despite the fact that *C. botulinum* E strains lack *botR* in close proximity to *bont/E*, the kinetics of growth, *bont/E* expression, and BoNT/E production are similar to those in *C. botulinum* A and proteolytic and non-proteolytic strains of *C. botulinum* B [33,43,44], supporting the notion that additional regulatory genes are involved in the control of neurotoxin genes. The non-proteolytic *C. botulinum* E, F6, non-proteolytic *C. baratii* F7, and non-proteolytic *C. butyricum* E contain an *orfX-p47-ntnh-bont* locus without *botR*. However, *botR* is present just upstream of the *ha* operon in non-proteolytic *C. botulinum* B4 and *C. argentinense*, which both have a *ha-ntnh-bont* botulinum locus, similar to *C. botulinum* C and D. In *C. baratii* F7, a gene encoding an UviA-like protein belonging to the same subgroup of sigma factors as BotR lies upstream of the *orfX* operon. This gene is also found in non-proteolytic *C. botulinum* E and F6, but distantly located of the botulinum locus [45]. The UviA-like protein in non-proteolytic strains might have a similar regulatory function of botulinum locus genes to BotR.

5. Two-Component Systems

Two-component systems (TCSs) consist of two proteins which coordinately control gene transcription in response to extracellular signals. TCSs regulate various bacterial physiological processes required for adaptation to environmental changes, such as development, cell division, metabolism, pathogenicity, and antibiotic/bacteriocin resistance. One component is a transmembrane protein with an extracellular sensor domain and, more rarely, a cytoplasmic domain, called sensor histidine kinase (SHK), which senses environmental stimuli including small molecules, ions, toxics, dissolved gases, pH, temperature, osmotic pressure, redox potential, or other yet unknown factors. SHK communicates with a corresponding response regulator (RR) by a phosphorelay. Most of the SHK and RR genes are closely located and organized in operons. Signal sensing by the N-terminal region of SHK induces the phosphorylation of a conserved histidine (His) in the C-terminal part. SHKs are in dimeric form and retain the phosphorylated His in a dimeric helical domain. The His phosphoryl group is then transferred to a conserved aspartate residue in the receiver domain of the corresponding RR. This results in a conformational change of the RR and enhanced affinity for specific promoter(s) [34,46–48].

C. botulinum and *C. tetani* contain numerous TCS genes in their genomes. In *C. botulinum* strain Hall, 39 genes have been found to encode RR proteins based on the presence of conserved motifs such as the signal receiver domain and DNA binding domain. Thirty RR genes are located in close proximity to a SHK gene, thus corresponding to 30 TCS operons. An additional nine RR genes and nine SHK genes are considered as orphan regulators [34,37]. *C. tetani* strain E88 possesses 30 TCS genes, 19 of which are homologous to related genes of *C. botulinum* A strain Hall, with $\geq$45% identity at the amino acid level (Figure 2) [32]. Most of the *C. botulinum* and *C. tetani* TCS genes have homologs in other clostridia, indicating that these clostridia might share similar regulatory pathways [32,34].

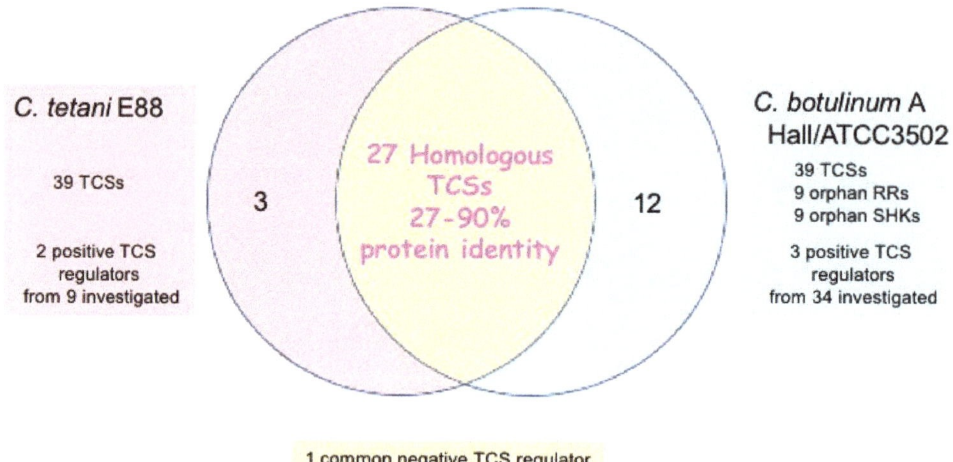

Figure 2. Two-component systems (TCSs) involved in neurotoxin gene regulation in *Clostridium botulinum* strain Hall and ATCC3502 and in *Clostridium tetani* strain E88.

In *C. botulinum* A strain Hall, 34 regulatory genes (29 TCSs and five RR orphan regulatory genes) and an additional TCS gene in strain ATCC3502 (a derivative of strain Hall), as well as nine TCS genes in *C. tetani* E88, have been investigated for their possible regulatory contribution in toxinogenesis [32,34,49]. In *C. botulinum* A strain Hall, three TCSs have been reported as positive regulators of BoNT synthesis in a *bontR*-independent manner (as tested by an antisense mRNA strategy, yielding recombinant strains with 10- to 100-fold decreased BoNT/A production, which was determined by ELISA) [50]. Among them, one TCS belonging to the OmpR family shows 65% protein identity with a *C. tetani* TCS and shares similarity with VirI/VirJ of *Clostridium perfringens*, which has been reported as a regulator of toxin synthesis (GenBank BAA78773, BAA78774). Although this TCS is a positive regulator of BoNT synthesis, the *C. tetani* homolog is not involved in TeNT synthesis. The two other TCSs that are positive regulators of BoNT synthesis have no homolog in *C. tetani*. Two additional TCSs in *C. botulinum* show an indirect effect on BoNT production as they have pleiotropic effects, notably by impairing cell wall synthesis or assembly. The two *C. tetani* TCSs that positively regulate the TeNT synthesis (as tested by an antisense mRNA strategy, yielding recombinant strains with two- to five-fold decreased TeNT production, which was determined by ELISA) have ineffective homologs in BoNT synthesis in *C. botulinum* A [32] (Table 1). Only one TCS that is conserved (100% protein identity) in *C. botulinum* A strain ATCC3502 and in *C. tetani* E88 is a negative regulator of neurotoxin synthesis in both microorganisms (as evaluated by the ClosTron (http://www.clostron.com (accessed on 13 April 2022), Nottingham, UK) strategy in strain ATCC3502, where BoNT/A levels were monitored by ELISA; in strain E88, the antisense mRNA strategy was applied and TeNT levels were determined by ELISA) [32,49]. In *C. botulinum*, this TCS binds to both promoters of the *ntnh-bont* and *ha* operons and prevents their transcription by impairing the binding of BotR [49]. In *C. tetani*, the corresponding TCS binds also to the *tent* promoter and likely retains the same mechanism of action as in *C. botulinum* ATCC3502. However, the apparent TCS counterpart in *C. botulinum* strain Hall, which shows only 58% protein identity, is apparently not involved in the regulation of BoNT synthesis, as judged from the lack of BoNT level alterations assayed by ELISA in the recombinant Hall strain using the antisense RNA strategy targeting this TCS (Table 1) [32,50]. Thereby, various TCSs control the neurotoxin synthesis in *C. botulinum* and *C. tetani*. Although *C. botulinum* and *C. tetani* share homologous TCSs, most of them have distinct functional roles in the control of neurotoxin synthesis in the two microorganisms.

Table 1. Two-component systems involved **in toxin** gene regulation in *C. botulinum* Hall and *C. tetani* E88.

	C. tetani E88					*C. botulinum* Hall			
Genetic Localization	Locus Tag	Role	Family RR	Regulation of TeNT Synthesis	Ref.	Homolog	Protein Identity (RR)	Regulation of BoNT Synthesis	Refs.
chr	CTC_RS10150 CTC_RS10155	SHK RR	LytR/AlgR	Positive	[32]	CLC_3250 CLC_3251	55%	None	[50]
plasmid	CTC_RS13810 CTC_RS13805	SHK RR	OmpR	Positive	[32]	CLC_1431 CLC_1432	56%	None	[50]
	No homolog		OmpR			CLC_1093 CLC_1094		Positive	[50]
	No homolog		OmpR			CLC_1913 CLC_1914		Positive	[50]
chr	CTC_RS02080 CTC_RS02085	RR SHK	OmpR	None	[32]	CLC_0661 CLC_0663	65%	Positive	[50]
chr	CTC_RS10030 CTC_RS10035	SHK RR	OmpR	None	[32]	CLC_0410 CLC_0411	68%	Cell wall alteration	[50]
	No homolog		OmpR			CLC_3293 CLC_3294		Cell wall alteration	[50]
chr	CTC_RS07310 CTC_RS07315	SHK RR	OmpR	Negative	[32]	strain Hall CLC_0842 CLC_0843	58%	None	[50]
						strain ATCC3502 CBO_0786 CBO_0787	100%	Negative	[49]

chr, chromosome; RR, response regulator; SHK, sensor histidine kinase. positive effects are in green, and negative effects are in pink.

6. Metabolism and Toxin Gene Regulation

Toxin synthesis, as with protein synthesis in general, is dependent on the metabolic activity of the bacteria. How is neurotoxin synthesis linked to the general metabolism?

6.1. CodY

In Gram-positive bacteria, CodY (control of dciA (decoyinine induced operon) Y) is a master regulator of metabolism, sporulation, and virulence. In *Bacillus subtilis*, CodY controls more than 100 genes involved in the adaptation to nutrient restriction and transition from the exponential growth phase to the stationary growth phase. Typically, CodY acts by binding to the promoter of target genes in a GTP(guanosine triphosphate)- and branched-chain amino acid-dependent manner; these are indicators of the general metabolism status of the bacterium [51]. Thereby, in *B. subtilis*, CodY senses intracellular levels of GTP and branched amino acids such as isoleucine, whose levels are high during the exponential growth and decrease in mostly repressed gene transcription. In contrast, at low GTP or isoleucine levels, CodY induces the de-repression of genes which are involved in adaptive responses to nutrient limitation, such as those coding for extracellular degradative enzymes, transport systems, and catabolic pathways [51,52]. CodY is conserved in clostridia, including the toxigenic species *C. botulinum*, *C. tetani*, *C. perfringens*, and *C. difficile*. In *C. botulinum* A strain ATCC3502, CodY binds to the promoter of the *ntnh-bont* operon at high GTP levels, whereas isoleucine is ineffective, and stimulates toxin gene transcription and BoNT/A synthesis (as tested with the ClosTron system, and determining BoNT/A levels by ELISA and *bont/A* transcription by qPCR) [53]. The precise role of CodY in *C. botulinum* is still elusive: does CodY directly regulate *ntnh-bont* transcription or interfere with *botR* or a TCS gene such as by repressing the negative TCS regulator? CodY is also a positive regulator of TeNT synthesis in *C. tetani* (as tested by the antisense mRNA strategy, and determining TeNT levelsby ELISA, and *tent* transcription by qPCR) [32]. CodY binds to the *tent* promoter but not to that of *tetR* [32]. BoNT and TeNT synthesis is dependent on the availability of a carbon source such as glucose [54–56]. CodY controls carbon metabolism in *B. subtilis*. Notably, under glucose-rich conditions in culture medium, CodY and CcpA (catabolite control protein A), a regulator of carbon catabolism, facilitate the conversion of excess pyruvate resulting from glycolysis into excretable overflow compounds such as acetate, lactate, and acetoin [57]. A similar mechanism of CodY in glucose/pyruvate metabolism has been suggested in *C. botulinum* A [53]. CcpA is conserved in *C. botulinum* and *C. tetani*. However, the role of CcpA in these microorganisms remains to be elucidated. In contrast, in *C. difficile*, CodY and CcpA are negative regulators of toxin A (TcdA) and toxin B (TcdB). Glucose and rapidly metabolizable carbohydrates inhibit toxin synthesis in *C. difficile*. CodY and CcpA, which are activated by glucose and rapidly metabolizable carbohydrates, bind to the promoter of TcdR and repress its transcription and subsequently that of *tcdA* and *tcdB* [58–60]. The opposite regulatory pathways of toxin synthesis linked to carbohydrate metabolism controlled by CodY and CcpA between *C. botulinum*/*C. tetani* and *C. difficile* are intriguing. This indicates that toxin synthesis in *C. botulinum* and *C. tertani* requires energy from carbohydrate metabolism, mainly glucose, the main carbohydrate fermented by these bacteria, while *C. difficile* mainly uses amino acid metabolism as an energy source, notably through the Stickland reaction, for toxin production [60–63]. These divergent regulatory pathways might have evolved during bacterial adaptation to different environments: soil for *C. botulinum*/*C. tetani* and the intestine for *C. difficile*.

6.2. Spo0A

Spo0A is a master regulator of the initial steps of sporulation in *Bacillus* and clostridia. However, the mode of activation of Spo0A differs in the two classes of bacteria. Nutrient limitation, notably carbohydrate, nitrogen, and phosphorus limitation, is the major signal leading to Spo0A activation through a phosphorelay including five sensor kinases and subsequent positive transcriptional regulation of critical sporulation-essential genes [64]. The kinases that activate Spo0A in *Bacillus* are not conserved in clostridia. Orphan Spo0A-activating histidine kinases have been identified in clostridia, such as *Clostridium acetobutylicum*, *C. perfringens*, *C. difficile*, *C. botulinum*, and *C. tetani*. Clostridia sense different environmental stimuli to initiate sporulation, including external pH resulting from fermentation, with the subsequent release of acidic end-products (acetate, butyrate) and unknown factors [65–69]. In clostridia, Spo0A displays additional functions apart from sporulation initiation. In *C. acetobutylicum*, Spo0A is activated at the end of the exponential growth phase and controls the shift between acidogenesis that occurs during the exponential growth, and solventogenesis that is coupled to the onset of sporulation [70,71]. In *C. perfringens*, Spo0A controls the production of toxins (*C. perfringens* enterotoxin and TpeL), which are synthesized during the sporulation process [72,73]. The role of Spo0A in the regulation of TcdA and TcdB synthesis is variable according to the genetic background of *C. difficile* strains [62,74]. Spo0A coordinates the expression of a large number of *C. difficile* genes involved in multiple additional functions, such as nutrient transport, metabolic pathways including the production of butyrate, surface protein assembly, and flagellar biosynthesis [75].

Spo0A is highly conserved in all *C. botulinum* genomes: an orphan sensor histidine kinase that is able to phosphorylate Spo0A has been identified [69]. In *C. botulinum* ATCC3502, Spo0A is expressed during the exponential growth and its expression decreases during the entry into the stationary phase, while the subsequent transcription of sigma factors essential for sporulation increases [76,77]. It is not known whether Spo0A affects the expression of *bont* in group I *C. botulinum*. Adaptation to cultivation at high temperatures (45 °C) represses both *bont*/A and sporulation genes in strain ATCC3502, but no co-regulation of these genes has been evidenced [78]. No correlation between sporulation and the production of BoNT/A has been observed in two other *C. botulinum* A strains [33]. Moreover, the strain Hall A-*hyper* produces high levels of BoNT/A and is unable to sporulate [79]. Similarly, the highly TeNT-producing *C. tetani* strain used for vaccine production is a non-sporulating strain [80], and Spo0A has not been found to control TeNT synthesis (as tested by the antisense mRNA strategy, TeNT monitoring by ELISA, and *tent* transcription by qPCR) [32]. In contrast, in group II *C. botulinum* E, Spo0A is a positive regulator of BoNT/E synthesis and sporulation (as tested with the ClosTron system, toxin monitoring by ELISA, and gene transcription by qPCR). Spo0A binds to a conserved motif in the promoters of the *ntnh-bont/A* operon together with CodY, AbrB (putative repressor of *bont/E*), sigma K (belonging to the sigma factor cascade of sporulation), and an UviA-like regulator [43]. Thus, Spo0A might directly and indirectly regulate the transcription of *bont/E*. Thereby, group II *C. botulinum* E strains that have an UviA-like regulator instead of BotR likely use specific and common regulatory pathways of *bont* expression compared to *C. botulinum* A strains which belong to the distinct physiological and genetic group 1.

In group III *C. botulinum* C and D, the production of the C2 toxin, which is an ADP-ribosyltransferase targeting monomeric actin, is linked to sporulation [81]. However, the regulatory pathway of C2 toxin genes, and the possible involvement of *spo0A* and/or other sporulation genes, has not yet been elucidated.

6.3. Amino Acid/Peptide Metabolism

C. botulinum and *C. tetani* produce high levels of toxins in complex media rich in peptones and other nutrients, whereas chemically defined media even containing almost all amino acids and vitamins as well as a carbon source usually yield 10- to 100-fold lower toxin titers [55,82–84]. Licona-Cassani et al. showed that, although *C. tetani* grew in a chemically defined medium, toxin production was obtained only when casein-derived peptides were added to the medium [82]. In addition to variations in toxin production according to different media, variations in BoNT or TeNT yields are often observed from batch to batch of the same culture medium, even using the same bacterial strain. The transcription of neurotoxin genes and toxin synthesis occur mainly within a short time interval between the late exponential growth and early stationary growth phase [33,85,86]. Thus, nutritional and environmental factors influence the regulation of toxin synthesis in *C. botulinum* and *C. tetani*, which takes place in a restricted phase of bacterial growth. Peptides and amino acids appear to be important regulatory factors at the transcriptional and posttranscriptional levels. Indeed, large amounts (0.8–0.9 g/L) of amino acids (aspartate, glutamate, serine, histidine, threonine) downregulate *tetR* and *tent* by a yet non-identified regulatory pathway [86]. Arginine is an essential amino acid for *C. botulinum* growth, but an excess of arginine represses BoNT production in proteolytic group I *C. botulinum* [87]. Arginine deiminase leads to arginine catabolites that increase the pH and induce subsequent BoNT degradation by not yet characterized proteases. BoNT and botulinum complexes are stable at acidic pH in media without excess of arginine [88]. Supplementation with a high amount of glucose (50 g/L) that induces acidification counteracts the effect of arginine. Interestingly, BoNT synthesis is coupled to protease production [89]. Likely, proteases that are active at alkaline pH induce BoNT degradation. Secreted proteases are required for protein substrate degradation, resulting in peptides and amino acids that are taken up into the bacteria through transport systems and used for protein synthesis, including neurotoxin synthesis. Indeed, the *C. botulinum* A and *C. tetani* genomes contain numerous protease/peptidase and transport system genes [37,80].

Peptides in culture media were found to be critical for TeNT synthesis by *C. tetani*. Since culture media containing casein pancreatic digests support high levels of TeNT production, peptides derived from casein tryptic digestion were investigated. Histidine-containing peptides as well as hydrophobic peptides containing the motif proline–aromatic acid–proline were the most effective in promoting TeNT production [84,90,91]. It is noteworthy that genome analysis of *C. tetani* shows the presence of numerous peptidases and amino acid degradation pathways [92]. The kinetics of *C. tetani* growth in a complex medium show rapid exponential growth (stage I, around 10–12 h), then a slower linear growth (stage II, around 30 h), followed by a stationary phase and subsequent autolysis. During stage I, the amino acids are consumed and the genes involved in amino acid degradation pathways are overexpressed, corroborating amino acid catabolism that provides energy used for the rapid biomass formation during this growth phase. The pH decreases due to organic acid production, *tetR* and *tent* are not expressed, and TeNT is not synthesized. Once free amino acids are depleted in the culture medium, *C. tetani* uses peptides, whose metabolism requires transporters that are more energy-costly, and enters the linear growth phase II. The transition from free amino acid to peptide consumption is associated with increased pH due to the reduction of organic acids to alcohols and solvents, and the production of ammonia from peptide metabolism. During phase II, *tetR* and *tent* are highly expressed, as well as *codY*, the TCS that positively regulates *tent*, and two additional sigma factors located on the large plasmid containing *tent*, resulting in TeNT synthesis [54,82,86,93]. In complex media, glucose is consumed during the first phase of growth, leading to rapid bacterial multiplication and pH decrease. Then, the nitrogen source is used and the pH increases. TeNT is synthesized only during this second phase, when glucose is no longer or weakly available and when peptides are used for energy production. Thus, as shown by Fratelli et al., the balance between the nitrogen and carbon sources, as well as the subsequent pH of culture media, are critical factors [54,93]. *C. botulinum* and *C. tetani* likely share common metabolic pathways and subsequent toxin gene regulatory networks. In both microorganisms, the transition from amino acid utilization to peptides that are more common substrates in the environment seems to elicit the production of proteases. BoNT and TeNT are metalloproteases, but which recognize specific substrates in host neuronal cells. BoNT and TeNT possibly evolve from ancestor metalloproteases with a broader substrate range that were used by the bacteria for nutrient acquisition and that were regulated as the other proteases. Thus, the regulation of toxin synthesis in *C. botulinum* and *C. tetani* might represent a reminiscent common regulatory circuit controlling protease synthesis.

6.4. Other Nutritional and Environmental Factors

In addition to nutrients required for growth and protein synthesis, some nutritional and environmental factors might influence, directly or indirectly, toxin synthesis.

CO_2—A high concentration of CO_2 in the gas phase increases *bont* expression and BoNT synthesis in non-proteolytic group II *C. botulinum* B and E, although the growth rate is decreased. Indeed, a 70% CO_2 atmosphere versus 10% stimulates 2- to 5-fold greater toxin gene expression and BoNT formation. In high and low CO_2 concentrations, toxin gene expression occurs in the same growth phase, mainly in the late exponential growth and early stationary phase [94,95]. The signaling pathways in the regulation by CO_2 are not known. CO_2 can dissolve in the liquid medium and generate bicarbonate, which influences protein synthesis through carboxylation reactions. CO_2 (35%) in the gas phase of *C. tetani* culture versus nitrogen atmosphere (unpublished) or the addition of sodium carbonate (100 mM) in the culture medium increases the production of TeNT approximately two-fold [32], despite reduced growth in the CO_2-rich atmosphere (approximately three-fold). In contrast, elevated CO_2 in the gas phase of proteolytic group I *C. botulinum* B and E has no effect on toxin gene expression [96], suggesting that CO_2 triggers a signaling pathway controlling toxin synthesis in non-proteolytic strains.

Inorganic phosphate—Inorganic phosphate has been found to control TeNT synthesis in *C. tetani*. Supplementation of culture medium with inorganic phosphate (optimum concentration 40 mM) stimulates *tent* expression and TeNT production approximately three-fold without impairing the growth rate [32]. Inorganic phosphate is involved in multiple biochemical reactions; its effect on toxin gene transcription might be mediated by TCSs. *C. tetani* genome contains two TCSs putatively involved in phosphate uptake, one of which has been found to negatively regulate TeNT synthesis [32]. Inorganic phosphate is apparently not involved in BoNT production in *C. botulinum* A, as tested by supplementation of the TGY (trypticase-glucose-yeast extract) culture medium with 20 to 150 mM Na_2HPO_4 and monitoring BoNT/A production (strain Hall) in the culture supernatant by titration of the mouse lethal activity (unpublished). Control of the virulence mechanism by inorganic phosphate and TCS from the PhoP/PhoR family has been found in several pathogens [97,98]. TCSs control the homeostasis of phosphate according to the availability in the environment. However, the precise subsequent phosphate-dependent signaling pathways controlling virulence remain largely unknown.

pH—*C. botulinum* and *C. tetani* grow and produce the neurotoxins in a wide range of pH (pH 4.5–5 to 9). The initial pH of the growth medium was found to influence the autolysis of *C. tetani*. An initial pH of 6.1 seems optimal for TeNT production [84]. In a complex medium, high pH (7.8) downregulates *tent* [99]. The mechanism of toxin gene regulation by pH is not yet identified.

The culture pH of proteolytic *C. botulinum* A, B grown in complex media typically drops (pH 6–6.3 with initial glucose concentration up to 1%, and until pH 5.5 with glucose 1.5%) during the exponential growth phase, and then stabilizes and slightly increases during the stationary phase [56,85,88,100]. Maintaining an acidic pH (pH 5.7–6) during the culture does not modify the BoNT yield in the culture supernatant, whereas an alkaline pH (pH 7.2 and above, manually adjusted or by supplementation of the culture medium with 2% arginine) decreases the BoNT level [56,88]. The pH does not influence BoNT synthesis at the transcriptional level, but affects BoNT stability by activating a BoNT degrading metalloprotease in alkaline conditions [88].

Temperature—In contrast to *C. difficile*, in which a high temperature (42 °C) prevents *tcdR* and toxin gene expression, temperatures of 37–44 °C have no influence on *botR* and *bont* transcription in group I *C. botulinum* A. However, a high temperature induces the production of protease(s), which inactivate BoNT/A [33]. TeNT production is usually obtained by *C. tetani* culture at 33–35 °C [82,84,99].

Group II *C. botulinum* has an optimum temperature of 25 °C for growth and toxin production, but the strains of this group can grow and form toxins at temperatures as low as 3.0–3.3 °C in 5 to 7 weeks [101]. Investigation with *C. botulinum* E showed that growth and toxin production are lower at 10 °C than at 30 °C. However, *bontE* transcription relative to growth was similar at 10 °C and 30 °C [102]. A TCS is involved in the cold adaptation of *C. botulinum* E [103,104]. Similarly, cold tolerance of growth at 15 °C in *C. botulinum* A strain ATCC3502 requires the contribution of a TCS [105]. This TCS is conserved in *C. botulinum* A strain Hall, but it has not been identified as a regulator of BoNT/A synthesis [50]. Thus, temperature is important for growth and toxin production, but temperature seems to have no direct role in the regulation of toxin synthesis in *C. botulinum* and *C. tetani*.

7. Small RNA

In addition to regulatory proteins, bacteria use regulatory RNAs to modulate gene transcription or translation initiation for adaptive responses to environmental changes. Environmental bacteria have to adapt their physiology and metabolism to various hostile conditions and pathogenic bacteria have to cope with adverse host interactions; thus, they have to adapt rapidly their gene expression, notably that of virulence genes [106,107]. Regulatory RNAs are small molecules, typically between 50 and 500 nucleotides (small RNAs or sRNAs), and are non-protein coding. The main advantage of sRNAs versus regulatory proteins is their speed in controlling gene expression based on the faster availability of sRNAs due to the lower energy needed for their production by transcription and not by translation, as in the case of regulatory proteins, the faster turnover of sRNAs since RNAs are less stable than proteins, the rapid control of mRNA function by pairing with specific motifs in the untranslated region (UTR), or, in some cases, by acting at a posttranscriptional level [108,109]. Although most sRNAs are inhibitors of gene expression, some are activators. sRNAs lie in or overlap with the 5′ or 3′ UTRs of target genes, in intergenic regions, or in the opposite DNA strand and are transcribed as antisense sRNAs. Based on their genomic localization and mechanism of action, sRNAs are divided into several classes: *cis*-encoded sRNAs are transcribed from the DNA strand opposite to the target sequence and interact by perfect base pairing with mRNAs; *trans*-encoded sRNAs are distantly transcribed from target mRNA genes and recognize their target mRNAs by multiple and discontinuous short contacts. Clustered, regularly interspaced short palindromic repeat (CRISPR) RNAs interact with foreign DNA or RNA. Another class of sRNAs bind to regulatory proteins and antagonize their function. RNA riboswitches sense metabolites or environmental cues; they are usually located in the 5′ UTR of their target genes. Certain sRNAs interact with regulatory proteins, notably by promoting protein sequestration, but most sRNAs interfere with mRNA by inhibition of their translation, such as by blocking the ribosome binding site and/or by impairment of mRNA stability, resulting in subsequent degradation [108–111]. Numerous sRNAs act at a posttranscriptional level, whereas regulatory proteins preferentially act at transcriptional steps. In fact, numerous bacterial gene regulations involve mixed regulatory networks, including both transcriptionally acting regulatory proteins and posttranscriptionally acting sRNAs [112].

sRNAs are widespread in Gram-positive bacteria and notably in clostridia. In environmental clostridia such as *C. acetobutylicum*, sRNAs (159 predicted) have a crucial role in solventogenesis, growth, and the response to toxic metabolites [113–115]. In pathogenic clostridia, sRNAs have been described in *C. perfringens* to be involved in toxin production, as well as in *Clostridioides difficile* in adaptation to host and anti-phage defenses [116–118]. More than 200 sRNAs are predicted in the genomes of groups I and II *C. botulinum*, and 137 in the genome of *C. tetani* E88 [119]. A sRNA has been identified in the 3′ UTR of *tent* in the *C. tetani* E88 strain that is conserved in all toxigenic *C. tetani* strains [120]. However, no such sRNA has been reported in *C. botulinum*. This sRNA is expressed concomitantly with *tent* and negatively regulates *tent* expression and TeNT synthesis. The sRNA (approximately 140 nucleotides) contains a predicted junction-loop-exposed 14-nucleotide-long sequence that perfectly matches to a complementary sequence in the 5′ region of *tent* mRNA. Thus, the sRNA-mediated inhibitory regulatory activity is likely based on the sequestration of *tent* mRNA. In addition, this sRNA impairs *C. tetani* growth, notably by reducing the exponential growth phase [120]. Pleiotropic effects of sRNAs have also been found for the regulatory RNA VR-RNA of *C. perfringens* that controls 147 genes, including genes of toxins and virulence factors (alpha-toxin, kappa-toxin, hyaluronidase, sialidases) as well as genes involved in capsule synthesis [116,121].

8. Quorum Sensing

Quorum sensing is a cell-to-cell communication that bacteria use to adapt their physiology and behavior in response to cell densities. Bacteria produce and secrete extracellular signaling molecules called autoinducers. When a certain threshold of bacterial density is reached, accumulated autoinducers are detected by the bacteria, leading to coordinated changes in gene expression and behavior. Thus, quorum sensing allows a synchronized adaptation to environmental conditions in nonclonal bacterial populations [122,123]. A well-characterized quorum-sensing system is the Agr system of *Staphylococcus aureus* [124]. The autoinducer is a small peptide (autoinducing peptide, AIP) produced by *agrD* as a precursor, which is processed by the membrane endoprotease AgrB. When AIP reaches a threshold concentration in the environment, it is sensed by the TCS AgrA/AgrC, which in turn upregulates the expression of a small RNA and subsequently stimulates toxin synthesis. AgrC is the histidine kinase cell surface receptor that recognizes AIP and activates the response regulator AgrA [122]. Quorum sensing has been evidenced in *C. botulinum* [125], and an Agr-like system has been identified in group I *C. botulinum* strains called *agr-1/agr-2*; *agr-1* and *agr-2* are homologs of *agrB*, and *agrD*, respectively [126]. However, the Agr system in *C. botulinum* plays a different role than in *S. aureus*. Agr-1 seems to be involved in sporulation, while Agr-2 possibly controls toxin production [126]. Homologous genes of *agrA* and *agrC* have also been identified in *C. botulinum* A Hall strain, but silencing *agrA* did not impact BoNT/A production [34]. Thus, the regulatory quorum-sensing pathways in *C. botulinum* remain to be defined. A computational model of group I *C. botulinum* A growth and toxin production based on nutrient availability, cell density, and quorum-sensing signaling has been proposed in agreement with experimental data [127].

It is noteworthy that a cultivation technique was developed in the 1950s and 1960s to obtain high levels of toxin [128]. This method consists of dialyzed cultures. The bacteria are inoculated inside a dialysis bag (usually a cellophane bag) containing saline, which is immersed in a culture medium. This technique was used for the production of BoNT and TeNT, with a 5–50-fold increase in toxin yields compared to cultures in a flask with the same culture medium [128–131]. A very high bacterial density is obtained by culturing clostridia in dialysis bags and likely a quorum-sensing-mediated regulation is involved in the production of high toxin levels. Moreover, the high bacterial density leads to increased autolysis, which contributes to toxin release into the extracellular medium.

9. Concluding Remarks

The toxigenic environmental bacteria *C. botulinum* and *C. tetani* contain multiple and complex regulatory networks to control neurotoxin production. These only partially deciphered networks include alternative sigma factors, TCSs, sRNAs, quorum-sensing systems, and regulators of bacterial metabolism that interlock the bacterial growth with toxin production (Figure 3). Both microbial species share some common regulatory mechanisms, notably an alternative sigma factor, the gene of which is located upstream of the neurotoxin gene, and an inhibitory TCS. They retain a similar kinetic pattern of toxin production, mainly occurring at the transition between the exponential growth and early stationary growth phases. This is possibly linked to the alternative sigma factors BotR and TetR, which are expressed concomitantly with the neurotoxin genes. However, *C. botulinum* and *C. tetani* use distinct signaling pathways, notably distinct TCSs, probably reflecting the recognition of different nutritional and/or environmental signals. Although toxinogenesis is dependent on the general metabolism in *C. botulinum* and *C. tetani*, the nutritional requirements for toxinogenesis seem different between these microorganisms. However, only a few nutritional and environmental factors controlling the toxinogenesis have so far been identified, such as CO_2 in group II *C. botulinum* and inorganic phosphate in *C. tetani*. Differences in nutritional factors, particularly in the nature and composition of peptides and amino acids required for toxinogenesis, seem a major hallmark between *C. botulinum* and *C. tetani*. This likely reflects the different ecological niches used by these bacteria. Group I *C. botulinum* prefers neutral to slightly alkaline soils with low organic content, while group II *C. botulinum* are mostly found in more acidic soils with high levels of organic matter or as commensals in the intestines of certain animals. *C. tetani* is found primarily in neutral or alkaline soils at sufficient temperatures (>20 °C) and levels of moisture (15%) [41,132]. Moreover, group I but not group II *C. botulinum* strains can colonize the intestines of humans and produce BoNT in situ, leading to infant botulism or adult intestinal toxemia botulism [133]. *C. tetani* has not been reported to colonize the digestive tract and to induce intestinal tetanus [134]. Thus, *C. botulinum* groups and *C. tetani* have adapted to particular environments, notably through complex and specific regulatory systems that sense extracellular signals, leading to adapted gene expression. In addition to regulatory proteins and sRNAs, clostridia sense environmental factors by specific arrays of surface-associated proteins [135]. Do BoNTs and TeNT represent adaptive factors? These neurotoxins that attack specifically the nervous systems of vertebrates seem not to be involved in environmental adaptation. Indeed, non-toxigenic strains of *C. botulinum* and *C. tetani* can multiply, sporulate, and survive in the environment in the same manner as their toxigenic counterparts. BoNTs and TeNT, which likely evolved from a common protease ancestor, possibly retain common regulatory mechanisms with other proteases/peptidases required for the utilization of specific nutrient sources [20,136]. This is further supported by the observation that toxin synthesis is initiated at the transition from amino acid/carbohydrate to peptide metabolism in *C. Tetani* [54,93], and possibly in *C. botulinum*.

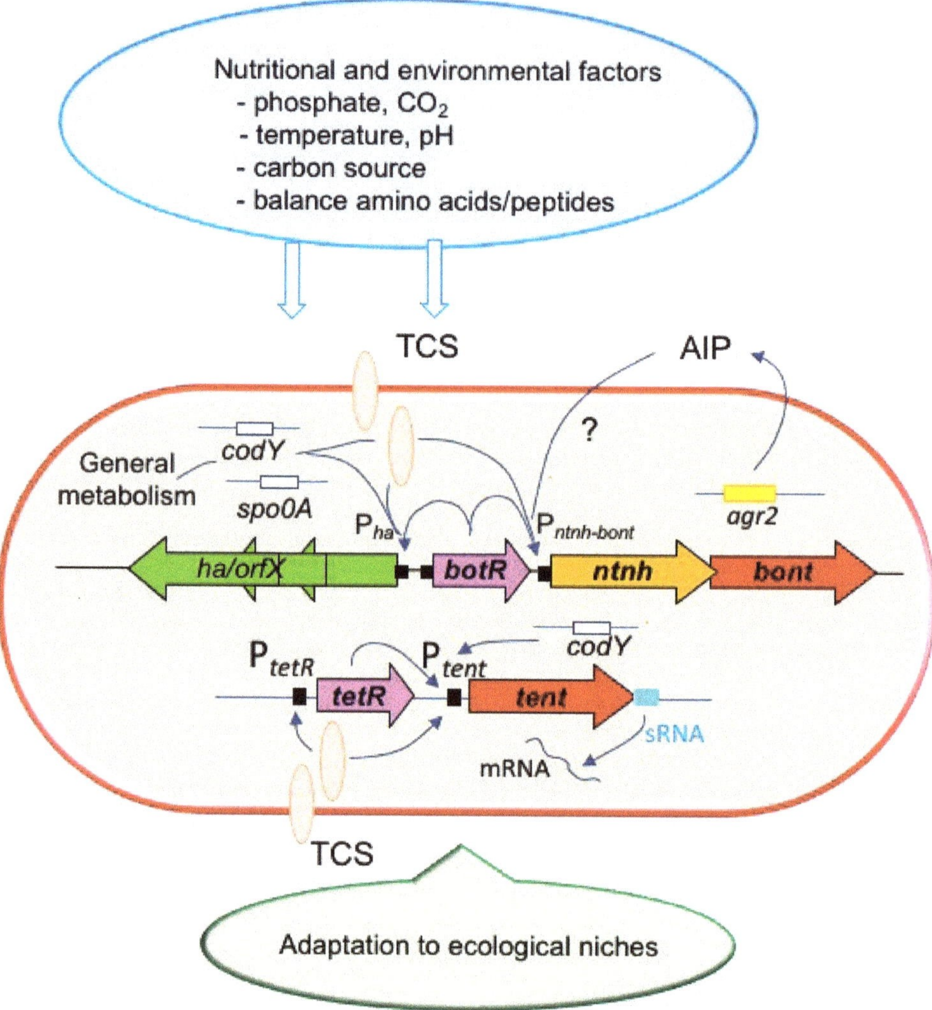

Figure 3. Schematic representation of the regulatory pathways in *Clostridium botulinum* and *Clostridium tetani*. TCS, two-component system; AIP, autoinducing peptide. External factors act through TCSs and/or other unknown receptors or transporters (blue arrows).

Author Contributions: Conceptualization, writing and editing, M.R.P.; review, H.B. All authors have read and agreed to the published version of the manuscript.

Funding: This review received no external funding.

Institutional Review Board Statement: Not applicable.

Informed Consent Statement: Not applicable.

Data Availability Statement: Not applicable.

Conflicts of Interest: The authors declare no conflict of interest.

References

1. Dong, M.; Masuyer, G.; Stenmark, P. Botulinum and Tetanus Neurotoxins. *Annu. Rev. Biochem.* **2019**, *88*, 811–837. [CrossRef] [PubMed]
2. Pirazzini, M.; Montecucco, C.; Rossetto, O. Toxicology and pharmacology of botulinum and tetanus neurotoxins: An update. *Arch. Toxicol.* **2022**, *96*, 1521–1539. [CrossRef] [PubMed]
3. Dover, N.; Barash, J.R.; Hill, K.K.; Xie, G.; Arnon, S.S. Molecular characterization of a novel botulinum neurotoxin type H gene. *J. Infect. Dis.* **2014**, *209*, 192–202. [CrossRef] [PubMed]
4. Zhang, S.; Masuyer, G.; Zhang, J.; Shen, Y.; Lundin, D.; Henriksson, L.; Miyashita, S.I.; Martinez-Carranza, M.; Dong, M.; Stenmark, P. Identification and characterization of a novel botulinum neurotoxin. *Nat. Commun.* **2017**, *8*, 14130. [CrossRef]
5. Peck, M.W.; Smith, T.J.; Anniballi, F.; Austin, J.W.; Bano, L.; Bradshaw, M.; Cuervo, P.; Cheng, L.W.; Derman, Y.; Dorner, B.G.; et al. Historical Perspectives and Guidelines for Botulinum Neurotoxin Subtype Nomenclature. *Toxins* **2017**, *9*, 38. [CrossRef] [PubMed]
6. Contreras, E.; Masuyer, G.; Qureshi, N.; Chawla, S.; Dhillon, H.S.; Lee, H.L.; Chen, J.; Stenmark, P.; Gill, S.S. A neurotoxin that specifically targets Anopheles mosquitoes. *Nat. Commun.* **2019**, *10*, 2869. [CrossRef]
7. Zornetta, I.; Azarnia Tehran, D.; Arrigoni, G.; Anniballi, F.; Bano, L.; Leka, O.; Zanotti, G.; Binz, T.; Montecucco, C. The first non *Clostridial botulinum*-like toxin cleaves VAMP within the juxtamembrane domain. *Sci. Rep.* **2016**, *6*, 30257. [CrossRef]
8. Brunt, J.; Carter, A.T.; Stringer, S.C.; Peck, M.W. Identification of a novel botulinum neurotoxin gene cluster in *Enterococcus*. *FEBS Lett.* **2018**, *592*, 310–317. [CrossRef]
9. Zhang, S.; Lebreton, F.; Mansfield, M.J.; Miyashita, S.I.; Zhang, J.; Schwartzman, J.A.; Tao, L.; Masuyer, G.; Martinez-Carranza, M.; Stenmark, P.; et al. Identification of a Botulinum Neurotoxin-like Toxin in a Commensal Strain of *Enterococcus faecium*. *Cell Host Microbe* **2018**, *23*, 169–176.e6. [CrossRef]
10. Mansfield, M.J.; Wentz, T.G.; Zhang, S.; Lee, E.J.; Dong, M.; Sharma, S.K.; Doxey, A.C. Bioinformatic discovery of a toxin family in *Chryseobacterium piperi* with sequence similarity to botulinum neurotoxins. *Sci. Rep.* **2019**, *9*, 1634. [CrossRef]
11. Wentz, T.G.; Muruvanda, T.; Lomonaco, S.; Thirunavukkarasu, N.; Hoffmann, M.; Allard, M.W.; Hodge, D.R.; Pillai, S.P.; Hammack, T.S.; Brown, E.W.; et al. Closed Genome Sequence of *Chryseobacterium piperi* Strain CTM(T)/ATCC BAA-1782, a Gram-Negative Bacterium with Clostridial Neurotoxin-Like Coding Sequences. *Genome Announc.* **2017**, *5*, e01296-17. [CrossRef] [PubMed]
12. Dong, M.; Stenmark, P. The Structure and Classification of Botulinum Toxins. In *Handbook of Experimental Pharmacology*; Springer: Berlin/Heidelberg, Germany, 2019; pp. 1–23.
13. Chapeton-Montes, D.; Plourde, L.; Bouchier, C.; Ma, L.; Diancourt, L.; Criscuolo, A.; Popoff, M.R.; Bruggemann, H. The population structure of *Clostridium tetani* deduced from its pan-genome. *Sci. Rep.* **2019**, *9*, 11220. [CrossRef] [PubMed]
14. Cohen, J.E.; Wang, R.; Shen, R.F.; Wu, W.W.; Keller, J.E. Comparative pathogenomics of *Clostridium tetani*. *PLoS ONE* **2017**, *12*, e0182909. [CrossRef] [PubMed]
15. Singh, B.R.; Wang, T.; Kukreja, R.; Cai, S. The botulinum neurotoxin complex and the role of ancillary proteins. In *Molecular Aspects of Botulinum Neurotoxin*; Foster, K.A., Ed.; Springer: New York, NY, USA, 2014; Volume 4, pp. 68–101.
16. Hauser, D.; Eklund, M.W.; Boquet, P.; Popoff, M.R. Organization of the botulinum neurotoxin C1 gene and its associated non-toxic protein genes in *Clostridium botulinum* C468. *Mol. Gen. Genet.* **1994**, *243*, 631–640. [CrossRef] [PubMed]
17. Gu, S.; Rumpel, S.; Zhou, J.; Strotmeier, J.; Bigalke, H.; Perry, K.; Shoemaker, C.B.; Rummel, A.; Jin, R. Botulinum neurotoxin is shielded by NTNHA in an interlocked complex. *Science* **2012**, *335*, 977–981. [CrossRef]
18. Doxey, A.C.; Lynch, M.D.; Muller, K.M.; Meiering, E.M.; McConkey, B.J. Insights into the evolutionary origins of clostridial neurotoxins from analysis of the *Clostridium botulinum* strain A neurotoxin gene cluster. *BMC Evol. Biol.* **2008**, *8*, 316. [CrossRef]
19. Popoff, M.R.; Marvaud, J.C. Structural and genomic features of clostridial neurotoxins. In *The Comprehensive Sourcebook of Bacterial Protein Toxins*, 2nd ed.; Alouf, J.E., Freer, J.H., Eds.; Academic Press: London, UK, 1999; Volume 2, pp. 174–201.
20. Popoff, M.R.; Bouvet, P. Genetic characteristics of toxigenic Clostridia and toxin gene evolution. *Toxicon* **2013**, *75*, 63–89. [CrossRef]
21. Hill, K.K.; Smith, T.J. Genetic diversity within *Clostridium botulinum* serotypes, botulinum neurotoxin gene clusters and toxin subtypes. *Curr. Top. Microbiol. Immunol.* **2013**, *364*, 1–20.
22. Marvaud, J.C.; Gibert, M.; Inoue, K.; Fujinaga, V.; Oguma, K.; Popoff, M.R. *botR* is a positive regulator of botulinum neurotoxin and associated non toxic protein genes in *Clostridium botulinum* A. *Mol. Microbiol.* **1998**, *29*, 1009–1018. [CrossRef]
23. Poulain, B.; Molgo, J.; Popoff, M.R. Clostridial neurotoxins: From the cellular and molecular mode of action to their therapeutic use. In *The Comprehensive Sourcebook of Bacterial Protein Toxins*, 4th ed.; Alouf, J., Ladant, D., Popoff, M.R., Eds.; Elsevier: Amsterdam, The Netherlands, 2015; pp. 287–336.
24. Fujinaga, Y.; Popoff, M.R. Translocation and dissemination of botulinum neurotoxin from the intestinal tract. *Toxicon* **2018**, *147*, 13–18. [CrossRef]
25. Lee, K.; Zhong, X.; Gu, S.; Kruel, A.M.; Dorner, M.B.; Perry, K.; Rummel, A.; Dong, M.; Jin, R. Molecular basis for disruption of E-cadherin adhesion by botulinum neurotoxin A complex. *Science* **2014**, *344*, 1405–1410. [CrossRef] [PubMed]
26. Matsumura, T.; Sugawara, Y.; Yutani, M.; Amatsu, S.; Yagita, H.; Kohda, T.; Fukuoka, S.; Nakamura, Y.; Fukuda, S.; Hase, K.; et al. Botulinum toxin A complex exploits intestinal M cells to enter the host and exert neurotoxicity. *Nat. Commun.* **2015**, *6*, 6255. [CrossRef] [PubMed]
27. Marvaud, J.C.; Eisel, U.; Binz, T.; Niemann, H.; Popoff, M.R. *tetR* is a positive regulator of the Tetanus toxin gene in *Clostridium tetani* and is homologous to *botR*. *Infect. Immun.* **1998**, *66*, 5698–5702. [CrossRef] [PubMed]

28. Bruggemann, H. Genomics of clostridial pathogens: Implication of extrachromosomal elements in pathogenicity. *Curr. Opin. Microbiol.* **2005**, *8*, 601–605. [CrossRef] [PubMed]

29. Raffestin, S.; Dupuy, B.; Marvaud, J.C.; Popoff, M.R. BotR/A and TetR are alternative RNA polymerase sigma factors controlling the expression of the neurotoxin and associated protein genes in *Clostridium botulinum* type A and *Clostridium tetani*. *Mol. Microbiol.* **2005**, *55*, 235–249. [CrossRef]

30. Dupuy, B.; Raffestin, S.; Matamouros, S.; Mani, N.; Popoff, M.R.; Sonenshein, A.L. Regulation of toxin and bacteriocin gene expression in *Clostridium* by interchangeable RNA polymerase sigma factors. *Mol. Microbiol.* **2006**, *60*, 1044–1057. [CrossRef]

31. Helmann, J.D. The extracytoplasmic function (ECF) sigma factors. *Adv. Microbial. Physiol.* **2002**, *46*, 47–110.

32. Chapeton-Montes, D.; Plourde, L.; Deneve, C.; Garnier, D.; Barbirato, F.; Colombié, V.; Demay, S.; Haustant, G.; Gorgette, O.; Schmitt, C.; et al. Tetanus Toxin Synthesis is Under the Control of A Complex Network of Regulatory Genes in *Clostridium tetani*. *Toxins* **2020**, *12*, 328. [CrossRef]

33. Couesnon, A.; Raffestin, S.; Popoff, M.R. Expression of botulinum neurotoxins A and E, and associated non-toxin genes, during the transition phase and stability at high temperature: Analysis by quantitative reverse transcription-PCR. *Microbiology* **2006**, *152*, 759–770. [CrossRef]

34. Connan, C.; Popoff, M.R. Two-component systems and toxinogenesis regulation in *Clostridium botulinum*. *Res. Microbiol.* **2015**, *166*, 332–343. [CrossRef]

35. Ghosh, T.; Bose, D.; Zhang, X. Mechanisms for activating bacterial RNA polymerase. *FEMS Microbiol. Rev.* **2010**, *34*, 611–627. [CrossRef] [PubMed]

36. Roberts, M.; Rowley, G.; Kormanec, J.; Zalm, M.E.J. The Role of Alternative Sigma Factors in Pathogen Virulence. In *Foodborne Pathogens*; Gurtler, J., Doyle, M., Kornacki, J., Eds.; Springer: Cham, Switzerland, 2017; pp. 229–303. [CrossRef]

37. Sebaihia, M.; Peck, M.W.; Minton, N.P.; Thomson, N.R.; Holden, M.T.; Mitchell, W.J.; Carter, A.T.; Bentley, S.D.; Mason, D.R.; Crossman, L.; et al. Genome sequence of a proteolytic (Group I) *Clostridium botulinum* strain Hall A and comparative analysis of the clostridial genomes. *Genome Res.* **2007**, *17*, 1082–1092. [CrossRef] [PubMed]

38. Cook, H.; Ussery, D.W. Sigma factors in a thousand *E. coli* genomes. *Environ. Microbiol.* **2013**, *15*, 3121–3129. [CrossRef] [PubMed]

39. Gruber, T.M.; Gross, C.A. Multiple Sigma Subunits and the Partitioning of Bacterial Transcription Space. *Annu. Rev. Microbiol.* **2003**, *57*, 441–466. [CrossRef] [PubMed]

40. Espelund, M.; Klaveness, D. Botulism outbreaks in natural environments—An update. *Front. Microbiol.* **2014**, *5*, 287. [CrossRef] [PubMed]

41. Popoff, M.R. Ecology of neurotoxigenic strains of Clostridia. In *Clostridial Neurotoxins: The Molecular Pathogenesis of Tetanus and Botulism*; Montecucco, C., Ed.; Springer: Berlin/Heidelberg, Germany, 1995; Volume 195, pp. 1–29.

42. Woudstra, C.; Skarin, H.; Anniballi, F.; Fenicia, L.; Bano, L.; Drigo, I.; Koene, M.; Bayon-Auboyer, M.H.; Buffereau, J.P.; De Medici, D.; et al. Neurotoxin gene profiling of clostridium botulinum types C and D native to different countries within Europe. *Appl. Environ. Microbiol.* **2012**, *78*, 3120–3127. [CrossRef]

43. Mascher, G.; Mertaoja, A.; Korkeala, H.; Lindstrom, M. Neurotoxin synthesis is positively regulated by the sporulation transcription factor Spo0A in *Clostridium botulinum* type E. *Environ. Microbiol.* **2017**, *19*, 4287–4300. [CrossRef] [PubMed]

44. Lovenklev, M.; Holst, E.; Borch, E.; Radstrom, P. Relative neurotoxin gene expression in *Clostridium botulinum* type B, determined using quantitative reverse transcription-PCR. *Appl. Environ. Microbiol.* **2004**, *70*, 2919–2927. [CrossRef]

45. Dover, N.; Barash, J.R.; Burke, J.N.; Hill, K.K.; Detter, J.C.; Arnon, S.S. Arrangement of the *Clostridium baratii* F7 toxin gene cluster with identification of a sigma factor that recognizes the botulinum toxin gene cluster promoters. *PLoS ONE* **2014**, *9*, e97983. [CrossRef]

46. Casino, P.; Rubio, V.; Marina, A. The mechanism of signal transduction by two-component systems. *Curr. Opin. Struct. Biol.* **2010**, *20*, 763–771. [CrossRef]

47. Jacob-Dubuisson, F.; Mechaly, A.; Betton, J.M.; Antoine, R. Structural insights into the signalling mechanisms of two-component systems. *Nat. Rev. Microbiol.* **2018**, *16*, 585–593. [CrossRef] [PubMed]

48. Zschiedrich, C.P.; Keidel, V.; Szurmant, H. Molecular Mechanisms of Two-Component Signal Transduction. *J. Mol. Biol.* **2016**, *428*, 3752–3775. [CrossRef] [PubMed]

49. Zhang, Z.; Korkeala, H.; Dahlsten, E.; Sahala, E.; Heap, J.T.; Minton, N.P.; Lindstrom, M. Two-component signal transduction system CBO0787/CBO0786 represses transcription from botulinum neurotoxin promoters in *Clostridium botulinum* ATCC 3502. *PLoS Pathog.* **2013**, *9*, e1003252. [CrossRef] [PubMed]

50. Connan, C.; Brueggemann, H.; Mazuet, C.; Raffestin, S.; Cayet, N.; Popoff, M.R. Two-Component Systems Are Involved in the Regulation of Botulinum Neurotoxin Synthesis in Clostridium botulinum Type A Strain Hall. *PLoS ONE* **2012**, *7*, e41848.

51. Sonenshein, A.L. CodY, a global regulator of stationary phase and virulence in Gram-positive bacteria. *Curr. Opin. Microbiol.* **2005**, *8*, 203–207. [CrossRef]

52. Ratnayake-Lecamwasam, M.; Serror, P.; Wong, K.W.; Sonenshein, A.L. *Bacillus subtilis* CodY represses early-stationary-phase genes by sensing GTP levels. *Genes Dev.* **2001**, *15*, 1093–1103. [CrossRef]

53. Zhang, Z.; Dahlsten, E.; Korkeala, H.; Lindström, M. Positive Regulation of Botulinum Neurotoxin Gene Expression by CodY in *Clostridium botulinum* ATCC 3502. *Appl. Environ. Microbiol.* **2014**, *80*, 7651–7658. [CrossRef]

54. Fratelli, F.; Siquini, T.J.; de Abreu, M.E.; Higashi, H.G.; Converti, A.; de Carvalho, J.C. Fed-batch production of tetanus toxin by *Clostridium tetani*. *Biotechnol. Prog.* **2010**, *26*, 88–92. [CrossRef]

55. Shone, C.C.; Tranter, H.S. Growth of Clostridia and preparation of their neurotoxins. In *Clostridial Neurotoxins*; Montecucco, C., Ed.; Springer: Berlin/Heidelberg, Germany, 1995; Volume 195, pp. 143–160.

56. Siegel, L.S.; Metzger, J.F. Toxin production by *Clostridium botulinum* type A under various fermentation conditions. *Appl. Environ. Microbiol.* **1979**, *38*, 606–611. [CrossRef]

57. Sonenshein, A.L. Control of key metabolic intersections in *Bacillus subtilis*. *Nat. Rev. Microbiol.* **2007**, *5*, 917–927. [CrossRef]

58. Antunes, A.; Martin-Verstraete, I.; Dupuy, B. CcpA-mediated repression of *Clostridium difficile* toxin gene expression. *Mol. Microbiol.* **2011**, *79*, 882–899. [CrossRef] [PubMed]

59. Dineen, S.S.; Villapakkam, A.C.; Nordman, J.T.; Sonenshein, A.L. Repression of *Clostridium difficile* toxin gene expression by CodY. *Mol. Microbiol.* **2007**, *66*, 206–219. [CrossRef] [PubMed]

60. Richardson, A.R.; Somerville, G.A.; Sonenshein, A.L. Regulating the Intersection of Metabolism and Pathogenesis in Gram-positive Bacteria. *Microbiol. Spectr.* **2015**, *3*. [CrossRef]

61. Antunes, A.; Camiade, E.; Monot, M.; Courtois, E.; Barbut, F.; Sernova, N.V.; Rodionov, D.A.; Martin-Verstraete, I.; Dupuy, B. Global transcriptional control by glucose and carbon regulator CcpA in *Clostridium difficile*. *Nucleic Acids Res.* **2012**, *40*, 10701–10718. [CrossRef]

62. Bouillaut, L.; Dubois, T.; Sonenshein, A.L.; Dupuy, B. Integration of metabolism and virulence in *Clostridium difficile*. *Res. Microbiol.* **2015**, *166*, 375–383. [CrossRef]

63. Girinathan, B.P.; DiBenedetto, N.; Worley, J.N.; Peltier, J.; Arrieta-Ortiz, M.L.; Immanuel, S.R.C.; Lavin, R.; Delaney, M.L.; Cummins, C.K.; Hoffman, M.; et al. In vivo commensal control of *Clostridioides difficile* virulence. *Cell Host Microbe* **2021**, *29*, 1693–1708.e1697. [CrossRef]

64. Sonenshein, A.L. Control of sporulation initiation in *Bacillus subtilis*. *Curr. Opin. Microbiol.* **2000**, *3*, 561–566. [CrossRef]

65. Dürre, P. Ancestral sporulation initiation. *Mol Microbiol.* **2011**, *80*, 584–587. [CrossRef]

66. Dürre, P. Physiology and Sporulation in *Clostridium*. *Microbiol. Spectr.* **2014**, *2*, TBS-0010-2012. [CrossRef]

67. Paredes, C.J.; Alsaker, K.V.; Papoutsakis, E.T. A comparative genomic view of clostridial sporulation and physiology. *Nat. Rev. Microbiol.* **2005**, *3*, 969–978. [CrossRef]

68. Talukdar, P.K.; Olguín-Araneda, V.; Alnoman, M.; Paredes-Sabja, D.; Sarker, M.R. Updates on the sporulation process in *Clostridium* species. *Res. Microbiol.* **2015**, *166*, 225–235. [CrossRef] [PubMed]

69. Wörner, K.; Szurmant, H.; Chiang, C.; Hoch, J.A. Phosphorylation and functional analysis of the sporulation initiation factor Spo0A from *Clostridium botulinum*. *Mol. Microbiol.* **2006**, *59*, 1000–1012. [CrossRef] [PubMed]

70. Ravagnani, A.; Jennert, K.C.; Steiner, E.; Grunberg, R.; Jefferies, J.R.; Wilkinson, S.R.; Young, D.I.; Tidswell, E.C.; Brown, D.P.; Youngman, P.; et al. Spo0A directly controls the switch from acid to solvent production in solvent-forming clostridia. *Mol. Microbiol.* **2000**, *37*, 1172–1185. [CrossRef] [PubMed]

71. Dürre, P. Metabolic networks in *Clostridium acetobutylicum*: Interaction of sporulation, solventogenesis and toxin formation. In *Clostridia, Molecular Biology in the Postgenomic Era*; Brüggemann, H., Gottschalk, G., Eds.; Caister Academic Press: Norfolk, UK, 2009; pp. 215–227.

72. Harry, K.H.; Zhou, R.; Kroos, L.; Melville, S.B. Sporulation and enterotoxin (CPE) synthesis are controlled by the sporulation-specific sigma factors SigE and SigK in *Clostridium perfringens*. *J. Bacteriol.* **2009**, *191*, 2728–2742. [CrossRef] [PubMed]

73. Paredes-Sabja, D.; Sarker, N.; Sarker, M.R. *Clostridium perfringens* tpeL is expressed during sporulation. *Microb. Pathog.* **2011**, *51*, 384–388. [CrossRef]

74. Mackin, K.E.; Carter, G.P.; Howarth, P.; Rood, J.I.; Lyras, D. Spo0A differentially regulates toxin production in evolutionarily diverse strains of *Clostridium difficile*. *PLoS ONE* **2013**, *8*, e79666. [CrossRef]

75. Pettit, L.J.; Browne, H.P.; Yu, L.; Smits, W.K.; Fagan, R.P.; Barquist, L.; Martin, M.J.; Goulding, D.; Duncan, S.H.; Flint, H.J.; et al. Functional genomics reveals that *Clostridium difficile* Spo0A coordinates sporulation, virulence and metabolism. *BMC Genom.* **2014**, *15*, 160. [CrossRef]

76. Kirk, D.G.; Palonen, E.; Korkeala, H.; Lindstrom, M. Evaluation of normalization reference genes for RT-qPCR analysis of spo0A and four sporulation sigma factor genes in *Clostridium botulinum* Group I strain ATCC 3502. *Anaerobe* **2014**, *26*, 14–19. [CrossRef]

77. Kirk, D.G.; Zhang, Z.; Korkeala, H.; Lindstrom, M. Alternative sigma factors SigF, SigE, and SigG are essential for sporulation in *Clostridium botulinum* ATCC 3502. *Appl. Environ. Microbiol.* **2014**, *80*, 5141–5150. [CrossRef]

78. Selby, K.; Mascher, G.; Somervuo, P.; Lindström, M.; Korkeala, H. Heat shock and prolonged heat stress attenuate neurotoxin and sporulation gene expression in group I *Clostridium botulinum* strain ATCC 3502. *PLoS ONE* **2017**, *12*, e0176944. [CrossRef]

79. Dineen, S.S.; Bradshaw, M.; Johnson, E.A. Neurotoxin gene clusters in *Clostridium botulinum* type A strains: Sequence comparison and evolutionary implications. *Curr. Microbiol.* **2003**, *46*, 342–352. [CrossRef] [PubMed]

80. Brüggemann, H.; Bäumer, S.; Fricke, W.F.; Wiezr, A.; Liesagang, H.; Decker, I.; Herzberg, C.; Martinez-Arias, R.; Henne, A.; Gottschalk, G. The genome sequence of *Clostridium tetani*, the causative agent of tetanus disease. *Proc. Natl. Acad. Sci. USA* **2003**, *100*, 1316–1321. [CrossRef] [PubMed]

81. Nakamura, S.; Serikawa, T.; Yamakawa, K.; Nishida, S.; Kozaki, S.; Sakaguchi, S. Sporulation and C2 toxin production by *Clostridium botulinum* type C strains producing no C1 toxin. *Microbiol. Immunol.* **1978**, *22*, 591–596. [CrossRef] [PubMed]

82. Licona-Cassani, C.; Steen, J.A.; Zaragoza, N.E.; Moonen, G.; Moutafis, G.; Hodson, M.P.; Power, J.; Nielsen, L.K.; Marcellin, E. Tetanus toxin production is triggered by the transition from amino acid consumption to peptides. *Anaerobe* **2016**, *41*, 113–124. [CrossRef] [PubMed]

83. Whitmer, M.E.; Johnson, E.A. Development of improved defined media for *Clostridium botulinum* serotypes A, B, and E. *Appl. Environ. Microbiol.* **1988**, *54*, 753–759. [CrossRef] [PubMed]

84. Garrigues, L.; Do, T.D.; Bideaux, C.; Guillouet, S.E.; Meynial-Salles, I. Insights into *Clostridium tetani*: From genome to bioreactors. *Biotechnol. Adv.* **2021**, *54*, 107781. [CrossRef] [PubMed]

85. Bradshaw, M.; Dineen, S.S.; Maks, N.D.; Johnson, E.A. Regulation of neurotoxin complex expression in *Clostridium botulinum* strains 62A, Hall A-*hyper*, and NCTC2916. *Anaerobe* **2004**, *10*, 321–333. [CrossRef]

86. Orellana, C.A.; Zaragoza, N.E.; Licona-Cassani, C.; Palfreyman, R.W.; Cowie, N.; Moonen, G.; Moutafis, G.; Power, J.; Nielsen, L.K.; Marcellin, E. Time-course transcriptomics reveals that amino acids catabolism plays a key role in toxinogenesis and morphology in Clostridium tetani. *J. Ind. Microbiol. Biotechnol.* **2020**, *47*, 1059–1073. [CrossRef]

87. Fredrick, C.M.; Lin, G.; Johnson, E.A. Regulation of Botulinum Neurotoxin Synthesis and Toxin Complex Formation by Arginine and Glucose in *Clostridium botulinum* ATCC 3502. *Appl. Environ. Microbiol.* **2017**, *83*, e00642-17. [CrossRef]

88. Inzalaco, H.N.; Tepp, W.H.; Fredrick, C.; Bradshaw, M.; Johnson, E.A.; Pellett, S. Posttranslational Regulation of Botulinum Neurotoxin Production in *Clostridium botulinum* Hall A-hyper. *mSphere* **2021**, *6*, e0032821. [CrossRef]

89. Patterson-Curtis, S.I.; Johnson, E.A. Regulation of neurotoxin and protease formation in *Clostridium botulinum* Okra B and Hall A by arginine. *Appl. Environ. Microbiol.* **1989**, *55*, 1544–1548. [CrossRef] [PubMed]

90. Mueller, J.H.; Miller, P.A. Essential role of histidine peptides in tetanus toxin production. *J. Biol. Chem.* **1956**, *223*, 185–194. [CrossRef]

91. Porfirio, Z.; Prado, S.M.; Vancetto, M.D.C.; Fratelli, F.; Alves, E.W.; Raw, I.; Fernandes, B.L.; Camargo, A.C.M.; Lebrun, I. Specific peptides of casein pancreatic digestion enhance the production of tetanus toxin. *J. Appl. Microbiol.* **1997**, *83*, 678–684. [CrossRef] [PubMed]

92. Brüggemann, H.; Gottschalk, G. Insights in metabolism and toxin production from the complete genome sequence of *Clostridium tetani*. *Anaerobe* **2004**, *10*, 53–68. [CrossRef]

93. Fratelli, F.; Siquini, T.J.; Prado, S.M.; Higashi, H.G.; Converti, A.; de Carvalho, J.C. Effect of medium composition on the production of tetanus toxin by *Clostridium tetani*. *Biotechnol. Prog.* **2005**, *21*, 756–761. [CrossRef]

94. Artin, I.; Carter, A.T.; Holst, E.; Lovenklev, M.; Mason, D.R.; Peck, M.W.; Radstrom, P. Effects of carbon dioxide on neurotoxin gene expression in nonproteolytic *Clostridium botulinum* Type E. *Appl. Environ. Microbiol.* **2008**, *74*, 2391–2397. [CrossRef]

95. Lovenklev, M.; Artin, I.; Hagberg, O.; Borch, E.; Holst, E.; Radstrom, P. Quantitative interaction effects of carbon dioxide, sodium chloride, and sodium nitrite on neurotoxin gene expression in nonproteolytic *Clostridium botulinum* type B. *Appl. Environ. Microbiol.* **2004**, *70*, 2928–2934. [CrossRef]

96. Artin, I.; Mason, D.R.; Pin, C.; Schelin, J.; Peck, M.W.; Holst, E.; Radstrom, P.; Carter, A.T. Effects of carbon dioxide on growth of proteolytic *Clostridium botulinum*, its ability to produce neurotoxin, and its transcriptome. *Appl. Environ. Microbiol.* **2010**, *76*, 1168–1172. [CrossRef]

97. Chekabab, S.M.; Harel, J.; Dozois, C.M. Interplay between genetic regulation of phosphate homeostasis and bacterial virulence. *Virulence* **2014**, *5*, 786–793. [CrossRef]

98. Lamarche, M.G.; Wanner, B.L.; Crépin, S.; Harel, J. The phosphate regulon and bacterial virulence: A regulatory network connecting phosphate homeostasis and pathogenesis. *FEMS Microbiol. Rev.* **2008**, *32*, 461–473. [CrossRef]

99. Pennings, J.L.A.; Abachin, E.; Esson, R.; Hodemaekers, H.; Francotte, A.; Claude, J.B.; Vanhee, C.; Uhlrich, S.; Vandebriel, R.J. Regulation of *Clostridium tetani* Neurotoxin Expression by Culture Conditions. *Toxins* **2022**, *14*, 31. [CrossRef] [PubMed]

100. Siegel, L.S.; Metzger, J.F. Effect of fermentation conditions on toxin production by Clostridium botulinum type B. *Appl. Environ. Microbiol.* **1980**, *40*, 1023–1026. [CrossRef] [PubMed]

101. Peck, M.W.; Stringer, S.C.; Carter, A.T. *Clostridium botulinum* in the post-genomic era. *Food Microbiol.* **2011**, *28*, 183–191. [CrossRef] [PubMed]

102. Chen, Y.; Korkeala, H.; Linden, J.; Lindström, M. Quantitative real-time reverse transcription-PCR analysis reveals stable and prolonged neurotoxin cluster gene activity in a *Clostridium botulinum* type E strain at refrigeration temperature. *Appl. Environ. Microbiol.* **2008**, *74*, 6132–6137. [CrossRef]

103. Mascher, G.; Derman, Y.; Kirk, D.G.; Palonen, E.; Lindström, M.; Korkeala, H. The CLO3403/CLO3404 two-component system of *Clostridium botulinum* E1 Beluga is important for cold shock response and growth at low temperatures. *Appl. Environ. Microbiol.* **2014**, *80*, 399–407. [CrossRef]

104. Derman, Y.; Isokallio, M.; Lindström, M.; Korkeala, H. The two-component system CBO2306/CBO2307 is important for cold adaptation of *Clostridium botulinum* ATCC 3502. *Int. J. Food Microbiol.* **2013**, *167*, 87–91. [CrossRef]

105. Dahlsten, E.; Zhang, Z.; Somervuo, P.; Minton, N.P.; Lindström, M.; Korkeala, H. The cold-induced two-component system CBO0366/CBO0365 regulates metabolic pathways with novel roles in group I *Clostridium botulinum* ATCC 3502 cold tolerance. *Appl. Environ. Microbiol.* **2014**, *80*, 306–319. [CrossRef]

106. Papenfort, K.; Vogel, J. Regulatory RNA in bacterial pathogens. *Cell Host Microbe* **2010**, *8*, 116–127. [CrossRef]

107. Westermann, A.J. Regulatory RNAs in Virulence and Host-Microbe Interactions. *Microbiol. Spectr.* **2018**, *6*. [CrossRef]

108. Gripenland, J.; Netterling, S.; Loh, E.; Tiensuu, T.; Toledo-Arana, A.; Johansson, J. RNAs: Regulators of bacterial virulence. *Nat. Rev. Microbiol.* **2010**, *8*, 857–866. [CrossRef]

109. Waters, L.S.; Storz, G. Regulatory RNAs in bacteria. *Cell* **2009**, *136*, 615–628. [CrossRef] [PubMed]

110. Cheah, H.L.; Raabe, C.A.; Lee, L.P.; Rozhdestvensky, T.S.; Citartan, M.; Ahmed, S.A.; Tang, T.H. Bacterial regulatory RNAs: Complexity, function, and putative drug targeting. *Crit. Rev. Biochem. Mol. Biol.* **2018**, *53*, 335–355. [CrossRef] [PubMed]

111. Romby, P.; Charpentier, E. An overview of RNAs with regulatory functions in gram-positive bacteria. *Cell Mol. Life Sci.* **2010**, *67*, 217–237. [CrossRef] [PubMed]

112. Brosse, A.; Guillier, M. Bacterial Small RNAs in Mixed Regulatory Networks. *Microbiol. Spectr.* **2018**, *6*. [CrossRef]

113. Jones, A.J.; Fast, A.G.; Clupper, M.; Papoutsakis, E.T. Small and Low but Potent: The Complex Regulatory Role of the Small RNA SolB in Solventogenesis in *Clostridium acetobutylicum*. *Appl. Environ. Microbiol.* **2018**, *84*, e00597-18. [CrossRef]

114. Venkataramanan, K.P.; Jones, S.W.; McCormick, K.P.; Kunjeti, S.G.; Ralston, M.T.; Meyers, B.C.; Papoutsakis, E.T. The Clostridium small RNome that responds to stress: The paradigm and importance of toxic metabolite stress in *C. acetobutylicum*. *BMC Genom.* **2013**, *14*, 849. [CrossRef]

115. Yang, Y.; Zhang, H.; Lang, N.; Zhang, L.; Chai, C.; He, H.; Jiang, W.; Gu, Y. The Small RNA sr8384 Is a Crucial Regulator of Cell Growth in Solventogenic Clostridia. *Appl. Environ. Microbiol.* **2020**, *86*, e00665-20. [CrossRef]

116. Ohtani, K.; Shimizu, T. Regulation of Toxin Production in Clostridium perfringens. *Toxins* **2016**, *8*, 207. [CrossRef]

117. Soutourina, O. RNA-based control mechanisms of Clostridium difficile. *Curr. Opin. Microbiol.* **2017**, *36*, 62–68. [CrossRef]

118. Kreis, V.; Soutourina, O. Clostridioides difficile—Phage relationship the RNA way. *Curr. Opin. Microbiol.* **2021**, *66*, 1–10. [CrossRef]

119. Chen, Y.; Indurthi, D.C.; Jones, S.W.; Papoutsakis, E.T. Small RNAs in the genus Clostridium. *mBio* **2010**, *2*, e00340-10. [CrossRef] [PubMed]

120. Brüggemann, H.; Chapeton-Montes, D.; Plourde, L.; Popoff, M.R. Identification of a non-coding RNA and its putative involvement in the regulation of tetanus toxin synthesis in *Clostridium tetani*. *Sci. Rep.* **2021**, *11*, 4157. [CrossRef] [PubMed]

121. Ohtani, K.; Hirakawa, H.; Tashiro, K.; Yoshizawa, S.; Kuhara, S.; Shimizu, T. Identification of a two-component VirR/VirS regulon in *Clostridium perfringens*. *Anaerobe* **2010**, *16*, 258–264. [CrossRef] [PubMed]

122. Abisado, R.G.; Benomar, S.; Klaus, J.R.; Dandekar, A.A.; Chandler, J.R. Bacterial Quorum Sensing and Microbial Community Interactions. *mBio* **2018**, *9*, e02331-17. [CrossRef] [PubMed]

123. Wang, S.; Payne, G.F.; Bentley, W.E. Quorum Sensing Communication: Molecularly Connecting Cells, Their Neighbors, and Even Devices. *Annu. Rev. Chem. Biomol. Eng.* **2020**, *11*, 447–468. [CrossRef] [PubMed]

124. Jenul, C.; Horswill, A.R. Regulation of *Staphylococcus aureus* Virulence. *Microbiol. Spectr.* **2019**, *7*. [CrossRef]

125. Zhao, L.; Montville, T.J.; Schaffner, D.W. Evidence for quorum sensing in *Clostridium botulinum* 56A. *Lett. Appl. Microbiol.* **2006**, *42*, 54–58. [CrossRef] [PubMed]

126. Cooksley, C.M.; Davis, I.J.; Winzer, K.; Chan, W.C.; Peck, M.W.; Minton, N.P. Regulation of neurotoxin production and sporulation by a Putative agrBD signaling system in proteolytic *Clostridium botulinum*. *Appl. Environ. Microbiol.* **2010**, *76*, 4448–4460. [CrossRef]

127. Ihekwaba, A.E.; Mura, I.; Walshaw, J.; Peck, M.W.; Barker, G.C. An Integrative Approach to Computational Modelling of the Gene Regulatory Network Controlling *Clostridium botulinum* Type A1 Toxin Production. *PLoS Comput. Biol.* **2016**, *12*, e1005205. [CrossRef] [PubMed]

128. Sterne, M.; Wentzel, L.M. A new method for the large-scale production of high-titre botulinum formol-toxoid types C and D. *J. Immunol.* **1950**, *65*, 175–183.

129. Koch, W.; Kaplan, D. A simple method for obtaining highly potent tetanus toxin. *J. Immunol.* **1953**, *70*, 1–5. [PubMed]

130. Vraný, B.; Hnátková, Z.; Lettl, A. Production of toxic antigens in dialyzed cultures of microorganisms. *Folia Microbiol.* **1988**, *33*, 148–154. [CrossRef] [PubMed]

131. Wentzel, L.M.; Sterne, M.; Polson, A. High toxicity of pure botulinum type D toxin. *Nature* **1950**, *166*, 739–740. [CrossRef] [PubMed]

132. Rasetti-Escargueil, C.; Lemichez, E.; Popoff, M.R. Public Health Risk Associated with Botulism as Foodborne Zoonoses. *Toxins* **2019**, *12*, 17. [CrossRef] [PubMed]

133. Dabritz, H.A.; Hill, K.K.; Barash, J.R.; Ticknor, L.O.; Helma, C.H.; Dover, N.; Payne, J.R.; Arnon, S.S. Molecular Epidemiology of Infant Botulism in California and Elsewhere, 1976-2010. *J. Infect. Dis.* **2014**, *210*, 1711–1722. [CrossRef] [PubMed]

134. Singh, B.R.; Li, B.; Read, D. Botulinum versus tetanus neurotoxins: Why is botulinum neurotoxin but not tetanus neurotoxin a food poison? *Toxicon* **1995**, *33*, 1541–1547. [CrossRef]

135. Bruggemann, H.; Gottschalk, G. Comparative genomics of clostridia: Link between the ecological niche and cell surface properties. *Ann. N. Y. Acad. Sci.* **2008**, *1125*, 73–81. [CrossRef]

136. Mansfield, M.J.; Doxey, A.C. Genomic insights into the evolution and ecology of botulinum neurotoxins. *Pathog. Dis.* **2018**, *76*, 4978416. [CrossRef]

Article

A Robust and Sensitive Spectrophotometric Assay for the Enzymatic Activity of Bacterial Adenylate Cyclase Toxins

Marilyne Davi [1], Mirko Sadi [1,2], Irene Pitard [3,4], Alexandre Chenal [1] and Daniel Ladant [1,*]

[1] Biochemistry of Macromolecular Interactions Unit, Department of Structural Biology and Chemistry, Institut Pasteur, Université Paris Cité, CNRS UMR 3528, 75015 Paris, France
[2] Université Paris Cité, 75014 Paris, France
[3] Structural Bioinformatic Unit, Department of Structural Biology and Chemistry, Institut Pasteur, Université Paris Cité, CNRS UMR 3528, 75015 Paris, France
[4] Université Paris Sorbonne, 75231 Paris, France
* Correspondence: daniel.ladant@pasteur.fr; Tel.: +33-1-4568-8400

Abstract: Various bacterial pathogens are producing toxins that target the cyclic Nucleotide Monophosphate (cNMPs) signaling pathways in order to facilitate host colonization. Among them, several are exhibiting potent nucleotidyl cyclase activities that are activated by eukaryotic factors, such as the adenylate cyclase (AC) toxin, CyaA, from *Bordetella pertussis* or the edema factor, EF, from *Bacillus anthracis*. The characterization of these toxins frequently requires accurate measurements of their enzymatic activity in vitro, in particular for deciphering their structure-to-function relationships by protein engineering and site-directed mutagenesis. Here we describe a simple and robust in vitro assay for AC activity based on the spectrophotometric detection of cyclic AMP (cAMP) after chromatographic separation on aluminum oxide. This assay can accurately detect down to fmol amounts of *B. pertussis* CyaA and can even be used in complex media, such as cell extracts. The relative advantages and disadvantages of this assay in comparison with other currently available methods are briefly discussed.

Keywords: adenylate cyclase toxin; *Bordetella pertussis*; cyclic nucleotide; cAMP; spectrophotometric enzymatic assay

Key Contribution: A simple and robust in vitro spectrophotometric assay for adenylate cyclase activity is described and shown to be capable of detecting fmol amounts of *Bordetella pertussis* CyaA toxin.

Citation: Davi, M.; Sadi, M.; Pitard, I.; Chenal, A.; Ladant, D. A Robust and Sensitive Spectrophotometric Assay for the Enzymatic Activity of Bacterial Adenylate Cyclase Toxins. *Toxins* **2022**, *14*, 691. https://doi.org/10.3390/toxins14100691

Received: 12 September 2022
Accepted: 28 September 2022
Published: 8 October 2022

Publisher's Note: MDPI stays neutral with regard to jurisdictional claims in published maps and institutional affiliations.

1. Introduction

Cyclic Nucleotides Monophosphates (cNMPs) are key messengers in most cell types that have been implicated in the modulation of numerous physiological processes [1]. Many pathogens have evolved sophisticated strategies to disrupt cNMP metabolism in the hosts they attempt to colonize [2]. In particular, several bacterial pathogens produce toxins that are endowed with endogenous nucleotidyl cyclase activity [3]. These toxins are able to invade eukaryotic cells where they are stimulated by endogenous co-factors to produce large amounts of cNMP, thus disrupting cell signaling and altering cell physiology. To date, the best characterized types are the adenylate cyclase (AC) toxin, CyaA, from *Bordetella pertussis* [4,5] and the edema factor, EF, from *Bacillus anthracis* [6]. They are key virulence factors that target the immune cells to promote efficient bacterial colonization of the hosts [7–9]. Both CyaA and EF are potently activated by the eukaryotic protein, calmodulin (CaM), and have very high catalytic efficiency (with kcat > 1000 s^{-1}). Their catalytic moieties share remarkable structural similarity, while these two toxins differ markedly in their sequences, their secretion mechanisms, and their modes of invasion of eukaryotic target cells [3]. A distinct sub-family of virulent factors with nucleotidyl cyclase activity [10,11] has been recently characterized as being activated by eukaryotic actin [12]. It includes the ExoY toxin that is a type-3 virulent effector produced by *P. aeruginosa* and various ExoY-like modules present in different MARTX (Multifunctional-Autoprocessing Repeats-in-Toxin) toxins produced by certain *Vibrio* species as well as by other Gram-negative bacteria [3,13].

The characterization of these virulent factors frequently requires the determination of their enzymatic activities, which is the conversion of adenosine triphosphate (ATP) into cyclic AMP (cAMP) and pyrophosphate (PPi). Indeed, numerous studies have been devoted to the identification of critical residues in the catalytic reaction and/or the binding of the toxins to their eukaryotic co-factors [2,3]. Simple and rapid assays for determining enzymatic activities of recombinant enzymes harboring specific mutations would greatly facilitate such investigations.

One of the most widely applied in vitro techniques uses a radioactive substrate [α-^{32}P or α-^{33}P]-ATP that is converted into a radioactive cAMP product that is subsequently isolated by chromatography on neutral alumina and quantified with a scintillation counter. A main drawback of this approach is that it requires specific equipment and laboratory premises to safely handle radioactivity. To avoid this major hurdle, a variety of alternative enzymatic assays have been set up, including the measurement of cAMP with ELISA, use of fluorescent ATP substrates, measurement of the pyrophosphate with colorimetric of fluorimetric coupled assays, or measurement of the reverse reaction (eg cAMP + PPi -> ATP) with colorimetric assays (see Table 1).

Here we describe a simple and robust in vitro assay for AC activity based on the spectrophotometric detection of cAMP after chromatographic separation on aluminum oxide. This assay is cost-effective and straightforward for routine monitoring of AC toxin activity and sensitive enough to detect down to fmol amounts of *B. pertussis* AC. Moreover, with proper controls, it can also be applied to quantify AC in complex mixtures, such as cell extracts. The relative advantages and disadvantages of this assay in comparison with other methods (Table 1) are presented in the Discussion section.

Table 1. Comparison of AC assay techniques.

Assay Principle	Characteristics	Drawbacks	References
Monitor production of cNMP product			
Radioactive: conversion of radioactive NTP into radioactive cNMP separated by Al_2O_3 chromatography and quantified in a scintillation counter	highly sensitive [a] (0.1 to 10 fmol) highly specific straightforward	Legal and environmental constraints of using radioactivity Expensive	[4,6,14]
Spectrometric: conversion of NTP into cNMP separated by Al_2O_3 chromatography and quantified by absorption at 260 nm	Sensitive (1 to 100 fmol of AC) straightforward inexpensive,	Possible application in complex media but requires proper controls	Present study
ELISA: Conversion of NTP into cNMP which is quantified by immunodetection	Highly Sensitive (0.01 to 10 fmol) Highly specific	Immunodetection (ELISA) assays are cumbersome and expensive Not high throughput	[15]
LC/MS: Conversion of NTP into cNMP which is quantified by liquid chromatography and Mass spectrometry	Highly sensitive (0.01 to 10 fmol) Highly specific Application to all cNMP	Special equipment (LC/MS) required Not high throughput	[16,17]
Monitor production of PPi product [b]			
Colorimetric: detection of PPi product after hydrolysis into $2 \times$ Pi quantified by colorimetry	Sensitive (30 to 100 fmol) Mid to high throughput	Not applicable in complex media or cell extracts [b] Requirement of coupled enzyme (PPiase)	[18,19]
Fluorometric: detection of PPi product after hydrolysis into $2 \times$ Pi and coupling to production of fluorescent resorufin	Sensitive (1 to 100 fmol) Mid to high throughput Kinetic mode	Not applicable in complex media Requirement of coupled enzymes (PPiase, PNP, HRP)	[20]
Conductimetry: detection of PPi product after hydrolysis into $2 \times$ Pi quantified by conductimetry	Low sensitivity (0.1 to 10 pmol) Kinetic mode	Not applicable in complex media Requirement of coupled enzyme (PPiase)	[21]
Monitor disappearance of ATP substrate [c]			
Fluorescent: monitoring of ATP decrease with fluorescent probe Tb(III)–norfloxacin	Sensitive (10 to 100 fmol) Kinetic mode Mid to high throughput inexpensive, straightforward	Applicability in complex media? Possible interference with ATP consumption or production by other enzymes.	[22]

Table 1. *Cont.*

Assay Principle	Characteristics	Drawbacks	References
Luminescence: monitoring of ATP decrease with luciferase (d)	Sensitive (10 to 100 fmol) Mid to high throughput	Applicability in complex media? Requirement of coupled enzyme (luciferase)	[23]
Reverse reaction: conversion of cAMP and PPi into ATP			
Spectrophotometric: ATP produced from cAMP and PPi-coupled by hexokinase and glucose-6-phosphate dehydrogenase-to the formation of NADPH measured by absorption at 334 nm	Low sensitivity (0.1 to 1 pmol) Kinetic mode Mid to high throughput	Interference with ATP production or consumption by other enzymes. Applicability in complex media? Requirement of coupled enzymes (hexokinase; glucose-6-phosphate dehydrogenase)	[24,25]

(a) Sensitivity: the values correspond to the lowest levels of CyaA (AC) or EF enzymes robustly detected in standard assays; (b) assays monitoring production of PPi product are usually not applicable in complex media or extracts due to presence of either Pi or Pi-generating reactions (any NTP hydrolyzing enzymes); (c) assays monitoring the consumption of ATP substrate will require that at least 20% of initial ATP is converted into cAMP in order to obtain reliable measurements. Moreover, these assays are usually not applicable in complex media or extracts due to the potential presence of ATP hydrolyzing or producing enzymes; and (d) in the luminescent assay described by Israeli et al. [23], there is an apparent discrepancy between the consumption of ATP (monitored by decrease of luminescence) and the appearance of cAMP which remains unclarified.

2. Results

2.1. Spectrophotometric Detection of AC Activity

Chromatography over aluminum oxide has been widely used for the separation of cyclic nucleotides from other nucleotides (NTP, NDP, or NMP), as these latter nucleotides are selectively adsorbed on Al_2O_3 [26,27]. In the following assay procedure, ATP is converted into cAMP by AC in a first step, and in a second step, cAMP is separated from ATP by chromatography on Al_2O_3; at pH 7.2–7.6, deprotonated ATP (ATP^{4-}, as well as ADP^{3-} and AMP^{2-}) is retained on the alumina, whereas $cAMP^-$ is not (provided the ionic strength of the elution buffer is higher than 0.1 M NaCl—data not shown). Thus, the alumina eluent exclusively contains cAMP, which can be easily determined by measuring the absorption at 260 nm, given the high absorption coefficient of cAMP (molar extinction coefficient, ε_{260nm} = 15,000 $M^{-1}\cdot cm^{-1}$). When using relatively pure AC preparations or highly-active ACs like bacterial AC toxins, there are no other contaminating molecules that absorb light at 260 nm, so the assay is very straightforward and robust. In standard conditions, the reaction is carried out at a 0.1 mL final volume with 2 mM ATP for appropriate time (from 5 min up to several hrs). Incubation is arrested by directly adding 0.3 g of Al_2O_3 dry powder to the reaction mixture (alternatively, an excess of EDTA can be added to chelate Mg^{2+}, an essential co-factor, before transferring the reaction mixture into pre-weighted Al_2O_3 dry powder). After the addition of 0.9 mL of HBS buffer (50 mM Hepes, pH 7.5, 0.2 M NaCl), the mixture is incubated for a few minutes with agitation, the tube is centrifuged (5 min at 10–13,000× *g*), and the absorption at 260 nm (A_{260}) of supernatant is recorded. When multiple assays are carried out in parallel, it is convenient to transfer supernatant (e.g., 0.3 mL) into a UV-transparent microplate to record the absorption with a microplate reader. Raw data are directly available in a spreadsheet file, which facilitates downstream calculations. In addition, the absorption of supernatant at another wavelength (e.g., at 340 nm, A_{340}) is also recorded in order to correct for the potential light scattering that may arise from the contamination of the supernatant by Al_2O_3 particles, as illustrated in Figure 1A. Figure 1B shows the correspondence between the absorption at 260 nm minus the absorption at 340 nm (A_{260}–A_{340}) of the Al_2O_3 supernatant as a function of when the cAMP spiked into the standard assay medium. An excellent linearity is observed for cAMP, which varies from 25 to 400 µM. These values correspond to the conversion of 1.25% to 20% of the initial ATP substrate (2 mM), a range that is suitable for standard enzymatic assays. The time and concentration dependence of cAMP synthesis by the purified catalytic domain of CyaA (encoded by the first 384 residues of toxin, AC384 [28]) is shown Figure 1C. Given the high catalytic efficiency of AC384 (with k_{cat} in the range of 1500 to 2000 s^{-1}) and its excellent time stability, this simple assay allows for an accurate determination (i.e., with robust absorption values) of its AC activity down to fmol amounts of purified AC enzyme.

A

	Assay Conditions	Raw absorption values				Corrected absorption values			
		260 nm		340 nm		A260 - A340		(A260 - A340) - Blank	
1	Hepes buffer (= Blank)	0.0833	0.12	0.0573	0.0955	0.026	0.0245	0	0
2		0.0721	0.0891	0.0433	0.0642	0.0288	0.0249	0.0028	0.0004
3	ATP 2 mM	0.0646	0.0698	0.0339	0.0389	0.0307	0.0309	0.0047	0.0064
4		0.0653	0.0824	0.0328	0.0527	0.0325	0.0297	0.0065	0.0052
5	cAMP 0.1 mM	0.1935	0.23	0.0504	0.0826	0.1431	0.1474	0.1171	0.1229
6		0.1845	0.1887	0.0435	0.0474	0.141	0.1413	0.115	0.1168
7	cAMP 0.2 mM	0.2996	0.3374	0.0406	0.0815	0.259	0.2559	0.233	0.2314
8		0.2977	0.3324	0.0516	0.0878	0.2461	0.2446	0.2201	0.2201
9	cAMP 0.3 mM	0.4188	0.4238	0.0526	0.0522	0.3662	0.3716	0.3402	0.3471
10		0.5011	0.4453	0.0771	0.1176	0.424	0.3277	0.398	0.3032
11	cAMP 0.4 mM	0.5572	0.5838	0.0761	0.1088	0.4811	0.475	0.4551	0.4505
12		0.5447	0.5816	0.0497	0.0838	0.495	0.4978	0.469	0.4733

Figure 1. *Cont.*

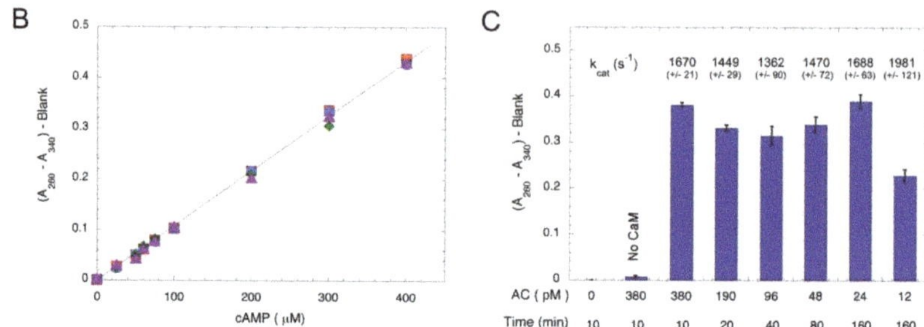

Figure 1. Photometric detection of AC. Panel (**A**): Spectrophotometric measurements of cAMP. Each line corresponds to an independent assay with the indicated nucleotide addition. Two 0.3 mL samples taken from the Al_2O_3 supernatant for each assay were transferred to a microplate and the absorption at 260 and 340 nm were recorded. The absorption at 340 nm provides a good estimate of the light scattering that arises from contaminating Al_2O_3 particles that are inadvertently aspirated in the 0.3 mL supernatant samples. Subtracting the absorption at 340 nm from the absorption at 260 nm yields a very reliable quantification of the cAMP content. Panel (**B**): Correspondence between the absorption at 260 nm minus absorption at 340 nm (A_{260}–A_{340}) of the Al_2O_3 supernatant as a function of cAMP (4 independent data points for each concentration) when it is added to the standard assay medium containing 2 mM ATP. Subtraction of the absorption at 340 nm allows for the correction of the potential light scattering arising from residual Al_2O_3 particles in the supernatant (see Material and methods for detail). Panel (**C**): Time and concentration dependence of *B. pertussis* AC activity. Reactions were carried out for the indicated times with the indicated final concentrations of AC384 in the presence of 2 mM ATP and 1 µM CaM (except in bar labeled "No CaM"). The calculated k_{cat} is indicated above the bar (mean of 4 independent experiments).

2.2. CaM-Dependent Activation of AC and EF Activities

Figure 2 shows the CaM-dependence of the AC enzymatic activity of *B. pertussis* AC384 as well as that of a purified edema factor, EF, from the *B. cereus* (G9241 strain; this protein differs from the well-studied B. anthracis EF toxin by three amino acids, Pitard et al. manuscript in preparation). Both enzymes show high catalytic activities and high CaM-affinities in accordance with prior studies [25,29,30]. These data also indicate that the three amino acid replacements in *B. cereus* EF (I318T, V694A, and N789K, using amino acid numbers from *B anthracis* EF) have no impact on the catalytic efficiency or CaM affinity of the enzyme. This is consistent with the fact that they are located at the protein surface, away from the catalytic site and the CaM binding interface [31].

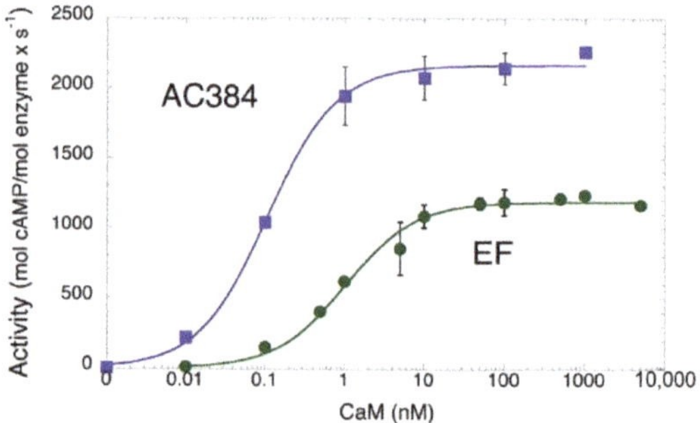

Figure 2. CaM-dependent AC activity of *B. pertussis* AC384 (blue) and *B. cereus* EF enzymes (green). AC activities (expressed in mol of cAMP produced per mol of enzyme per sec) at the indicated CaM concentrations were fitted to equation: A = (A_{Max} × [CaM])/(K_D + [CaM]). Maximal activities (A_{Max}) of 2165 ± 30 and 1183 ± 22 s^{-1} were calculated for AC384 and EF, respectively, while the CaM concentration at half-maximal activation ($\approx K_D$) were 0.11 ± 0.01 and 1.04 ± 0.13 nM for AC384 and EF, respectively.

2.3. Enzymatic Assays in Complex Media

One advantage of this spectrophotometric assay is that it does not require any coupled enzyme to quantify the formation of the cAMP product. Therefore, it can be applied to measure AC activity in conditions that might affect the activity of many enzymes, like the presence of high salt or chaotropic agent concentrations. Figure 3A,B show the AC384 activity in the presence of an increasing concentration of NaCl or urea, respectively. A gradual decrease of activity is observed in both cases, although AC384 retains a significant catalytic efficiency (i.e., higher than 50% of its maximal k_{cat}) even in the presence of up to 0.5 M NaCl or 0.2 M urea. We further examined the CaM-dependency of AC384 activity in the presence of 0.2 M NaCl or 0.2 M urea (Figure 3C) and found that although the maximal activity was reduced, the affinity for CaM was only marginally affected by these compounds.

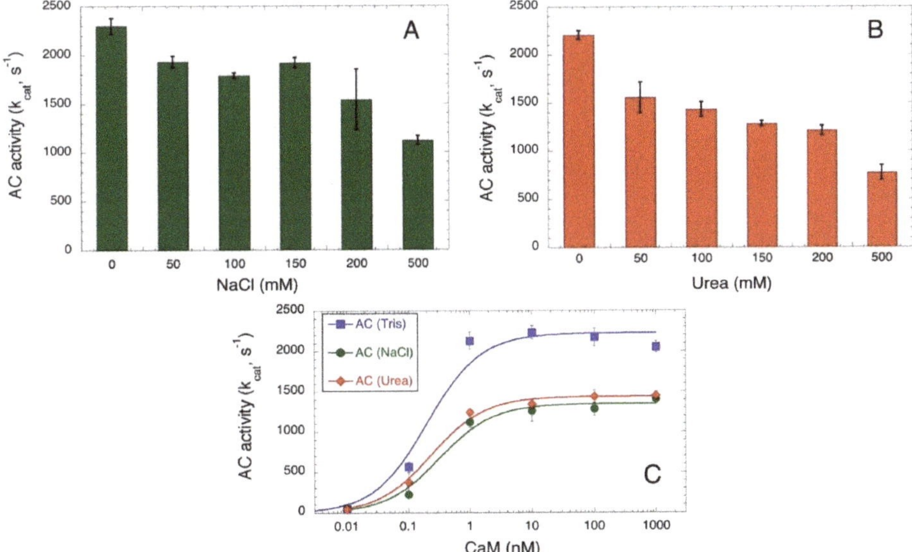

Figure 3. Salt and urea concentration dependence of AC384 activity. NaCl (Panel (**A**)) and urea (Panel (**B**)) concentration dependence of *B. pertussis* AC activity. Reactions were carried out for 10 min with 0.32 nM AC384 in the presence of 2 mM ATP and 1 μM CaM (mean and STD of 4 independent experiments). Control experiments (not shown) showed that ATP and cAMP binding to the Al_2O_3 were not modified by the presence of NaCl or urea at the tested concentrations. Panel (**C**): *B. pertussis* AC384 activities were measured at the indicated CaM concentrations in standard conditions (i.e., in Tris buffer, blue square) or in the presence of 0.2 M NaCl (green circle) or 0.2 M urea (red diamond). Data were fitted as in Figure 2. Maximal activities (in mol of cAMP per mol of AC384 per sec) were 2225 ± 95, 1351 ± 50, and 1440 ± 32 s^{-1} for assays in Tris buffer, in 0.2 M NaCl, and 0.2 M urea, respectively, while the CaM concentrations at half-maximal activation ($\approx K_D$) were 0.20 ± 0.06, 0.32 ± 0.08, and 0.24 ± 0.04 nM, respectively.

2.4. Characterization of a Detoxified CyaAE5-OVA Vaccine

The non-radioactive AC assay can be useful for the characterization of CyaA-based vaccines. Indeed, earlier works have shown that detoxified recombinant CyaAs can be used as efficient antigen delivery vehicles. Various CyaA-based vaccines have been designed to trigger immune responses against infectious agents and/or cancer-specific antigens [32–34] (for a review [35]). The demonstration of the lack of AC activity (that is responsible for CyaA toxicity) of the detoxified CyaA vaccine is key in the process of GMP (Good Manufacturing Practice) preparation of protein lots for clinical evaluation. The spectrophotometric AC assay can be easily implemented for this, as illustrated in Figure 4, where a detoxified CyaA protein (CyaAE5, resulting from a specific mutation in catalytic site) carrying a model epitope (OVA, derived from ovalbumin), CyaAE5-OVA [36,37], was assayed at various concentrations. No traces of cAMP synthesis could be evidenced even with the highest concentration tested (200 nM CyaAE5-OVA, corresponding to 20 pmol per assay). Given that CyaAE5-OVA (as all CyaA proteins) was stored in a buffer containing ≈6.6 M urea to prevent the irreversible aggregation of the protein [5,38,39], we checked that the residual urea added to each assay (as indicated in Figure 4) did not affect the enzymatic detection by spiking sub-pmol amounts of active AC384. In all cases, high amounts of cAMP were easily detected. These experiments thus unambiguously demonstrate the total lack of AC activity of the purified CyaAE5-OVA preparation and indicate that this simple spectrophotometric AC assay could be easily implemented for control of GMP preparations of detoxified CyaA vaccines.

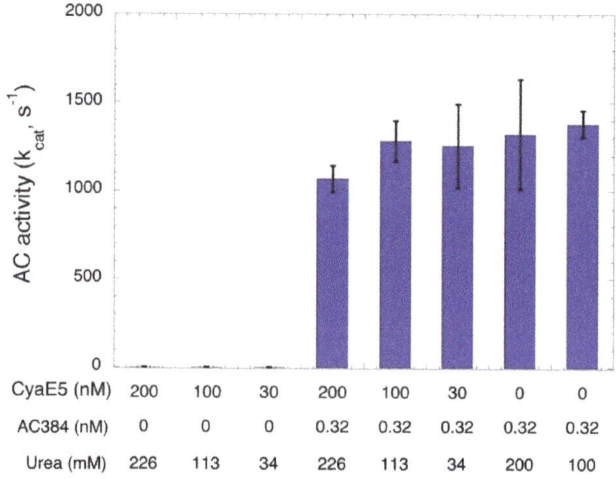

CyaE5 (nM)	200	100	30	200	100	30	0	0
AC384 (nM)	0	0	0	0.32	0.32	0.32	0.32	0.32
Urea (mM)	226	113	34	226	113	34	200	100

Figure 4. Characterization of a detoxified CyaAE5-OVA vaccine. AC activity of the purified CyaAE5-OVA protein (stored at 5.8 µM in 6.6 M urea, 20 mM Hepes-Na) at the indicated concentrations was monitored at 30 °C in the presence of 2 mM ATP and 1 µM CaM for 10 min (bars 1 to 6). As a control, in bars 4–6, 0.32 nM AC384 were spiked in the reaction mixtures in addition to the CyaAE5-OVA protein, in order to check that the residual urea concentration in each assay (indicated on the bottom line) did not affect the enzymatic activity. Results are the mean and STD of four independent experiments.

2.5. AC Assays in Crude Cell Extracts

To determine if the spectrophotometric assay could also be applied to unpurified or partially-purified AC preparations, or when the AC toxins would be diluted into complex cell extracts, we attempted to characterize CyaA binding to erythrocytes, which are frequently used as model target cells. Although these cells do not express the CyaA receptor (CD11b/CD18), CyaA can directly bind to the erythrocyte plasma membrane in a calcium- and temperature-dependent manner [40–43].

Purified CyaA was diluted in erythrocyte suspensions and incubated for 20 min at 30 °C in the presence of calcium (or in the presence of EDTA, or at 4 °C as controls). The cells were then extensively washed, resuspended in a small volume, and lysed with non-ionic detergent (e.g., Tween 20). The AC activities of the lysed samples were then assayed for various incubation times (0, 10, 20, and 30 min), and the absorption of the Al_2O_3 supernatants of all samples were recorded at different wavelengths (ie 260, 340, 405, and 595 nm). As anticipated, the cell lysates showed high absorption at 260, 340, or 405 nm (Figure 5). Yet, in control erythrocytes not incubated with CyaA, the absorption values remained constant over all the incubation periods (Figure 5A). In lysates of cells incubated with CyaA at 30 °C and in the presence of calcium (Figure 5B), the absorption at 260 nm (A_{260}) increased linearly with time while the absorption at other wavelengths remained constant. By independently monitoring the cAMP content in the Al_2O_3 supernatants with an ELISA assay, we confirmed that the A_{260} increase precisely corresponded to the synthesis of cAMP due to the CyaA enzyme bound to cells (Figure 5B). The binding of CyaA to erythrocytes at 4 °C in the presence of calcium or at 30 °C in the absence of calcium (Figure 5C) was reduced in accordance with prior studies [41–45]. Hence, this spectrophotometric assay may be adapted to monitor AC toxin activity in complex cell extracts despite high absorption at useful wavelengths.

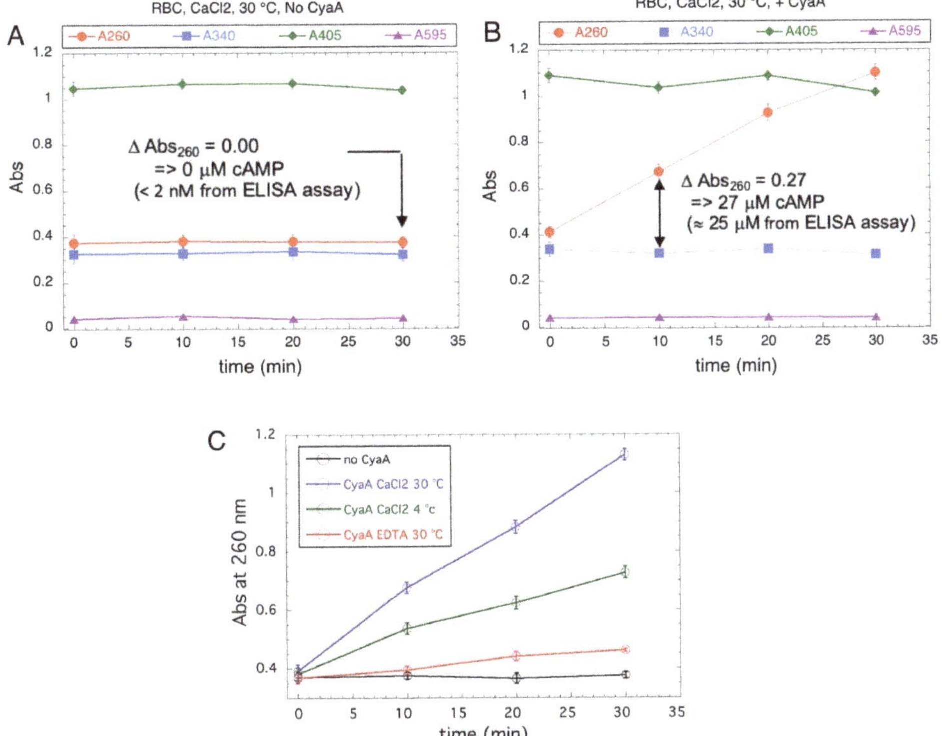

Figure 5. Assay of CyaA binding to erythrocyte. Panels A and B: Erythrocytes (RBC) were incubated at 30 °C with 2 mM $CaCl_2$, for 30 min without CyaA (Panel (**A**)) or with 11.2 nM CyaA (panel (**B**)), and then extensively washed as described in the Methods section. After lysis, 20 µL of RBC lysates were tested in standard AC assays for the indicated times and the absorption of the Al_2O_3 supernatants were recorded at 260, 340, 405, and 595 nm. The cyclic AMP content in the Al_2O_3 supernatants was determined for sample A after 30 min of incubation and for sample B after 10 min of incubation by a specific ELISA assay and compared to the cAMP quantification deduced from the absorption at 260 nm (Abs_{260}). Panel (**C**): Erythrocytes were incubated without (no CyaA, black) or with 11.2 nM CyaA at 4 or 30 °C with 2 mM $CaCl_2$ or 2 mM EDTA for 30 min (as indicated). After extensive washing and cell lysis, 20 µL of RBC lysates were tested in standard AC assays as above for the indicated times. The relative fractions of "bound" CyaA activity, as compared to the total CyaA added to each cell suspension, correspond to about 2.4% for the incubation at 30 °C in the presence of $CaCl_2$ (blue), about 1.1% for the incubation at 4 °C in the presence of $CaCl_2$ (green), and less than 0.15% for the incubation at 30 °C in the presence of EDTA (red).

3. Discussion

We show here that the direct spectrophotometric detection of cAMP after chromatographic separation on aluminum oxide can be used to provide a robust in vitro assay for bacterial AC toxin activities. This simple assay allows for an accurate determination (i.e., with robust absorption values) of the AC activity of *B. pertussis* CyaA. Thanks to the high catalytic efficiency ($k_{cat} > 1000$ s^{-1}) and excellent time stability of this enzyme, we could detect down to the fmol amounts of purified AC. The assay is cost-effective and straightforward for routine monitoring of AC activity in safe conditions (i.e., without requiring manipulation of radioactivity). It can be easily set up in any laboratory without requiring expensive equipment or facilities (e.g., for manipulating radioactivity). It is perfectly appropriate for most standard assays of the bacterial nucleotidyl cyclase toxins that exhibit a high catalytic activity, such as the *B. pertussis* CyaA and *B. anthracis* EF—as shown here—and could be adapted as well for *P. aeruginosa* ExoY and other Gram- ExoY-like enzymes, which have turnover numbers in the range of 50 to 500 s^{-1} (depending on enzymes and/or substrates) [10,11,46,47]. Moreover, this spectrophotometric assay can also be applied to quantify the AC activity in complex mixtures, such as cell extracts, by shifting to a kinetic mode where the cAMP synthesis is measured via the specific increase of absorption at 260 nm (A_{260}) as a function of the incubation time. Of course, proper controls should be performed to reliably establish that the A_{260} increase corresponds only to cAMP synthesis. It might also be adapted to characterize other less active nucleotidyl cyclases found in numerous bacterial or eukaryotic species—again provided that careful controls are applied.

Table 1 presents a comparison of the diverse assays that have been used to characterize bacterial AC toxins with their main advantages and drawbacks. The spectrometric assay described here lies in between the more sensitive and specific, but rather cumbersome, techniques that rely on cNMP detection (e.g., via radioactivity, ELISA, or LC/MS), and more high-throughput methods that might be less sensitive and more susceptible to interference by contaminating NTP- and/or pyrophosphate-metabolizing enzymes. Given its robustness and simplicity to set up, this in vitro spectrometric assay may thus become a method of choice for the routine characterization of the enzymatic activities of purified or partially-purified bacterial cyclase toxins.

4. Materials and Methods

4.1. Materials

Aluminum oxide 90 active neutral (activity stage I) for column chromatography was obtained from Merck (ref # 101077, 70–230 mesh ASTM, Al$_2$O$_3$). UV-transparent 96-well microplates (Nunc™ UV 96 well) were purchased from Thermo Scientific (Waltham, MA, USA). ATP, cAMP, Bovine Serum Albumin, and Tween 20 were from Sigma-Aldrich (Saint-Louis, MO, USA). All recombinant proteins were expressed in *E. coli* and purified as described previously: CyaA protein [39,42], CyaAE5-OVA [36,37], AC384 (catalytic domain of CyaA, corresponding to residues 1 to 384), and CaM [28]. The EF enzyme tested here was from the *Bacillus cereus* G9241 strain (kind gift from P. L. Goossens, Pitard et al. manuscript in preparation) and comprises the adenylyl cyclase domain without the protective antigen-binding domain (i.e., residues 291 to 798 of full-length EF). This protein differs from the wild-type EF toxin from *B anthracis* by three amino acid changes: I318T, V694A, N789K (aa numbering from *B anthracis* EF shown in 1XFV.pdb, [48]). It was expressed in *E. coli* and purified as described by Drum et al. [31].

4.2. Enzymatic AC Assay

The enzymatic reactions were carried out in 1.5 mL Eppendorf tubes by sequentially adding: 50 µL of AC×2 buffer (Adenylate Cyclase assay buffer 2-times concentrated: 100 mM Tris-HCl, pH 8.0, 15 mM MgCl$_2$, 0.2 mM CaCl$_2$, 1 mg/mL bovine serum albumin (BSA), 10 µL of CaM (calmodulin) diluted in buffer D (dilution buffer: 10 mM Tris-HCl, pH 8.0, 0.1% Tween 20, or other non-ionic detergent such NP40 or Triton ×100) to appropriate concentrations (final concentration of 2 µM for standard assays), 10 to 30 µL of enzyme samples to be tested, diluted in buffer D to appropriate concentrations. Buffer D was added to a complete volume of 90 μL. The tubes were equilibrated at 30 °C for 3–5 min in a Thermomixer (Eppendorf, Montesson, France), and the enzymatic reactions were initiated by adding 10 μL of 20 mM ATP (stock solution in H_2O, adjusted to pH ≈ 7.5). The tubes were incubated at 30 °C for appropriate incubation times (usually 5–20 min, but it could be extended to several hrs when needed, see Figure 1). A blank assay containing no enzyme (or no CaM) was carried out in parallel. The enzymatic reactions were stopped by adding 0.3 g (conveniently measured with a 0.2 mL Eppendorf tube) of dried aluminum oxide 90 active powder (the reaction was stopped immediately as the ATP binds to alumina). Alternatively, the enzymatic reaction could be stopped by adding 100 μL of 50 mM HEPES pH 7.5 or 50 mM EDTA (stop solution) to chelate the Mg^{2+} essential co-factor. Then, 0.9 mL (or 0.8 mL if 100 μL of stop solution have been added) of 20 mM Hepes-Na, pH 7.5 or 0.15 M NaCl (Chromatography buffer) were added to elute cAMP from the alumina powder. The tubes were mixed by inversion a few times over the course of 3–5 min and then centrifuged for 5 min at 12–14,000 rpm (room temperature) to pellet the alumina powder. The supernatants were carefully collected and the absorption at 260 and 340 nm were recorded. The cAMP concentration was then determined using an absorption coefficient at 260 nm of 15.4 mM^{-1} after subtracting the background absorption of the blank tube. The measurement of the absorption at 340 nm (A_{340}) is very convenient for monitoring the potential light diffusion that may happen due to the presence of traces contaminating alumina powder in the supernatant solution. In our routine procedure, the OD_{260}–OD_{340} was used to measure the cAMP content in each fraction.

A convenient format to carry out multiple assays is to transfer 0.3 mL aliquots of alumina supernatants into wells of a UV-transparent acrylic microplate, which allows for easy sequential measurements of the absorption at 260 and 340 nm (as well as other wavelengths see below) with direct data being recorded into an excel sheet. This greatly facilitates the downstream calculations.

4.3. Assay of CyaA Binding to Erythrocytes

CyaA toxin binding to sheep erythrocytes was essentially assayed as previously described [42,43]. Sheep erythrocytes (from Charles River Laboratories, Wilmington, MA, USA) were washed and resuspended (≈5% of dry pellet) in HBS buffer (20 mM Hepes-Na, pH 7.5, 150 mM NaCl) supplemented with 5 mM glucose. The purified CyaA toxin (stored at 1 mg/mL in 8 M urea, 20 mM Hepes-Na) was directly diluted to 11.2 nM (2 μg/mL final concentration) into 1 mL of erythrocyte suspension supplemented with either 2 mM $CaCl_2$ or 2 mM EDTA. A 20 μL aliquot was removed to determine the total adenylate cyclase activity added to each sample. The mixtures were then incubated at 30 °C, or 4 °C, for 30 min. The cell suspensions were chilled on ice, centrifuged at 4 °C, and cell pellets were washed two times with 1 mL cold HBS buffer (transferring the resuspended cells into new tubes to eliminate potential non-specific absorption of CyaA to the tube wall). Finally, the pelleted erythrocytes were resuspended in 100 μL of 10 mM Tris-HCl, pH 8.0, containing 0.1% Triton X−100 in order to lyse the cells. A total of 20 μL of lysates were tested in standard AC assays as described above and incubated for 0, 10, 20, or 30 min. The supernatants of aluminum oxide were then collected in a microplate and the absorption at 260, 340, 405, and 595 nm (the build-in filters available in our microplate reader–other wavelengths could be used as well in addition to the 260 nm) were successively recorded. Independent quantification of cAMP was performed on Al_2O_3 supernatants with an ELISA assay using a cAMP-biotinylated-BSA conjugate coated on ELISA plates and detected with a specific rabbit anti-cAMP antiserum [49].

Author Contributions: Conceptualization: D.L.; Investigation: M.D., M.S., I.P., A.C. and D.L.; Writing: original draft preparation: D.L.; Writing: review and editing: M.D., M.S., I.P., A.C. and D.L.; Project administration: D.L.; Funding acquisition: A.C. and D.L. All authors have read and agreed to the published version of the manuscript.

Funding: This research was funded by Institut Pasteur, the Centre National de la Recherche Scientifique, CNRS UMR 3528 (Biologie Structurale et Agents Infectieux), and Agence Nationale de la Recherche (grant #: ANR 21-CE11-0014-01-TransCyaA). M.S. was supported by the Pasteur–Paris University (PPU) International PhD Program. I.P. was funded by DGA support 2017033 and the Ecole Doctorale Complexité du Vivant (ED515).

Institutional Review Board Statement: Not applicable.

Informed Consent Statement: Not applicable.

Data Availability Statement: Data is contained within the article.

Acknowledgments: We thank P. L. Goossens, Institut Pasteur, and the Production and Purification of Recombinant Proteins Facility from Institut Pasteur for their assistance in producing the *B. cereus* EF enzyme.

Conflicts of Interest: The authors declare no conflict of interest.

References

1. Seifert, R.; Schneider, E.H.; Bähre, H. From Canonical to Non-Canonical Cyclic Nucleotides as Second Messengers: Pharmacological Implications. *Pharmacol. Ther.* **2015**, *148*, 154–184. [CrossRef]
2. *The Comprehensive Sourcebook of Bacterial Protein Toxins*, 4th ed.; Alouf, J.E.; Ladant, D.; Popoff, M.R. (Eds.) Elsevier: Amsterdam, The Netherlands, 2015.
3. Teixeira-Nunes, M.; Retailleau, P.; Comisso, M.; Deruelle, V.; Mechold, U.; Renault, L. Bacterial Nucleotidyl Cyclases Activated by Calmodulin or Actin in Host Cells: Enzyme Specificities and Cytotoxicity Mechanisms Identified to Date. *IJMS* **2022**, *23*, 6743. [CrossRef] [PubMed]
4. Hewlett, E.L.; Urban, M.A.; Manclark, C.R.; Wolff, J. Extracytoplasmic Adenylate Cyclase of *Bordetella pertussis*. *Proc. Natl. Acad. Sci. USA* **1976**, *73*, 1926–1930. [CrossRef] [PubMed]
5. Rogel, A.; Schultz, J.E.; Brownlie, R.M.; Coote, J.G.; Parton, R.; Hanski, E. *Bordetella pertussis* Adenylate Cyclase: Purification and Characterization of the Toxic Form of the Enzyme. *EMBO J.* **1989**, *8*, 2755–2760. [CrossRef] [PubMed]
6. Leppla, S.H. Anthrax Toxin Edema Factor: A Bacterial Adenylate Cyclase That Increases Cyclic AMP Concentrations of Eukaryotic Cells. *Proc. Natl. Acad. Sci. USA* **1982**, *79*, 3162–3166. [CrossRef] [PubMed]
7. Masin, J.; Osicka, R.; Bumba, L.; Sebo, P. *Bordetella* Adenylate Cyclase Toxin: A Unique Combination of a Pore-Forming Moiety with a Cell-Invading Adenylate Cyclase Enzyme. *Pathog. Dis.* **2015**, *73*, ftv075. [CrossRef] [PubMed]
8. Guiso, N. *Bordetella* Adenylate Cyclase-Hemolysin Toxins. *Toxins* **2017**, *9*, 277. [CrossRef]
9. Moayeri, M.; Leppla, S.H.; Vrentas, C.; Pomerantsev, A.P.; Liu, S. Anthrax Pathogenesis. *Annu. Rev. Microbiol.* **2015**, *69*, 185–208. [CrossRef]
10. Yahr, T.L.; Vallis, A.J.; Hancock, M.K.; Barbieri, J.T.; Frank, D.W. ExoY, an Adenylate Cyclase Secreted by the *Pseudomonas Aeruginosa* Type III System. *Proc. Natl. Acad. Sci. USA* **1998**, *95*, 13899–13904. [CrossRef]
11. Ochoa, C.D.; Alexeyev, M.; Pastukh, V.; Balczon, R.; Stevens, T. *Pseudomonas Aeruginosa* Exotoxin Y Is a Promiscuous Cyclase That Increases Endothelial Tau Phosphorylation and Permeability. *J. Biol. Chem.* **2012**, *287*, 25407–25418. [CrossRef]
12. Belyy, A.; Raoux-Barbot, D.; Saveanu, C.; Namane, A.; Ogryzko, V.; Worpenberg, L.; David, V.; Henriot, V.; Fellous, S.; Merrifield, C.; et al. Actin Activates *Pseudomonas Aeruginosa* ExoY Nucleotidyl Cyclase Toxin and ExoY-like Effector Domains from MARTX Toxins. *Nat. Commun.* **2016**, *7*, 13582. [CrossRef] [PubMed]
13. Ziolo, K.J.; Jeong, H.-G.; Kwak, J.S.; Yang, S.; Lavker, R.M.; Satchell, K.J.F. *Vibrio vulnificus* Biotype 3 Multifunctional Autoprocessing RTX Toxin Is an Adenylate Cyclase Toxin Essential for Virulence in Mice. *Infect. Immun.* **2014**, *82*, 2148–2157. [CrossRef] [PubMed]
14. White, A.A. Separation and Purification of Cyclic Nucleotides by Alumina Column Chromatography. *Methods Enzymol.* **1974**, *38*, 41–46. [CrossRef]
15. Duriez, E.; Goossens, P.L.; Becher, F.; Ezan, E. Femtomolar Detection of the Anthrax Edema Factor in Human and Animal Plasma. *Anal. Chem.* **2009**, *81*, 5935–5941. [CrossRef]
16. Gottle, M.; Dove, S.; Kees, F.; Schlossmann, J.; Geduhn, J.; Konig, B.; Shen, Y.; Tang, W.-J.; Kaever, V.; Seifert, R. Cytidylyl and Uridylyl Cyclase Activity of *Bacillus anthracis* Edema Factor and *Bordetella pertussis* CyaA. *Biochemistry* **2010**, *49*, 5494–5503. [CrossRef]
17. Beste, K.Y.; Spangler, C.M.; Burhenne, H.; Koch, K.-W.; Shen, Y.; Tang, W.-J.; Kaever, V.; Seifert, R. Nucleotidyl Cyclase Activity of Particulate Guanylyl Cyclase A: Comparison with Particulate Guanylyl Cyclases E and F, Soluble Guanylyl Cyclase and Bacterial Adenylyl Cyclases CyaA and Edema Factor. *PLoS ONE* **2013**, *8*, e70223. [CrossRef]
18. Laine, E.; Goncalves, C.; Karst, J.C.; Lesnard, A.; Rault, S.; Tang, W.-J.; Malliavin, T.E.; Ladant, D.; Blondel, A. Use of Allostery to Identify Inhibitors of Calmodulin-Induced Activation of *Bacillus anthracis* Edema Factor. *Proc. Natl. Acad. Sci. USA* **2010**, *107*, 11277–11282. [CrossRef]

19. Selwa, E.; Davi, M.; Chenal, A.; Sotomayor-Perez, A.-C.; Ladant, D.; Malliavin, T.E. Allosteric Activation of *Bordetella pertussis* Adenylyl Cyclase by Calmodulin: Molecular Dynamics and Mutagenesis Studies. *J. Biol. Chem.* **2014**, *289*, 21131–21141. [CrossRef]

20. Stein, R.L. Kinetic Studies of the Activation of *Bordetella pertussis* Adenylate Cyclase by Calmodulin. *Biochemistry* **2022**, *61*, 554–562. [CrossRef]

21. Lawrence, A.J.; Coote, J.G.; Kazi, Y.F.; Lawrence, P.D.; MacDonald-Fyall, J.; Orr, B.M.; Parton, R.; Riehle, M.; Sinclair, J.; Young, J.; et al. A Direct Pyrophosphatase-Coupled Assay Provides New Insights into the Activation of the Secreted Adenylate Cyclase from *Bordetella pertussis* by Calmodulin. *J. Biol. Chem.* **2002**, *277*, 22289–22296. [CrossRef]

22. Spangler, C.M.; Spangler, C.; Göttle, M.; Shen, Y.; Tang, W.-J.; Seifert, R.; Schäferling, M. A Fluorimetric Assay for Real-Time Monitoring of Adenylyl Cyclase Activity Based on Terbium Norfloxacin. *Anal. Biochem.* **2008**, *381*, 86–93. [CrossRef] [PubMed]

23. Israeli, M.; Rotem, S.; Elia, U.; Bar-Haim, E.; Cohen, O.; Chitlaru, T. A Simple Luminescent Adenylate-Cyclase Functional Assay for Evaluation of *Bacillus anthracis* Edema Factor Activity. *Toxins* **2016**, *8*, 243. [CrossRef] [PubMed]

24. Munier, H.; Bouhss, A.; Krin, E.; Danchin, A.; Gilles, A.M.; Glaser, P.; Barzu, O. The Role of Histidine 63 in the Catalytic Mechanism of *Bordetella pertussis* Adenylate Cyclase. *J. Biol. Chem.* **1992**, *267*, 9816–9820. [CrossRef]

25. Guo, Q.; Shen, Y.; Zhukovskaya, N.L.; Florián, J.; Tang, W.-J. Structural and Kinetic Analyses of the Interaction of Anthrax Adenylyl Cyclase Toxin with Reaction Products CAMP and Pyrophosphate. *J. Biol. Chem.* **2004**, *279*, 29427–29435. [CrossRef] [PubMed]

26. White, A.A.; Zenser, T.V. Separation of Cyclic 3′,5′-Nucleoside Monophosphates from Other Nucleotides on Aluminum Oxide Columns. Application to the Assay of Adenyl Cyclase and Guanyl Cyclase. *Anal. Biochem.* **1971**, *41*, 372–396. [CrossRef]

27. Ramachandran, J. A New Simple Method for Separation of Adenosine 3′,5′-Cyclic Monophosphate from Other Nucleotides and Its Use in the Assay of Adenyl Cyclase. *Anal. Biochem.* **1971**, *43*, 227–239. [CrossRef]

28. Vougier, S.; Mary, J.; Dautin, N.; Vinh, J.; Friguet, B.; Ladant, D. Essential Role of Methionine Residues in Calmodulin Binding to *Bordetella pertussis* Adenylate Cyclase, as Probed by Selective Oxidation and Repair by the Peptide Methionine Sulfoxide Reductases. *J. Biol. Chem.* **2004**, *279*, 30210–30218. [CrossRef]

29. Shen, Y.; Lee, Y.-S.; Soelaiman, S.; Bergson, P.; Lu, D.; Chen, A.; Beckingham, K.; Grabarek, Z.; Mrksich, M.; Tang, W.-J. Physiological Calcium Concentrations Regulate Calmodulin Binding and Catalysis of Adenylyl Cyclase Exotoxins. *EMBO J.* **2002**, *21*, 6721–6732. [CrossRef]

30. O'Brien, D.P.; Durand, D.; Voegele, A.; Hourdel, V.; Davi, M.; Chamot-Rooke, J.; Vachette, P.; Brier, S.; Ladant, D.; Chenal, A. Calmodulin Fishing with a Structurally Disordered Bait Triggers CyaA Catalysis. *PLoS Biol.* **2017**, *15*, e2004486. [CrossRef]

31. Drum, C.L.; Yan, S.-Z.; Bard, J.; Shen, Y.-Q.; Lu, D.; Soelaiman, S.; Grabarek, Z.; Bohm, A.; Tang, W.-J. Structural Basis for the Activation of Anthrax Adenylyl Cyclase Exotoxin by Calmodulin. *Nature* **2002**, *415*, 396–402. [CrossRef]

32. Saron, M.F.; Fayolle, C.; Sebo, P.; Ladant, D.; Ullmann, A.; Leclerc, C. Anti-Viral Protection Conferred by Recombinant Adenylate Cyclase Toxins from *Bordetella pertussis* Carrying a CD8+ T Cell Epitope from Lymphocytic Choriomeningitis Virus. *Proc. Natl. Acad. Sci. USA* **1997**, *94*, 3314–3319. [CrossRef] [PubMed]

33. Guermonprez, P.; Fayolle, C.; Rojas, M.-J.; Rescigno, M.; Ladant, D.; Leclerc, C. In Vivo Receptor-Mediated Delivery of a Recombinant Invasive Bacterial Toxoid to CD11c + CD8 Alpha -CD11bhigh Dendritic Cells. *Eur. J. Immunol.* **2002**, *32*, 3071–3081. [CrossRef]

34. Preville, X.; Ladant, D.; Timmerman, B.; Leclerc, C. Eradication of Established Tumors by Vaccination with Recombinant *Bordetella pertussis* Adenylate Cyclase Carrying the Human Papillomavirus 16 E7 Oncoprotein. *Cancer Res.* **2005**, *65*, 641–649. [CrossRef] [PubMed]

35. Ladant, D. Bioengineering of *Bordetella pertussis* Adenylate Cyclase Toxin for Vaccine Development and Other Biotechnological Purposes. *Toxins* **2021**, *13*, 83. [CrossRef] [PubMed]

36. Guermonprez, P.; Ladant, D.; Karimova, G.; Ullmann, A.; Leclerc, C. Direct Delivery of the *Bordetella pertussis* Adenylate Cyclase Toxin to the MHC Class I Antigen Presentation Pathway. *J. Immunol.* **1999**, *162*, 1910–1916.

37. Guermonprez, P.; Fayolle, C.; Karimova, G.; Ullmann, A.; Leclerc, C.; Ladant, D. *Bordetella pertussis* Adenylate Cyclase Toxin: A Vehicle to Deliver CD8-Positive T-Cell Epitopes into Antigen Presenting Cells. *Methods Enzymol.* **2000**, *326*, 527–542. [PubMed]

38. Hewlett, E.L.; Gordon, V.M.; McCaffery, J.D.; Sutherland, W.M.; Gray, M.C. Adenylate Cyclase Toxin from *Bordetella pertussis*. Identification and Purification of the Holotoxin Molecule. *J. Biol. Chem.* **1989**, *264*, 19379–19384. [CrossRef]

39. Karst, J.C.; Ntsogo Enguene, V.Y.; Cannella, S.E.; Subrini, O.; Hessel, A.; Debard, S.; Ladant, D.; Chenal, A. Calcium, Acylation, and Molecular Confinement Favor Folding of *Bordetella pertussis* Adenylate Cyclase CyaA Toxin into a Monomeric and Cytotoxic Form. *J. Biol. Chem.* **2014**, *289*, 30702–30716. [CrossRef]

40. Hanski, E.; Farfel, Z. *Bordetella pertussis* Invasive Adenylate Cyclase. Partial Resolution and Properties of Its Cellular Penetration. *J. Biol. Chem.* **1985**, *260*, 5526–5532. [CrossRef]

41. Rogel, A.; Hanski, E. Distinct Steps in the Penetration of Adenylate Cyclase Toxin of *Bordetella pertussis* into Sheep Erythrocytes. Translocation of the Toxin across the Membrane. *J. Biol. Chem.* **1992**, *267*, 22599–22605. [CrossRef]

42. Karimova, G.; Fayolle, C.; Gmira, S.; Ullmann, A.; Leclerc, C.; Ladant, D. Charge-Dependent Translocation of *Bordetella pertussis* Adenylate Cyclase Toxin into Eukaryotic Cells: Implication for the in Vivo Delivery of CD8(+) T Cell Epitopes into Antigen-Presenting Cells. *Proc. Natl. Acad. Sci. USA* **1998**, *95*, 12532–12537. [CrossRef]

43. Voegele, A.; Sadi, M.; O'Brien, D.P.; Gehan, P.; Raoux-Barbot, D.; Davi, M.; Hoos, S.; Brûlé, S.; Raynal, B.; Weber, P.; et al. A High-Affinity Calmodulin-Binding Site in the CyaA Toxin Translocation Domain Is Essential for Invasion of Eukaryotic Cells. *Adv. Sci.* **2021**, *8*, 2003630. [CrossRef] [PubMed]
44. Hewlett, E.L.; Gray, L.; Allietta, M.; Ehrmann, I.; Gordon, V.M.; Gray, M.C. Adenylate Cyclase Toxin from *Bordetella pertussis*. Conformational Change Associated with Toxin Activity. *J. Biol. Chem.* **1991**, *266*, 17503–17508. [CrossRef]
45. Masin, J.; Osickova, A.; Sukova, A.; Fiser, R.; Halada, P.; Bumba, L.; Linhartova, I.; Osicka, R.; Sebo, P. Negatively Charged Residues of the Segment Linking the Enzyme and Cytolysin Moieties Restrict the Membrane-Permeabilizing Capacity of Adenylate Cyclase Toxin. *Sci. Rep.* **2016**, *6*, 29137. [CrossRef]
46. Beckert, U.; Wolter, S.; Hartwig, C.; Bahre, H.; Kaever, V.; Ladant, D.; Frank, D.W.; Seifert, R. ExoY from *Pseudomonas Aeruginosa* Is a Nucleotidyl Cyclase with Preference for cGMP and cUMP Formation. *Biochem. Biophys. Res. Commun.* **2014**, *450*, 870–874. [CrossRef] [PubMed]
47. Raoux-Barbot, D.; Belyy, A.; Worpenberg, L.; Montluc, S.; Deville, C.; Henriot, V.; Velours, C.; Ladant, D.; Renault, L.; Mechold, U. Differential Regulation of Actin-Activated Nucleotidyl Cyclase Virulence Factors by Filamentous and Globular Actin. *PLoS ONE* **2018**, *13*, e0206133. [CrossRef] [PubMed]
48. Shen, Y.; Zhukovskaya, N.L.; Guo, Q.; Florián, J.; Tang, W.-J. Calcium-Independent Calmodulin Binding and Two-Metal-Ion Catalytic Mechanism of Anthrax Edema Factor. *EMBO J.* **2005**, *24*, 929–941. [CrossRef]
49. Karimova, G.; Pidoux, J.; Ullmann, A.; Ladant, D. A Bacterial Two-Hybrid System Based on a Reconstituted Signal Transduction Pathway. *Proc. Natl. Acad. Sci. USA* **1998**, *95*, 5752–5756. [CrossRef] [PubMed]

Concept Paper

Effector-Triggered Trained Immunity: An Innate Immune Memory to Microbial Virulence Factors?

Cedric Torre and Laurent Boyer *

Université Côte d'Azur, Inserm, C3M, 06200 Nice, France
* Correspondence: laurent.boyer@univ-cotedazur.fr

Abstract: In the last decade, a major dogma in the field of immunology has been called into question by the identification of a cell autonomous innate immune memory. This innate immune memory (also named trained immunity) was found to be mostly carried by innate immune cells and to be characterized by an exacerbated inflammatory response with a heightened expression of proinflammatory cytokines, including TNF-α, IL-6 and IL-1β. Unlike the vast majority of cytokines, IL-1β is produced as a proform (pro-IL-1β) and requires a proteolytic cleavage to exert its biological action. This cleavage takes place mainly within complex molecular platforms named inflammasomes. These platforms are assembled upon both the infectious or sterile activation of NOD-like receptors (NLRs), thereby allowing for the recruitment and activation of caspases and the subsequent maturation of pro-IL-1β into IL-1β. The NLRP3 inflammasome has recently been implicated both in western diet-induced trained immunity, and in the detection of microbial virulence factors (effector-triggered immunity (ETI)). Here, we will attempt to link these two immune processes and provide arguments to hypothesize the existence of trained immunity triggered by microbial virulence factors (effector-triggered trained immunity (ETTI)).

Keywords: trained immunity; effector-triggered immunity; effector-triggered trained immunity

Key Contribution: In this article we examine how trained immunity and effector-triggered immunity are connected through the NLRP3 inflammasome, pointing for a possible trained immunity mediated by microbial effectors.

Citation: Torre, C.; Boyer, L. Effector-Triggered Trained Immunity: An Innate Immune Memory to Microbial Virulence Factors?. *Toxins* **2022**, *14*, 798. https://doi.org/10.3390/toxins14110798

Received: 29 September 2022
Accepted: 15 November 2022
Published: 17 November 2022

Publisher's Note: MDPI stays neutral with regard to jurisdictional claims in published maps and institutional affiliations.

1. Introduction

To survive host–pathogen battles, pathogens have evolved various strategies to blunt or escape host immune responses. Among them, effector proteins, also called virulence factors, could be considered as surgical weapons of microbes that precisely target the Achilles' heel of the immune system with the goal of surviving and multiplying. Thus, the expression of virulence factors by microbes confers them with pathogenic properties, distinguishing commensals from pathogens. On the other hand, hosts have multiplied sensors to detect and gauge the pathogenic potential of microbes. Pattern Recognition Receptors (PRRs) discovery revolutionized the way of thinking about pathogen detection by linking the microbial ligand to innate immune signaling. The innate immune response subsequent to the recognition of Microbe-Associated Molecular Patterns (MAMPs) by PRR signaling has been termed Pattern-Triggered Immunity (PTI). Since this seminal discovery [1], additional sensing mechanisms have been proposed to explain how the host could distinguish commensals from pathogens.

2. Effector-Triggered Immunity

An elegant strategy based on the recognition of virulence factor activities was proposed. As virulence factors are specifically expressed by pathogens, being able to detect their activities appears to be a smart strategy to limit the number of sensors, which detect specifically pathogens and not all microbes. This innate immune sensing mechanism has been termed Effector-Triggered Immunity (ETI). By functioning as an additional layer of innate immune responses, ETI adds up to PTI to allow the host to gauge both the quantity of microbes and their pathogenic potential. ETI was first identified during plant–pathogen interactions, and knowledge in the matter has since deepened [2–4]. More recently, proofs of the conservation of this innate immune mechanism have emerged in animals [5]. Several publications have extensively reviewed the ETI similarities and differences between plants and animals [6–8]. Here, we will focus on the mechanisms of ETI linking RhoGTPases, inflammasomes and trained immunity.

Most bacterial toxins hijack the regulation of crucial cellular functions, such as those performed by RhoGTPases [9]. This was previously referred to as a virulence activity until recent reports showing that the modifications of RhoGTPases activity by microbial effectors are detected by the host cell through NOD-like receptors (NLRs) and the assembly of inflammasomes [5]. The Pyrin inflammasome was first shown to be activated in macrophages treated by bacterial toxins inactivating the RhoA GTPase and this finding was later extended to numerous bacterial factors inactivating the Rho family members found in various pathogenic bacteria [5,10–13]. The involvement of the NLRP3 inflammasome in ETI has recently been demonstrated, implicating the CNF1 toxin of uropathogenic *Escherichia coli* [14]. This study showed that both mouse and human macrophages detect CNF1 activity via NLRP3 inflammasome activation. CNF1 displays deamidase activity toward RhoGTPases. This modification destroys the GTPase activity and locks the RhoGTPases into an active GTP-bound form. CNF1-triggered activation of the Rac2 GTPase in macrophages activates the PAK1/2 kinases that in turn phosphorylate the Threonine 659 of human NLRP3 leading to its activation and subsequent IL-1β secretion. This way, the NLRP3 inflammasome detects the activation of Rac2 GTPase, leading to the bacterial clearance of CNF1-expressing *E. coli* in a mouse model of bacteremia. Importantly, this sensing mechanism has been reused by the innate immune system to sense other virulence factors activating Rac2 encoded by other pathogenic bacteria such as SopE from *Salmonella* spp. or DNT from *Bordetella* spp. [14].

All these studies have recently highlighted that an increasing number of virulence factors are sensed by inflammasomes; this leads to the urgent question of whether this recognition can trigger immune memory.

3. Trained Immunity

The immune system is classically divided into the following two compartments: the innate immunity as the first line of defense in the host's response to aggression, and the adaptive immunity that confers immune memory to the host. The dogma that immune memory is conferred by adaptive immunity has been challenged in recent years. Moreover, the scientific literature is replete with examples of immune memory in the absence of adaptive immunity in invertebrate organisms that naturally lack adaptive immunity, a process called innate immune memory [15].

However, it was only a decade ago that these observations were made in mammals [16–18]. This concept of innate immune memory emerged in the early 2010s, originally referred to as an adaptive form of innate immunity before to be termed trained immunity [16,19,20].

Mammalian innate immune memory, or Trained Immunity (TI), is a mechanism initiated by the action of an inflammatory stimulus on cells with or linked to immune functions, leading to the epigenetic and metabolic reprogramming of these cells. Thus, following a first stimulation, the prototypical cells of the innate immune compartment are trained; when restimulated, they will display an exacerbated inflammatory response that can be either beneficial or detrimental depending on the context.

Evidence of the beneficial effects for the host has been described by numerous investigations in mice [20]. As a result of these studies, it was shown that training mice with numerous microbial ligands could provide non-specific protection against an ensuing fatal infection. Beyond their specific action, live attenuated vaccines such as Bacille Calmette–

Guerin (BCG), measles, smallpox or poliomyelitis vaccines have a non-specific protective action against secondary infections with other pathogens [21]. Furthermore, the BCG vaccine has proven its non-specific and beneficial action in bladder cancer, where it is used for bladder instillation treatment [22]. Conversely, immune training can lead to harmful effects, including chronic inflammatory diseases such as atherosclerotic cardiovascular disease and neurodegenerative diseases (Alzheimer disease and dementia), and tumor growth and metastasis [20].

Historically, BCG, *Candida albicans* and its β-glucan were the first triggers known to induce a TI program [17,18]; however, different types of pathological inflammation triggers were recently shown as potential TI triggers. Most of these triggers engage the activation of the IL-1 pathway (Table 1). This includes the western diet (a high fat and high sugar diet) that was shown to induce a deleterious TI program in mice and it was interestingly shown to rely on the NLRP3 inflammasome [23]. Christ and colleagues showed that feeding the western diet to $Ldlr^{-/-}$ mice (a mouse model that develops atherosclerotic lesions and hypercholesterolemia) induces systemic inflammation, the proliferation of myeloid precursors, and their reprogramming, while $Ldlr^{-/-}$ $Nlrp3^{-/-}$ mice do not, pointing towards the importance of the NLRP3 inflammasome. Interestingly, the induction of TI by β-glucan was recently shown to limit NLRP3 inflammasome activation, repressing the production of IL-1β [24]. This highlights the importance of better characterizing the involvement of the NLRP3 inflammasome in TI. Considering that, depending on the context, NLRP3 activation triggers IL-1β secretion with or without cell death (pyroptosis versus hyperactivation), studies should be carried out in relation to inflammasome inducers and the strength of signals.

The NLRP3 inflammasome maturates both pro-IL-1β and pro-IL-18 into their biologically active forms. Interestingly, both cytokines have been associated with TI. IL-18 has been implicated in the induction of TI in specific contexts and has been associated with the involvement of TI in rheumatic pathologies such as primary Sjögren's syndrome [25,26].

A collection of works agrees on the fact that the NLRP3-IL-1β axis is central in TI regulation, which raises the exciting possibility that TI may also be induced by the microbial effectors sensed by NLRP3 inflammasome during ETI.

Table 1. Evidence for the implication of the IL-1 pathway and/or NLRP3 inflammasome in trained immunity.

Model	Inducer	Challenge	Evidence for the Implication of		Reference
			IL-1 Pathway	**NLRP3 Inflammasome**	
Mouse	β-glucan	LPS	Yes	No	[27]
Mouse	Western diet	LPS	Yes	Yes	[23]
Mouse	β-glucan	*M. tuberculosis*	Yes	No	[28]
Mouse	Periodontitis/arthritis	LPS	Yes	Suggested	[29]
Human	BCG vaccination	*M. tuberculosis/S. aureus/C. albicans*	Yes	No	[18]
Human	BCG vaccination	LPS	Yes	No	[30]
Human	BCG vaccination	TLR ligands	Yes	No	[31]
Human	BCG vaccination	Yellow fever virus vaccine strain	Yes	Suggested	[32]
Human	BCG vaccination	TLR ligands	Suggested	No	[33]

4. Effector-Triggered Trained Immunity?

So far, the mechanisms of TI described in the literature involve the NOD receptor, NOD2, as well as the NLRP3 inflammasome. The NOD2 receptor has been identified as the molecular bridge between BCG vaccination and the increased production of proinflammatory cytokines, leading to the protection of severe combined immunodeficiency SCID mice from disseminated candidiasis [18]. As described earlier, the NLRP3 inflammasome is involved in TI induced by a "western diet", a diet rich in sugar and saturated fatty acids [23]. Conversely, pyrin and NLRP3 are two major inflammasomes that have been implicated in ETI. The involvement of the NLRP3 inflammasome in both ETI and TI suggests that it could be implicated in TI in an infectious context, induced by a virulence factor, through a process that can be termed Effector-Triggered Trained Immunity (ETTI). In this view, it could be applied to bacterial effectors recently discovered for their detection by the NLRP3 inflammasome activation pathway [14]. Indeed, not only the *E. coli* CNF1 toxin but also the *Bordetella* spp. DNT (dermonecrotic toxin) and the *Salmonella* spp. SopE toxin are triggers of the NLRP3 inflammasome, suggesting their potential role as triggers of NLRP3-mediated trained immune response. If this hypothesis is true, one can expect that other NLRs involved in ETI, such as the Pyrin-inflammasome-detecting RhoA GTPase-inactivating toxins, may also be involved in ETTI. Furthermore, TI is induced by vaccination with live attenuated viruses [21]. Considering this, viruses can act as inducers of TI. Previous studies described a mechanism specific to positive single-stranded RNA ((+)ssRNA) viruses, whose proteases can induce an ETI response by activating the NLRP1 inflammasome [34]. To date, the NLRP1 inflammasome has not been implicated in TI and it would be interesting to investigate whether NLRP1 may be involved in ETTI.

The activation of the NLRP3 inflammasome requires both a transcriptional priming and an activation step triggered by various stimuli including PAMPs and DAMPs (Pathogen- and Danger-Associated Molecular Patterns). The priming step required for the production of pro-IL-1β cytokine and components for inflammasome assembly relies on the NF-κB pathway, which has been involved in the induction of TI. As an example, low-dose LPS—a well-known inducer of the NF-κB pathway—has been identified as an inducer of TI [35]. The recent study of NLRP3 activation triggered by the CNF1 toxin revealed that transcriptional priming is not mandatory to measure NLRP3-dependent Caspase-1 activation [14]. Here, we wanted to pinpoint the possible role of virulence factors in activating the inflammasome in the trained immunity process that can be observed with or without the priming step.

With regard to innate immune memory, a distinction is made between peripheral memory and central memory [20]. Peripheral memory is the memory carried by myeloid cells in circulation in the body, while central memory is the memory carried by hematopoietic stem cells in the bone marrow [36]. In this context, NLRP3 inflammasome activation has been shown to be involved in myeloid precursor release [27,37], and BCG exposure in the bone marrow modifies the transcriptional program of the hematopoietic stem and progenitor cells, promoting local cell growth and improved myelopoiesis [38]. This supports the idea of the involvement of bacterial toxins sensed by inflammasomes in the induction of a central TI.

For TI to last over time, there must be a transfer of epigenetic modifications to cells that have a long lifespan. Previously, using the planarian model, we demonstrated that an innate immune memory is carried by the neoblasts (planarian stem cells) and the innate memory could be transferred from one organism to another after irradiation; furthermore, other cells are also capable of exhibiting innate immune memory [39,40]. Later, Christ et al. proposed that the induction of a central memory within the context of the western diet is made possible by myeloid precursors [23]. Similar to these previous examples, infection with a microbe producing a microbial effector should lead to immune imprinting of myeloid precursors for a long lasting memory.

5. Conclusions

In this paper, we explore how trained immunity (TI) and effector-triggered immunity (ETI) are linked through the NLRP3 inflammasome, suggesting the existence of a trained immunity mediated by microbial effectors (ETTI). The involvement of the NLRP3 inflammasome in TI has been demonstrated in the context of sterile inflammation; however, its involvement in an infectious context remains to be proven. Moreover, this phenomenon might not be exclusive to the NLRP3 inflammasome and should therefore be considered for other inflammasomes. Last but not least, an interesting trans-kingdom perspective would be to return to the roots of effector-triggered immunity by determining whether trained immunity in plants is activated in response to virulence factors and whether it requires plant NLRs.

Author Contributions: Conceptualization, C.T. and L.B.; writing, C.T. and L.B.; funding acquisition, C.T. and L.B. All authors have read and agreed to the published version of the manuscript.

Funding: This research was funded by Agence Nationale de la Recherche (ANR-20-CE14-0049, ANR-11-LABX-0028-01 and IDEX UCAJEDI ANR-15-IDEX-01). C.T. is supported by a fellowship from Ville de Nice (2020).

Institutional Review Board Statement: Not applicable.

Informed Consent Statement: Not applicable.

Data Availability Statement: Not applicable.

Acknowledgments: We thank Juan A. Garcia-Sanchez, Patrick Munro and Orane Visvikis for critical reading the manuscript.

Conflicts of Interest: The authors declare no conflict of interest.

References

1. Lemaitre, B.; Nicolas, E.; Michaut, L.; Reichhart, J.M.; Hoffmann, J.A. The Dorsoventral Regulatory Gene Cassette Spätzle/Toll/Cactus Controls the Potent Antifungal Response in Drosophila Adults. *Cell* **1996**, *86*, 973–983. [CrossRef]
2. Jones, J.D.G.; Dangl, J.L. The Plant Immune System. *Nature* **2006**, *444*, 323–329. [CrossRef] [PubMed]
3. Adachi, H.; Kamoun, S. NLR Receptor Networks in Plants. *Essays Biochem.* **2022**, *66*, 541–549. [CrossRef] [PubMed]
4. Chen, J.; Zhang, X.; Rathjen, J.P.; Dodds, P.N. Direct Recognition of Pathogen Effectors by Plant NLR Immune Receptors and Downstream Signalling. *Essays Biochem.* **2022**, *66*, 471–483. [CrossRef] [PubMed]
5. Dufies, O.; Boyer, L. RhoGTPases and Inflammasomes: Guardians of Effector-Triggered Immunity. *PLoS Pathog.* **2021**, *17*, e1009504. [CrossRef]
6. Stuart, L.M.; Paquette, N.; Boyer, L. Effector-Triggered versus Pattern-Triggered Immunity: How Animals Sense Pathogens. *Nat. Rev. Immunol.* **2013**, *13*, 199–206. [CrossRef]
7. Rajamuthiah, R.; Mylonakis, E. Effector Triggered Immunity. *Virulence* **2014**, *5*, 697–702. [CrossRef]
8. Lopes Fischer, N.; Naseer, N.; Shin, S.; Brodsky, I.E. Effector-Triggered Immunity and Pathogen Sensing in Metazoans. *Nat. Microbiol.* **2020**, *5*, 14–26. [CrossRef]
9. Boquet, P.; Lemichez, E. Bacterial Virulence Factors Targeting Rho GTPases: Parasitism or Symbiosis? *Trends Cell Biol.* **2003**, *13*, 238–246. [CrossRef]
10. Ng, J.; Hirota, S.A.; Gross, O.; Li, Y.; Ulke-Lemee, A.; Potentier, M.S.; Schenck, L.P.; Vilaysane, A.; Seamone, M.E.; Feng, H.; et al. *Clostridium difficile* Toxin-Induced Inflammation and Intestinal Injury Are Mediated by the Inflammasome. *Gastroenterology* **2010**, *139*, 542–552.e3. [CrossRef]
11. Aubert, D.F.; Xu, H.; Yang, J.; Shi, X.; Gao, W.; Li, L.; Bisaro, F.; Chen, S.; Valvano, M.A.; Shao, F. A *Burkholderia* Type VI Effector Deamidates Rho GTPases to Activate the Pyrin Inflammasome and Trigger Inflammation. *Cell Host Microbe* **2016**, *19*, 664–674. [CrossRef]
12. Gao, W.; Yang, J.; Liu, W.; Wang, Y.; Shao, F. Site-Specific Phosphorylation and Microtubule Dynamics Control Pyrin Inflammasome Activation. *Proc. Natl. Acad. Sci. USA* **2016**, *113*, E4857–E4866. [CrossRef]
13. Medici, N.P.; Rashid, M.; Bliska, J.B. Characterization of Pyrin Dephosphorylation and Inflammasome Activation in Macrophages as Triggered by the *Yersinia* Effectors YopE and YopT. *Infect. Immun.* **2019**, *87*, e00822-18. [CrossRef]
14. Dufies, O.; Doye, A.; Courjon, J.; Torre, C.; Michel, G.; Loubatier, C.; Jacquel, A.; Chaintreuil, P.; Majoor, A.; Guinamard, R.R.; et al. *Escherichia coli* Rho GTPase-Activating Toxin CNF1 Mediates NLRP3 Inflammasome Activation via P21-Activated Kinases-1/2 during Bacteraemia in Mice. *Nat. Microbiol.* **2021**, *6*, 401–412. [CrossRef]
15. Milutinović, B.; Kurtz, J. Immune Memory in Invertebrates. *Semin. Immunol.* **2016**, *28*, 328–342. [CrossRef]
16. Netea, M.G.; Quintin, J.; van der Meer, J.W.M. Trained Immunity: A Memory for Innate Host Defense. *Cell Host Microbe* **2011**, *9*, 355–361. [CrossRef]
17. Quintin, J.; Saeed, S.; Martens, J.H.A.; Giamarellos-Bourboulis, E.J.; Ifrim, D.C.; Logie, C.; Jacobs, L.; Jansen, T.; Kullberg, B.-J.; Wijmenga, C.; et al. *Candida albicans* Infection Affords Protection against Reinfection via Functional Reprogramming of Monocytes. *Cell Host Microbe* **2012**, *12*, 223–232. [CrossRef]
18. Kleinnijenhuis, J.; Quintin, J.; Preijers, F.; Joosten, L.A.B.; Ifrim, D.C.; Saeed, S.; Jacobs, C.; van Loenhout, J.; de Jong, D.; Stunnenberg, H.G.; et al. Bacille Calmette-Guerin Induces NOD2-Dependent Nonspecific Protection from Reinfection via Epigenetic Reprogramming of Monocytes. *Proc. Natl. Acad. Sci. USA* **2012**, *109*, 17537–17542. [CrossRef]
19. Bowdish, D.M.E.; Loffredo, M.S.; Mukhopadhyay, S.; Mantovani, A.; Gordon, S. Macrophage Receptors Implicated in the "Adaptive" Form of Innate Immunity. *Microbes Infect.* **2007**, *9*, 1680–1687. [CrossRef]

20. Netea, M.G.; Domínguez-Andrés, J.; Barreiro, L.B.; Chavakis, T.; Divangahi, M.; Fuchs, E.; Joosten, L.A.B.; van der Meer, J.W.M.; Mhlanga, M.M.; Mulder, W.J.M.; et al. Defining Trained Immunity and Its Role in Health and Disease. *Nat. Rev. Immunol.* **2020**, *20*, 375–388. [CrossRef]

21. Benn, C.S.; Netea, M.G.; Selin, L.K.; Aaby, P. A Small Jab—A Big Effect: Nonspecific Immunomodulation by Vaccines. *Trends Immunol.* **2013**, *34*, 431–439. [CrossRef]

22. Redelman-Sidi, G.; Glickman, M.S.; Bochner, B.H. The Mechanism of Action of BCG Therapy for Bladder Cancer—A Current Perspective. *Nat. Rev. Urol.* **2014**, *11*, 153–162. [CrossRef]

23. Christ, A.; Günther, P.; Lauterbach, M.A.R.; Duewell, P.; Biswas, D.; Pelka, K.; Scholz, C.J.; Oosting, M.; Haendler, K.; Baßler, K.; et al. Western Diet Triggers NLRP3-Dependent Innate Immune Reprogramming. *Cell* **2018**, *172*, 162–175.e14. [CrossRef]

24. Camilli, G.; Bohm, M.; Piffer, A.C.; Lavenir, R.; Williams, D.L.; Neven, B.; Grateau, G.; Georgin-Lavialle, S.; Quintin, J. β-Glucan-Induced Reprogramming of Human Macrophages Inhibits NLRP3 Inflammasome Activation in Cryopyrinopathies. *J. Clin. Investig.* **2020**, *130*, 4561–4573. [CrossRef]

25. Moorlag, S.J.C.F.M.; Röring, R.J.; Joosten, L.A.B.; Netea, M.G. The Role of the Interleukin-1 Family in Trained Immunity. *Immunol. Rev.* **2018**, *281*, 28–39. [CrossRef]

26. Badii, M.; Gaal, O.; Popp, R.A.; Crişan, T.O.; Joosten, L.A.B. Trained Immunity and Inflammation in Rheumatic Diseases. *Joint Bone Spine* **2022**, *89*, 105364. [CrossRef]

27. Mitroulis, I.; Ruppova, K.; Wang, B.; Chen, L.-S.; Grzybek, M.; Grinenko, T.; Eugster, A.; Troullinaki, M.; Palladini, A.; Kourtzelis, I.; et al. Modulation of Myelopoiesis Progenitors Is an Integral Component of Trained Immunity. *Cell* **2018**, *172*, 147–161.e12. [CrossRef]

28. Moorlag, S.J.C.F.M.; Khan, N.; Novakovic, B.; Kaufmann, E.; Jansen, T.; van Crevel, R.; Divangahi, M.; Netea, M.G. β-Glucan Induces Protective Trained Immunity against *Mycobacterium tuberculosis* Infection: A Key Role for IL-1. *Cell Rep.* **2020**, *31*, 107634. [CrossRef]

29. Li, X.; Wang, H.; Yu, X.; Saha, G.; Kalafati, L.; Ioannidis, C.; Mitroulis, I.; Netea, M.G.; Chavakis, T.; Hajishengallis, G. Maladaptive Innate Immune Training of Myelopoiesis Links Inflammatory Comorbidities. *Cell* **2022**, *185*, 1709–1727.e18. [CrossRef]

30. Kleinnijenhuis, J.; Quintin, J.; Preijers, F.; Benn, C.S.; Joosten, L.A.B.; Jacobs, C.; van Loenhout, J.; Xavier, R.J.; Aaby, P.; van der Meer, J.W.M.; et al. Long-Lasting Effects of BCG Vaccination on Both Heterologous Th1/Th17 Responses and Innate Trained Immunity. *J. Innate Immun.* **2014**, *6*, 152–158. [CrossRef]

31. Jensen, K.J.; Larsen, N.; Biering-Sørensen, S.; Andersen, A.; Eriksen, H.B.; Monteiro, I.; Hougaard, D.; Aaby, P.; Netea, M.G.; Flanagan, K.L.; et al. Heterologous Immunological Effects of Early BCG Vaccination in Low-Birth-Weight Infants in Guinea-Bissau: A Randomized-Controlled Trial. *J. Infect. Dis.* **2015**, *211*, 956–967. [CrossRef]

32. Arts, R.J.W.; Moorlag, S.J.C.F.M.; Novakovic, B.; Li, Y.; Wang, S.-Y.; Oosting, M.; Kumar, V.; Xavier, R.J.; Wijmenga, C.; Joosten, L.A.B.; et al. BCG Vaccination Protects against Experimental Viral Infection in Humans through the Induction of Cytokines Associated with Trained Immunity. *Cell Host Microbe* **2018**, *23*, 89–100.e5. [CrossRef]

33. Freyne, B.; Donath, S.; Germano, S.; Gardiner, K.; Casalaz, D.; Robins-Browne, R.M.; Amenyogbe, N.; Messina, N.L.; Netea, M.G.; Flanagan, K.L.; et al. Neonatal BCG Vaccination Influences Cytokine Responses to Toll-like Receptor Ligands and Heterologous Antigens. *J. Infect. Dis.* **2018**, *217*, 1798–1808. [CrossRef]

34. Tsu, B.V.; Fay, E.J.; Nguyen, K.T.; Corley, M.R.; Hosuru, B.; Dominguez, V.A.; Daugherty, M.D. Running With Scissors: Evolutionary Conflicts Between Viral Proteases and the Host Immune System. *Front. Immunol.* **2021**, *12*, 769543. [CrossRef]

35. Wendeln, A.-C.; Degenhardt, K.; Kaurani, L.; Gertig, M.; Ulas, T.; Jain, G.; Wagner, J.; Häsler, L.M.; Wild, K.; Skodras, A.; et al. Innate Immune Memory in the Brain Shapes Neurological Disease Hallmarks. *Nature* **2018**, *556*, 332–338. [CrossRef]

36. Domínguez-Andrés, J.; Dos Santos, J.C.; Bekkering, S.; Mulder, W.J.M.; van der Meer, J.W.; Riksen, N.P.; Joosten, L.A.B.; Netea, M.G. Trained Immunity: Adaptation within Innate Immune Mechanisms. *Physiol. Rev.* **2022**, *103*, 313–346. [CrossRef]

37. Lenkiewicz, A.M.; Adamiak, M.; Thapa, A.; Bujko, K.; Pedziwiatr, D.; Abdel-Latif, A.K.; Kucia, M.; Ratajczak, J.; Ratajczak, M.Z. The Nlrp3 Inflammasome Orchestrates Mobilization of Bone Marrow-Residing Stem Cells into Peripheral Blood. *Stem Cell Rev. Rep.* **2019**, *15*, 391–403. [CrossRef]

38. Kaufmann, E.; Sanz, J.; Dunn, J.L.; Khan, N.; Mendonça, L.E.; Pacis, A.; Tzelepis, F.; Pernet, E.; Dumaine, A.; Grenier, J.-C.; et al. BCG Educates Hematopoietic Stem Cells to Generate Protective Innate Immunity against Tuberculosis. *Cell* **2018**, *172*, 176–190.e19. [CrossRef]

39. Torre, C.; Abnave, P.; Tsoumtsa, L.L.; Mottola, G.; Lepolard, C.; Trouplin, V.; Gimenez, G.; Desrousseaux, J.; Gempp, S.; Levasseur, A.; et al. *Staphylococcus aureus* Promotes Smed-PGRP-2/Smed-Setd8-1 Methyltransferase Signalling in Planarian Neoblasts to Sensitize Anti-Bacterial Gene Responses During Re-Infection. *EBioMedicine* **2017**, *20*, 150–160. [CrossRef]

40. Hamada, A.; Torre, C.; Drancourt, M.; Ghigo, E. Trained Immunity Carried by Non-Immune Cells. *Front. Microbiol.* **2018**, *9*, 3225. [CrossRef]

Review

Superantigens, a Paradox of the Immune Response

Sofia Noli Truant, Daniela María Redolfi, María Belén Sarratea, Emilio Luis Malchiodi and Marisa Mariel Fernández *

Cátedra de Inmunología—IDEHU (UBA-CONICET), Facultad de Farmacia y Bioquímica, Universidad de Buenos Aires, Junín 956 4 Piso, Ciudad Autónoma de Buenos Aires 1113, Argentina
* Correspondence: mmfernan@ffyb.uba.ar; Tel.: +54-11-52874411

Abstract: Staphylococcal enterotoxins are a wide family of bacterial exotoxins with the capacity to activate as much as 20% of the host T cells, which is why they were called superantigens. Superantigens (SAgs) can cause multiple diseases in humans and cattle, ranging from mild to life-threatening infections. Almost all *S. aureus* isolates encode at least one of these toxins, though there is no complete knowledge about how their production is triggered. One of the main problems with the available evidence for these toxins is that most studies have been conducted with a few superantigens; however, the resulting characteristics are attributed to the whole group. Although these toxins share homology and a two-domain structure organization, the similarity ratio varies from 20 to 89% among different SAgs, implying wide heterogeneity. Furthermore, every attempt to structurally classify these proteins has failed to answer differential biological functionalities. Taking these concerns into account, it might not be appropriate to extrapolate all the information that is currently available to every staphylococcal SAg. Here, we aimed to gather the available information about all staphylococcal SAgs, considering their functions and pathogenicity, their ability to interact with the immune system as well as their capacity to be used as immunotherapeutic agents, resembling the two faces of Dr. Jekyll and Mr. Hyde.

Keywords: staphylococcal superantigen; enterotoxin; toxin pathogenicity; immunomodulation; molecular and cellular targets

Key Contribution: Bacterial superantigens manipulate the host immune response favoring the spread and colonization of *Staphylococcus aureus*, causing multiple affections. This review summarizes the current knowledge on every superantigen described, focusing on the dual nature of these toxins.

Citation: Noli Truant, S.; Redolfi, D.M.; Sarratea, M.B.; Malchiodi, E.L.; Fernández, M.M. Superantigens, a Paradox of the Immune Response. *Toxins* **2022**, *14*, 800. https://doi.org/10.3390/toxins14110800

Received: 14 September 2022
Accepted: 27 October 2022
Published: 18 November 2022

Publisher's Note: MDPI stays neutral with regard to jurisdictional claims in published maps and institutional affiliations.

1. Introduction

The term superantigen (SAg) was introduced by White et al. (1989) to describe a group of bacterial proteins targeting, as unprocessed molecules, the variable portion of the β chain of the T cell receptor (TCR) and the major histocompatibility complex type II molecules (MHC-II) expressed on antigen-presenting cells (APC). As a result of this simultaneous interaction, there is a massive activation of the immune system along with an intense proliferation of T cells, either CD4+ or CD8+ [1,2]. Although the term superantigen was introduced in 1989, these toxins had been described in the early 1980s as the causative agent of the highly lethal Toxic Shock Syndrome (TSS) associated with tampons [3–5], which was later explained by the deregulation of the immune system caused by SAgs characterized by generalized multiple organ failure caused by a pro-inflammatory cytokine storm. SAgs are also responsible for food poisoning and triggering autoimmune diseases in sensitive hosts, among other conditions. They can promote immunosuppression in the infected host, allowing bacterial spread and further sepsis [6–8].

SAgs are produced by Gram-positive bacteria such as *S. aureus* and *Streptococcus pyogenes*, but they have also been found in other species such as β-hemolytic streptococci, coagulase-negative staphylococci, Gram-negative *Yersinia pseudotuberculosis*, *Pseudomonas fluorescens* and cell wall-less bacteria *Mycoplasma arthritidis* [9–12]. Furthermore, SAgs encoded by murine and human retroviruses have also been described [13–15].

Until now, the best characterized SAgs were produced by *S. aureus* and *S. pyogenes*. Although these two bacteria are not phylogenetically related species, staphylococcal and streptococcal SAgs share sequence homology and biological similarities. Some SAgs from these two species are so close that they belong to the same group or family, as is very well described for the SE family or Group II, where the streptococcal superantigen SpeA is more similar to staphylococcal enterotoxin B (SEB) than to staphylococcal enterotoxin G (SEG). SpeA displays a low degree of similarity with the other streptococcal SAgs. This finding supports the horizontal transmission of genes between species. This theory is also reinforced by the encoding of *sags* in genetic mobile elements, such as plasmids or transposons [16].

Interestingly, the intense immune stimulation induced by SAgs could allow the spread and further colonization of the infected host by the pathogen instead of favoring its eradication. This characteristic would turn superantigen activity into a paradox of the immune response.

2. Staphylococcal SAgs

2.1. An Overall Description

S. aureus is one of the major pathogens responsible for human and veterinarian diseases causing mild to life-threatening infections. Human beings are the primary reservoir for this bacterium. It is well known that about 20–30% of the population persistently carries this microorganism in the anterior nares, while about 50–60% are intermittently colonized [17–20]. The colonization of other extra nasal sites, such as the skin and gastrointestinal tract [21–24] has also been reported. It should be mentioned that colonization is considered one of the main risk factors for *S. aureus* diseases [25].

This pathogen harbors several virulence factors among which some are surface proteins and many others are secreted. Staphylococcal enterotoxins (SEs) or SAgs are some of the most important and are associated with both colonization and pathogenicity of *S. aureus*.

The first staphylococcal SAgs described were the classical SAgs, including staphylococcal enterotoxins A to E and the non-emetic toxin TSST-1. Due to its strong association with TSS, the toxin formerly called SEF was later named Toxic Shock Syndrome Toxin number 1 or TSST-1 [26–33]. Since that first group was described, 24 more SAgs were identified, and were considered non-classical or new SAgs.

The first non-classical SAg, SEH, was described in 1994 [34]. Later, SE*I*J was acknowledged [35] at the same time that Munson et al. described two new SEs, G and I [36]. In 2001, Jarraud et al. defined the *egc* operon for the first time, as an enterotoxin gene nursery encoding for the proteins SEO, SEM, SEI, SEN and SEG (firstly named SEL, SEM, SEI, SEK and SEG) and two pseudogenes, ψ ent1 and ψ ent2 [37]. Moreover, SEK and SEL were simultaneously described [38–40]. In the early 2000s, genetic variations within the *egc* cluster were described, which allowed the identification of SE*I*U and SE*I*V [41,42].

Later on, SEP [43], SES and SET [44], SEQ [45], SER [46], SE*I*W [47], SE*I*X [48] SE*I*Y [49] and SE*I*Z [50] were described. Lastly, SE01 [51], SE02 [52], and SE*I*26 and SE*I*27 were identified and characterized among the complex form by *S. aureus*, *S. argenteus* and *S. schweitzeri* [53]. Table 1 summarizes the information described above. Phylogenetic group correspondence is explained below in Section 2.4.

According to the International Nomenclature for Staphylococcal Superantigens, *S. aureus* enterotoxins (SEs) are defined by their emetic ability when ingested, while the term SAg considers the effects on the immune system. Taking that into account, only the toxins that induce emesis after oral administration in a primate model are designated as SEs. Other related toxins that either lack emetic properties in this model or have not been tested are defined as "staphylococcal enterotoxin–like" (SE*I*) SAgs, to indicate that their potential role in staphylococcal food poisoning has not been confirmed [54].

To date, TSST-1 and SE*l*J are the only tested superantigens with non-emetic properties, along with a report of a truncated SE*l*X [55].

Table 1. Staphylococcal SAgs.

SAg	Type	Year of Discovery	Phylogenetic Group
SEA	Classical	1962	III
SEB	Classical	1962	II
SEC	Classical	1965	II
SED	Classical	1967	III
SEE	Classical	1971	III
TSST-1 (SEF)	Classical	1981	II
SEG	New	1998	I
SEH	New	1994	III
SEI	New	1998	V
SE*l*J	New	1998	III
SEK	New	2001	V
SEL	New	2001	V
SEM	New	2001	V
SEN	New	2001	III
SEO	New	2001	III
SEP	New	2005	III
SEQ	New	2002	V
SER	New	2004	II
SES	New	2008	III
SET	New	2008	I
SE*l*U	New	2003	II
SE*l*V	New	2006	V
SE*l*W	New	2012	III
SE*l*X	New	2011	I
SE*l*Y	New	2015	I
SE*l*Z	New	2015	I
SE*l*26	New	2018	V
SE*l*27	New	2018	II
SE01	New	2017	III
SE02	New	2020	II

Furthermore, some of the so-called SE*l* and SE proteins have already been tested for their emetic properties in a proposed model using house musk shrews *(suncus murinus)* [56,57]. In this model, in addition to all the other SEs, SE*l*Y has been tested and shown to have emetic properties [44,49,56,58,59]. Recently, a new emetic animal model was established using common marmosets and, in this case, SEA, SEB, SEC, SEI, SE*l*Y and SE02 showed emesis-inducing activity, while TSST-1 did not [52,60].

It has been shown that almost every isolated strain of *S. aureus* carries at least one *sag* gene in its genome [61–64]. Although there are some differences in the percentages of total gene prevalence reported, a lower prevalence of *sag* genes (70–85%) was found when a few *sag* genes were analyzed [61,65], whereas in other studies evaluating a larger number of genes, the percentages were almost 98% [66,67]. Noteworthy, as mentioned before, 30 SAgs have been hitherto described, which suggests some underestimation of SAg prevalence. Additionally, more than 50% of the isolates encode for the *egc* operon—with eight different forms described—being that the *egc* SAgs are the most prevalent genes [37,62,65,68–71]. However, there are some differences in the methodology used to define the presence of the operon; while some papers considered all *egc* SAgs, others picked two (such as *seg* and *sei* or *sem* and *seo*) without considering the operon variants, which in some cases may underestimate the prevalence of the operon itself. With regard to this issue, a consensus method should be established to assess the presence of the *egc* operon, which should definitely consider all the toxins involved in the operon, taking into account possible DNA mutations in their genes.

2.2. Production and Detection of Staphylococcal SAgs

Although there are some statements about the production of almost every SAg, in addition to the 6 classical ones (SEA-SEE and TSST-1), there is not much information regarding the individual gene expression and the mechanism of regulation involved in their production. However, many factors and pathways have been described to impact positively and negatively in the production of SAgs and other virulence factors [72–74].

Considering the classical SAgs, early reports using mainly single or double-gel diffusion indicated that SEB and SEA are produced in all the growing phases [75–78]. In contrast, SEC seems to be produced during the exponential growth phase or at the early stationary phase [79,80]. In regard to SED, mRNA expression appears to be higher in the stationary growth phase using qPCR [81], and evidence of its production with SEE at the stationary phase has also been provided [82].

With respect to TSST-1, although its production occurs principally at the stationary phase, it was found that several environmental elements are capable of triggering or inhibiting its production, including temperature, carbon dioxide atmosphere and antibiotic pressure [16,83–85].

Many authors rely on the conclusions reported by Derzelle and collaborators to describe the patterns of staphylococcal enterotoxins expression [86]. These authors found a differential expression behavior between classical and new SAgs. Their work is limited to mRNA, as they performed RT-qPCR assays; however, protein secretion was not evaluated. Furthermore, although their analysis was performed in a total of 28 strains, the number of strains used per gene was variable and dependable on the sag gene profile of each strain, between one and thirteen. In the case of the *egc* operon, only *seg* and *seu* were evaluated in all *egc* positive strains (thirteen), and the remaining genes (*sei*, *sen*, *sem* and *seo*) were only evaluated in three strains. Therefore, these results should be validated using other strategies for reliable conclusions. A similar analysis for SE02 showed the production of this SAg in vitro in the early exponential phase and to a lesser extent in the stationary phase, with these results confirmed by Western blotting [52].

In vitro assays indicate that *egc* superantigens are produced at much lower concentrations than classical SAgs TSST-1, SEB and SEC, which may explain the relevance of these toxins in pathogenesis [87–89]. Particularly, it was proved that TSST-1 was produced in a high level in biofilms, which could be transferred to production on mucosal and skin surfaces [89].

Some evidence suggests that the regulation of gene expression is not the same for *S. aureus* in vitro than during infection in vivo [90,91], showing differences in SAg expression even between tissues or isolation sources [82,92]. As a result, there is a different SAg expression despite the evaluation of DNAs encoding the same *sag* genes. In addition, it has been proposed that the detection of SAg mRNA does not correlate with the presence of toxins [93].

Proteomics approaches confirmed the production of some toxins such as SEB, SEC, SEH, SEK and SEQ both in exponential and stationary phases, and TTST-1, SEL, SE*l*U and SEP in the stationary phase [84,94–96], but the whole exo-proteome was analyzed in vitro; therefore, relevance in vivo has not been clarified [97]. In addition, a wide heterogeneity between the proteins identified amongst the different isolates analyzed in vitro [98,99] has been reported.

Although PCR is the chosen method to identify the presence of *sag* genes in bacterial DNA, it is not effective to evaluate the presence of SAgs in patients, food or in vitro-generated samples [100–102].

At present, ELISA-based methods are considered the best option for the detection and quantification of SEs/SE*l*s in several samples because of their robust and sensitive results, with reported detection limits between 0.25 and 1 ng of SAgs per g of sample [102]. However, there are no available ELISA kits for each of the 30 SAgs identified, which complicates the evaluation of the full profile of SAgs potentially produced by *S. aureus*. An approach based on a combination of chromatography and MS measures simultaneously a group of toxins; however, it is not easy to conduct and does not cover all existing SAgs [103].

In addition, Surface Plasmon Resonance (SPR) was also postulated as a method to detect SAgs in different biological samples, showing a limit of detection in the picomolar range, with the advantage to measure simultaneously three or more SAgs present in the same assay depending on the biosensor used [104–109].

All in all, each method has its advantages and disadvantages and there is no ideal technique to assess the production of all toxins; however, it is important to bear in mind the limitations of the method used at the time of reaching conclusions.

2.3. Superantigens and Human Diseases

2.3.1. Toxic Shock Syndrome

Superantigens have been associated with many illnesses caused by *S. aureus* infections. One of the most well-recognized diseases is Toxic Shock Syndrome (TSS), a potentially lethal febrile illness related to multiorgan dysfunction that occurs as a consequence of the cytokine storm produced by SAgs. TTS is characterized by hypotension; diarrhea; labored breathing; and changes in heart, liver and kidney function [110–112]. This pathology has a high incidence in women during their menstrual periods, and is particularly called menstrual TSS (mTSS) [112]. TSST-1 is associated with all menstrual cases, in some instances, coproduced with SEA or SEC-1; conversely, SEB production has been negatively related to this syndrome [111,113–116]. Regarding non-classical SAgs, Jarraud and collaborators suggested that the production of SEG and SEI by *S. aureus* strains that do not produce TSST-1, SEA to SEE and SEH and isolated from mTSS patients may explain these clinical manifestations [117].

Studies in porcine vaginal tissue have proved that, initially, the presence of *S. aureus* triggers the innate immune system activation and increases TSST-1 flux across the mucosa [118]. Furthermore, it has been demonstrated that TSST-1 is essential to stimulate systemic inflammation by inducing the production of IFN-γ and suppressing autophagy [119,120].

TSST-1 biochemical and biological properties have been studied, and among them, its capacity to interact with CD40 on epithelial cells. This interaction promotes chemokine production and attracts other components of the adaptive immune response. Consequently, local inflammation occurs and provokes mucosal disruption, facilitating TSST-1 penetration and enhancing mTSS [121].

Moreover, mTSS progress has been associated with oxygen introduction into the anaerobic vagina. This phenomenon was related to the use of medical devices to control menstrual flow, especially certain tampons whose composition enhanced their ability to trap oxygen within its fibers. Remarkably, device insertion does not cause significant vaginal oxygenation. In addition to the introduction of oxygen, the increase in body temperature, as well as protein, low glucose levels and neutral pH generate an environment that is favorable for TSST-1 production. In addition, the device's capability to increase the permeability of the mucosa promotes SAg penetration [114,122–125].

Moreover, it has been reported that tampon wearing time influences the development of mTSS [122]. Some studies propose that the tampon material also affects *S. aureus* proliferation and TSST-1 production, while others suggest that the tampon structure and fiber density have a higher impact on colonization and toxin production than the nature of its material [124,126]. It has been described that the application of menstrual cups could promote *S. aureus* growth and biofilm formation with toxin production. Although those conditions are affected by cup composition, toxin production induced by cup use is lower than that caused by tampon use. This fact has to be related to its incapacity to trap oxygen [124,125].

Non-menstrual TSS has also been associated with TSST-1 and other classical superantigens such as SEB, SEC1 and SEA [61,113–115,127–129]. A few studies suggest that SEB production is prevalent in *S. aureus* isolates that are negative for TSST-1 in non-menstrual

TSS cases [113,129]. Furthermore, in case-control investigations, it was demonstrated that *sea* gene is correlated with TSS, whereas the presence of *sem* and *seo* genes is correlated with sepsis [61].

2.3.2. Infection Endocarditis

Infection endocarditis (IE) is a disease of the endocardium and cardiac valves, that causes cardiac and multi-organic symptoms usually as a consequence of *S. aureus* bacteremia [130,131]. Classical superantigens, such as TSST-1, SEC, SEB, SEA, SED and SEE are related to IE, as well as non-classical, from the *egc* operon [87,132–135]. It has been shown that TSST-1 and *egc* operon gene deletion raised mortality in a rabbit model. Many studies have associated disease development with vegetation formation in aortic valves, whose formation decreases with the deletion of *egc*, SEC and TSST-1 genes [87,133].

Another action mechanism that could influence the inflammation process is the secretion of IL-8 by aortic endothelial cells induced by SEC [134]. In addition, it has been demonstrated that SEC inhibits the pro-angiogenic factor serpin E1 impeding reendothelialization and promoting disease [133]. On the other hand, TSST-1 also acts on the endothelium, producing upregulation of vascular and intercellular adhesion molecules (VCAM-1 and ICAM-1), raising permeability and affecting tissue re-composition [135].

2.3.3. Pneumonia

Pneumonia is an infectious respiratory disease frequently related to *S. aureus* infections in both community and hospital settings [136]. Clinical features include cough, fever, tachypnea, tachycardia and pulmonary crackles, leading to severe pulmonary infections [137,138]. Staphylococcal pneumonia caused by methicillin-resistant Staphylococcus aureus (MRSA) has been of growing concern due to its resistance to several antibiotics and its higher risk of morbidity and mortality [139,140]. It has been observed that several isolates obtained from patients with lung damage produced TSST-1, SEB and SEC in vitro [141,142]. Spauding et al. [33] proposed that SAgs contribute to pneumonia development once they are produced by *S. aureus* in the respiratory tract, generating an intrapulmonary inflammation. In addition, the cytotoxic effect on endothelial cells of the alveolar–capillary barrier may cause edema and respiratory distress. Furthermore, T cell activation and the release of cytokines may contribute to the pathogenesis and would be associated with the clinical symptoms [143,144]. However, there is no solid evidence proving that SAgs are the cause of pneumonia in humans.

2.3.4. Staphylococcal Food Poisoning

Staphylococcal food poisoning (SFP) is a foodborne disease that provokes abdominal pain, intense diarrhea and vomiting [2]. This pathology has been related to many SAgs in investigations from all over the world. The presence of classical SAg genes, such as *sea*, *seb*, *sec*, *sed* and *see*, in *S. aureus* isolates from dairy products, animals and infected patients has been associated with SFP development [101,145–150]. In those samples, not only classical SAg genes, but also *egc* genes such as *sei*, *seg*, *sem*, *sen*, *seo* and *seu* have been detected [101,145,147,148,150–153]. Moreover, other non-classical SAg genes, such as *seh*, *sej*, *sek*, *sep*, *seq* and *ser*, have been identified in isolations related to SFP [101,145,147,149,150,154–157].

While genetic studies and genomic patterns are relevant to the study of this illness, the confirmation of the physical presence of toxins in food products suspected of contamination or an analysis of isolation capability of expressing SEs is needed to verify their contribution to SFP. Notably, some isolations from SFP patients and milk samples containing *seg* and *sei* genes did not produce detectable levels of SEG and SEI. On the other hand, isolations harboring *seh* were able to produce significant levels of this SAg. Moreover, some positive isolations for *sea*, *seb* and *sec* did not express those toxins in vitro [93,158]. Other studies showed the presence of SEC, SED, SEA, SEB and SEE in samples related to SFP cases and their expression in isolates from sick people or food [159]. SEH and SEA expressions were also shown in isolates from foodstuffs causing poisoning events [154,155].

SEI and SEM production in isolates associated with food poisoning was also corroborated in several cases [160].

A variable distribution of *S. aureus* genotypes and SEs expression has been shown worldwide, suggesting that the distribution of SAg genotypes may be related to geographical distribution [156,161].

Some mechanisms of action for vomiting and diarrhea have been proposed in several animal models. Many studies suggested that enterotoxins stimulate the vagus nerve in the gut, which transmits the signal to the vomiting center in the brain [162,163]. Supporting this idea, a few studies have shown that SEA increases 5-hydroxytryptamine (5-HT) and histamine release in the intestinal tract, which would increase the discharge of vagal afferent fibers and trigger an emetic effect; it was also shown that this SAg needs vagal 5-HT3 receptors to stimulate vomiting [60,162]. Additionally, it has been described that some SAgs, such as SEA, SEB, SEE and TSST-1, can penetrate the gut lining. Interestingly, it is proposed that transcytosis is mediated by an amino acid sequence highly conserved across superantigens [164]. After transcytosis, the activation of the local immune system occurs. Using a mice model, it was shown that SEB administration resulted in an early expansion of peripheral T cells and Th1 cytokine secretion [165], and that CD4+ T cell stimulation by SEB provoked a disruption of the jejune tonic and stimulated ion transport, creating water movement and contributing to a diarrheal response [166]. Furthermore, SEA and SEB could interact with the MH C II of human subepithelial intestinal myofibroblasts, triggering the secretion of pro-inflammatory cytokines. These effects may play a role in inflammatory injury [167].

2.3.5. Autoimmune Pathologies

SAgs have been correlated with the onset of autoimmune pathologies [168,169]. SAgs may activate autoreactive T and B cells. In addition, APC activation may lead to alterations in antigen processing, with the production and presentation of autoantigens as a consequence [170].

Kawasaki Disease

Kawasaki disease (KD) is an autoimmune, febrile and inflammatory syndrome that involves coronary vasculitis. It is reported that this illness may have different etiologies such as viral and bacterial infections, imbalance in immune response and genetic factors [171–174].

Although, individuals that develop KD may have a genetic predisposition, some theories propose that SAgs could be implicated in massive immune stimulation, activating T cells, B cells and macrophages, and promoting a considerable pro-inflammatory cytokine secretion, allowing vasculitis development [173,174]. A proposed mechanism of action suggests that superantigens, secreted by *S. aureus* in the gut microbiota, may penetrate the mucosal membrane and activate Th1, T cytotoxic and B cells, generating autoreactive cells capable of enhancing illness development [173,175].

SAgs secretion by *S. aureus* isolations from patients with KD has been demonstrated in several cases [176–178]. Moreover, IgM antibodies against TSST-1, SEA, SEB and SEC were detected in KD patients' sera, and titers were significantly higher than in the control group [179,180].

In contrast, some authors state that there is no evidence of SAg involvement in the pathogenesis of KD, or that it remains unclear [171,181,182].

Diabetes Mellitus

Some animal models and in vitro experiments suggest that SAgs could have a role in diabetes mellitus II (DMII) progression. Chronic exposure to sublethal doses of TSST-1 in a rabbit model showed that this SAg induced glucose tolerance in vivo and systemic inflammation. In addition, SAg-treated adipocytes produced a higher level of pro-inflammatory cytokines (IL-6, IL-8 and TNF) in comparison to the control group. Moreover, TSST-1 induced lipolysis and insulin resistance in adipocytes. In addition, the liver damage shown in the TSST-1 group could lead to an increase in circulating endotoxin levels, which could impact the development of insulin resistance in DMII. Altogether, it was postulated that SAgs may promote insulin resistance and systemic inflammation, leading to disease worsening [183].

Other investigations on adipocyte metabolism showed a possible relationship between SAgs and insulin resistance and cytokine production. A study showed that SEA interacts with the human cytokine receptor gp130, which is present ubiquitously in human cells, and particularly in adipocytes, and consequently reduces insulin signaling and modulates glucose uptake. Binding was mediated by a structure that is well-conserved in several other SAgs [184]. In addition, it was demonstrated that *egc* SAgs, SEI, SEG, SEM and SEO could bind gp130, by SPR [62]. Furthermore, the effect of TSST-1 and SEB in adipocyte cytokine production was analyzed, showing that they provoke IL-6 and IL-8 secretion, which could contribute to inflammation and diabetes persistence [185].

Despite the previous reports, there are no significant studies in humans that demonstrate that there is a significant correlation between DMII development and chronic SAg exposure; thus, further investigations should be conducted.

Rheumatoid Arthritis

Rheumatoid arthritis (RA) is an autoimmune disease characterized by inflammation and infiltration of synovial tissues that can lead to important damage in articulations [186]. It has been suggested that SAgs could trigger the disease by activating autoreactive T cells, could have cross-reactivity with self-antigens and may contribute to long-term inflammation [186,187].

Numerous studies in RA patients demonstrated the presence of SEC, SEA and SED and their genes in blood samples, synovial fluids and DNA extracts from those samples [188–190]. The *see* gene was present in several samples of synovial fluid of RA patients, but the expression of the toxin was not analyzed [191]. SEA, SEB, SEC and TSST-1 were also identified in the synovial fluids of a group of pediatric RA patients [192]. All these studies have a limitation in common: they did not assess other SAgs.

In addition, some studies found antibodies against SEB and TSST-1, with significantly higher titers in RA patients than in healthy groups [193,194].

Several animal models were used to investigate the correlation between SAgs and RA development and, as mentioned earlier, it was proposed that the inflammatory response and the reactive cells generated by SAgs play an important role in the development of the illness and its reactivations and severity [195–199]. It was suggested that pro-inflammatory cytokine secretion induced by SAgs may enhance the expression of MHC II in synoviocytes, promoting the release of chemotactic factors and cell infiltration. In addition, SEA and SEB significantly increase the proliferation of T cells in co-culture with synoviocytes from RA patients, compared to the control group. These results suggest that the interaction between both cells and SAgs may be important in disease pathogenicity [200].

Atopic Dermatitis

Atopic dermatitis (AD) is a pruritic skin disorder that presents with eczema, inflammation and immune cell infiltration into the local skin lesion [201]. Chronic exposure to SAgs may be a factor that promotes disease incidence, as *S. aureus* colonization is usual in AD patients [202,203]. It is proposed that the disease is caused by inflammatory and allergic factors, and SAgs may be implicated in both. On the one hand, SAgs can activate T cells promoting pro-inflammatory cytokine release and contributing to skin injury, enhancing the Th1 profile. On the other hand, SAgs may activate B cells promoting SAg-specific IgE which can stimulate basophile and mast cell degranulation, leading to a Th2 response [204].

The relationship between SAgs and AD has been based on several factors, such as the presence of anti-SAgs IgE in sera from patients with AD. Anti-SEA, SEB and TSST-1 IgE has been documented in several AD patients with significantly higher titers than in the control groups [201,205–207]. Moreover, a total IgE increase has been positively correlated with a SAg-specific IgE increment in patient sera, and it was demonstrated that SAg-specific IgE titers increased with the severity of the condition [206].

In addition, SEA, SEB and TSST-1 provoked a SAg-specific histamine release by basophils from AD patients after being sensitized with IgE-SAg specific serum, in contrast to the control group. Notably, basophil samples from patients lacking these antibodies did not secrete histamine. These results showed that SAgs and their interaction with basophils mediated by IgE could exacerbate AD, promoting an allergic response [201].

Another important fact, which relates SAgs and AD development, is the production of SAgs by *S. aureus* isolated from patients. Compared with control groups, these isolates had a significantly higher production of SAgs and greater heterogeneity of encoded genes. Moreover, they were more likely to produce TSST-1, SEB and SEC [208]. In addition, *S. aureus* isolated from AD patients produced TSST-1, SEA, SEB, SEC and SED [201,209,210].

SAgs are able to induce these pathologies due to their ability to promote a pro-inflammatory environment by upregulating co-stimulatory molecules, which are usually expressed at low levels, preventing the stimulation of autoreactive lymphocytes. Another important point to note is the globular and compact structure shared by these toxins, which is the key to remaining as a whole and active molecule during the passage through the lysosome vesicle and reaching again the dendritic cell surface in the lymph nodes, where they are able to activate different lymphocyte T populations by the crosslinking of the TCRs in association with the MHC-II molecules and other recently described molecular targets, such as CD28 and B7 [211–214]. In addition, their ability to interact with diverse molecular targets in different types of cells opens up a new avenue to study their functions in new pathologies.

2.4. Staphylococcal Classification, Structure and Molecular Targets

Staphylococcal SAgs are homolog proteins that share from 20 to 89% similarity in their amino acid sequence (Supplementary Table S1). Based on their sequence of nucleotides and amino acids, SAgs were separated into five evolutive groups or families: Group I, which includes TSST-1, SET, SE*l*X, SE*l*Y and members of the superantigen-like toxins; Group II, including SEB, SECs, SEG, SER, SE*l*U, SE*l*W, SE*l*Z, SE*l*27 and SE02; Group III, comprising SEA, SED, SEE, SE*l*J, SEH, SEN, SEO, SEP, SES and SE01; Group V, gathering SEI, SEK, SEL, SEQ, SEM SE*l*V and the recently described SE*l*26. Group IV is represented by another group of toxins produced by *S. pyogenes* [33] (Table 1).

Staphylococcal and streptococcal superantigens share a common three-dimensional arrangement comprising an N-terminal β-barrel domain (OB-fold) and a C-terminal β-grasp domain with β-sheets that wrap around a long α-helix. TSST-1 is the only SE described that is less related to the other SEs from a structural point of view. Despite this fact, the basic tertiary structure present in TSST-1 is also found in the other SEs, even though these toxins have more structural complexity. This overall structure is represented in Figure 1. The loop known as the disulfide loop is located between the β4 and β5 strands. Due to its flexibility, no electron density from the crystallographic data is recorded. Experimental data suggest that this loop is responsible for the emetic properties of the SEs. The direct mutagenesis of the residues involved in the disulfide loop eliminates the emetic capacity in SEC1 [215,216].

The major and best-described molecular targets of SAgs are the MHC-II molecules and the TCR. Different regions of the toxins participate in the interaction with these targets. The interaction with the MHC-II is well documented for SAgs of different groups. SAgs mostly interact with the DR isoform of the MHC-II molecule, showing less interaction with DQ and DP. SAgs–MHC-II structures have been already solved, showing the binding surface between the SAg and the DR1 isoform of the MHC-II molecule (Figure 2).

SAgs can interact with the α and/or β chain of the MHC-II molecule, alone or in a complex with a peptide. The interaction with the polymorphic β chain is characterized by strong affinity which is mediated by zinc. SAgs involve highly conserved His and Asp residues, in positions 207 and 209, respectively, to establish coordination with this metal.

As Zn^{2+} needs four atoms to complete the interaction, the interaction could involve one more amino acid in position 169, mainly a His or Asp, and the β chain completes the interaction with a His in position 81. SEH, which also displays this kind of interaction with the MHC-II molecule, only contributes with two residues to the union. A water molecule contributes to completing the coordination required by Zn^{2+}. All the studied SAgs that contact the MHC-II β chain also contact the bound peptide, which provides more specificity to the binding. On the contrary, the interaction with the invariant α chain usually displays low affinity. SAgs such as SEB and SEC3 that bind to the low-affinity site on the MHC-II α-domain can recognize multiple HLA types (DR, DQ, DP). This is easily explained because the DR α-chain is nonpolymorphic and the SAg makes no contact with the bound peptide [217–219].

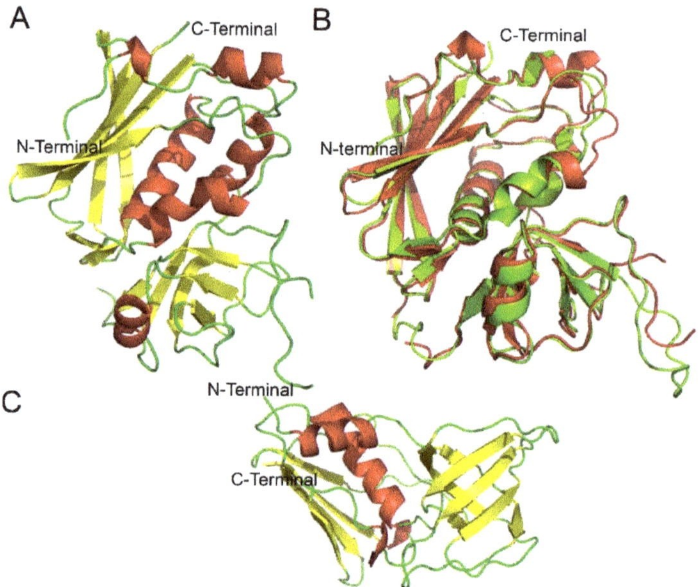

Figure 1. Structural features of Staphylococcal superantigens. (**A**) Overall structure of staphylococcal enterotoxin G (SEG) PDB accession number 1XXG. The general structure of SEG is displayed as a cartoon, and the secondary structures are colored yellow, β strands; red, α helixes; and flexible loops, green. The N-terminal domain and the C-terminal extreme are both located in the same domain of the molecule. (**B**) Superimposition of SEG and SEB. SEG overall structure (red) was superimposed over SEB structure (green) with an RMS of 0.714 Å, suggesting high similarity. (**C**) Overall structure of TSST-1, showing a simpler general structure than that of other staphylococcal enterotoxins. All the figures were performed using the PyMOL software.

The interaction with the TCR is very well studied for Group II SAgs. The SE members of this group, SEB, SEC1-3 and SEG are crystallized in complex with the murine Vβ8.2 chain. In addition, the streptococcal SAg, SpeA, which also belongs to this group, is crystallized in complex with this TCR. All the SAgs of this group involve the Asn 25 (SEB numbering) or Asn 24 (SEG numbering) as the hottest spot to interact with the TCR Vβ chain [220]. The SAgs contact the region CDR2, FR2 and 3 and the hypervariable region HV4 of the TCR, except SEG, which does not bind HV4. Despite this fact, SEG shows the highest affinity to the TCR already documented by wild-type SAg. SEG interacts with mVβ8.2 with an affinity in the nanomolar range (500 nM). An analysis of the mVβ8.2-SEG interface elucidates the higher affinity of this complex compared to others as it establishes the highest number of hydrogen bonds in the smallest binding surface [221] (Figure 3). This phenomenon may at least partially explain the early activation of T cells bearing mVβ8.2 by SEG compared to other members of Group II. SEG induces the increase in Vβ8.1+2 T cells in mice 48 h after inoculation, followed by the apoptosis of this cell population. The other members of this group cause this same effect at 96 h, showing a temporary delay compared with SEG biological function. Considering this early immunosuppression, SEG-mVβ8.2 interaction could be advantageous to the pathogen as it may facilitate the colonization of the host [222].

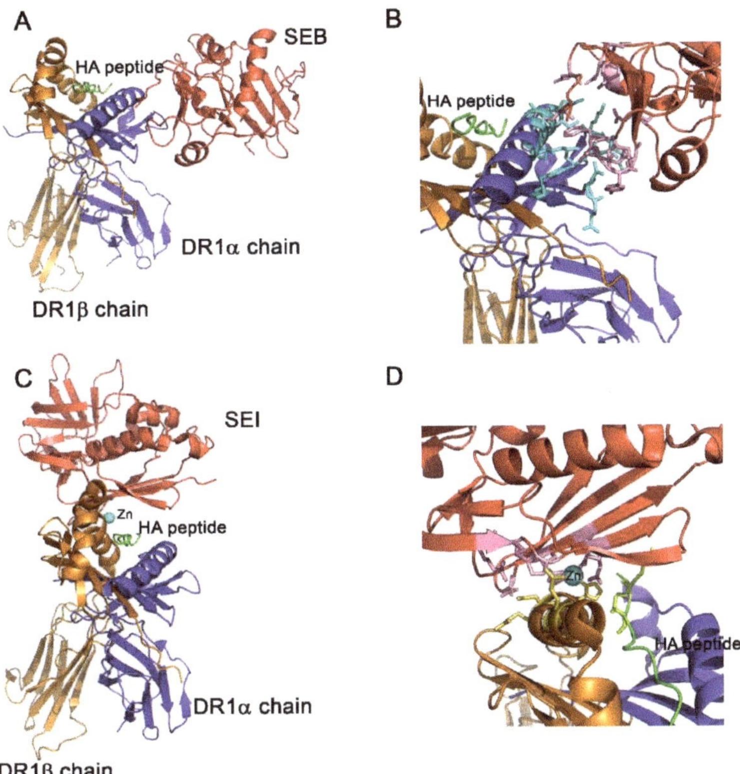

Figure 2. Structure of the SAg-HA-HLA DR1 complex. (**A**) Ribbon diagram of SEB-HA-HLA DR1 complex. (**B**) The interface of the interaction is shown in detail. SEB compromises residues in positions 43, 44, 45, 46, 47, 67, 89, 92, 94, 95, 115 and 209. HLA DR1α chain, involves the residues: 13, 17, 18, 36, 37, 39, 57, 60, 61, 63, 67 and 68. Non-contacts are found between the peptide and SEB or SEB with the DR1β chain. (**C**) Ribbon diagram of the SEI-HA-HLA DR1 complex. (**D**) The interface of the interaction is shown in detail. The interaction is coordinated by Zn^{2+}. This metal ion interacts with His81 of the DR1β chain and His169, His207 and Asp209 of SEI. SEI compromises residues in positions 98, 100, 105 and 211 to contact the residues 307 and 309 of the hemagglutinin (HA). No contacts are found between SEI and the DR1α chain. In all panels, the superantigen is colored red; the HLAD1α chain, blue; and the DR1β chain, orange. Zn^{2+} is represented as a sphere in cyan and the HA peptide, green. The residues conforming the interaction surface are represented as balls and sticks and colored pink (SEB or SEI); cyan, HLAD1α chain; yellow, DR1β chain; and light green, HA peptide. The figures were performed using PyMOL and the analysis of the structures was carried out using the CCP4i suite program.

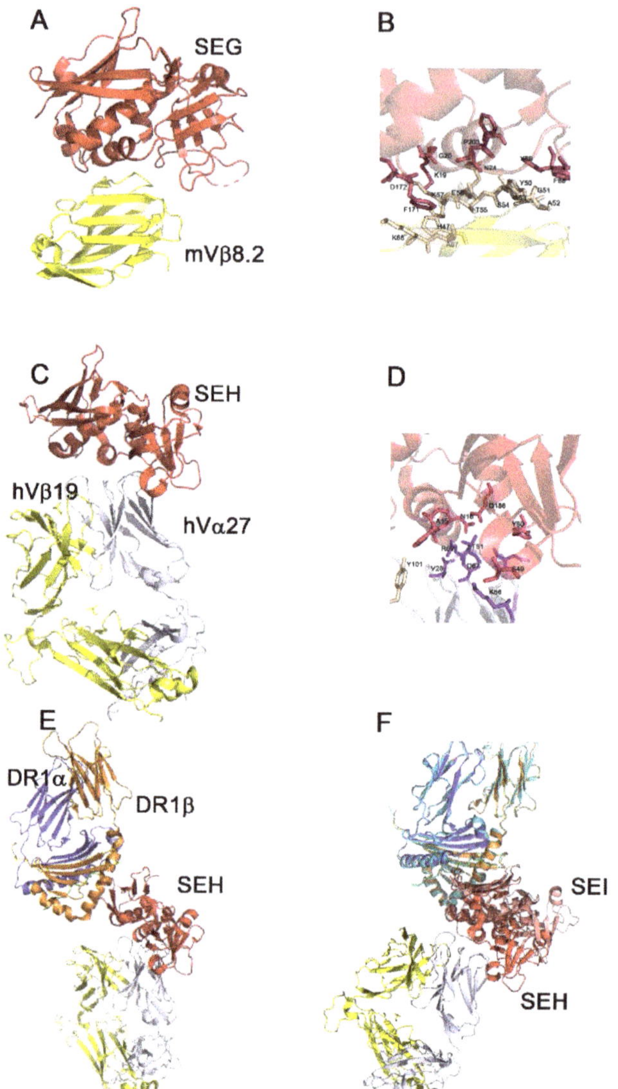

Figure 3. Crystallographic structures of the SAg-TCR interaction. (**A**) Ribbon diagram of SEG-mVβ8.2 complex. SEG is colored red and the TCR β chain, yellow (**B**) as shown in detail. SEG residues are colored pink and mVβ8.2 residues, wheat. Residues are indicated with a one letter code and numbered. (**C**) Ribbon diagram of the SEH–human TCR complex. SEH is colored red and the TCR, light blue (α chain) and yellow (β chain). (**D**) The interface of the interaction is shown in detail. SEH residues are colored hot pink, hVα27 chain residues are colored violet and hVβ19 residues are colored wheat. Residues are indicated with a one letter code and numbered. (**E**) Ribbon diagram of the SEH-TCR-MHC-II tri molecular complex. SEH is colored red, HLA-HA-DR1 is colored blue (α chain) and orange (β chain) and the TCR as indicated in C. (**F**) Superimposition of the SEI-HA-HLADR1 complex over the trimolecular complex using the DR1 as template. SEI shown in pink is clearly away from the interaction surface with the TCR α chain. The figures were performed using PyMOL and the analyses of the structures were carried out using the CCP4i suite program.

In 2010, Saline et al. solved the crystal structure of the ternary complex between the human MHC class II molecule DR1 loaded with the Influenza hemagglutinin (HA) (306–318) peptide (PKYVKQNTLKLAT), the human TCR (JM22:TRAV27/TRBV19) and the enterotoxin SEH at 2.3 Å of the resolution. The structure reveals that SEH mostly contacts the variable portion of the TCR α chain and only 6% of the variable portion of the β chain. Until now, SEH was the only SE described which presents this kind of binding [223].

With the aim to investigate if other staphylococcal enterotoxins interacting with the DR1 β chain and displaying a similar orientation than SEH over the MHC-II molecule could contact the TCR α chain, we superimposed the structure of SEI-DR1 over the structure of the ternary complex using the DR1 β chain as the template. SEI and SEH show a displacement that puts SEI away from the sphere of the interaction with the variable portion of the TCR α chain (Figure 3). Furthermore, a sequence analysis that took into account all the staphylococcal enterotoxins contacting the DR1 β chain [219,223] shows that none of them conserve the crucial residues to contact the variable portion of the TCR α chain, nor hydrogen bonds or Van der Waals contacts. Considering all the available structural information, SEH would be the only described SE that presents that kind of binding.

The interaction between staphylococcal SAgs and other molecular targets, such as CD1, B7, CD28 and gp130, is well documented. Even though these interactions were not tested in all the staphylococcal SAgs, the conserved residues in their sequence strongly suggest that all of them could bind these molecular targets. Nevertheless, more studies should be conducted to determine the physiological consequences of these bindings. Kaempfer et al. reported in deep detail the interaction between SAgs and B7/CD28 molecules, and as a consequence of this binding, SAgs would improve the contact between B7-2 and CD28, inducing T cell hyperactivation [211–213,224].

The interaction between CD1 and staphylococcal SAgs was described by Gregory and collaborators in 2000, and the engagement between SAgs and the CD1a isoform on the monocyte surface disturbs the intracellular calcium flux, altering the biological functions of the target cell [225].

The new molecular targets described increase the cell populations affected by SAgs, implying direct effects outside the immune system.

Latterly, the interaction between gp130, the interleukin 6 receptor and SEA was described by Banke [184]. Gp130 binds SEA with medium affinity inducing phosphorylation of STAT3 in adipocytes. They suggested that an Asp in position 227 would be necessary to stabilize the interaction. Nonetheless, more recently, Noli Truant and collaborators demonstrated that non-classical SAgs lacking Asp227 also interact with gp130 with micromolar affinity [62]. Later, it was shown that SEE can bind human gp130 with a similar affinity to SEA, whereas SEH displays a ten-fold lower affinity [226].

In recent years, Schlievert and collaborators showed that TSST-1, SEB and SEC induce pro-inflammatory chemokine production from human vaginal epithelial cells, being that TSST-1 is much more potent than the others. In addition, using CRISPR-Cas9 knockout CD40 cells, it was shown that this receptor is essential for the chemokine response and that the specificity of TSST-1 for menstrual TSS is in part dependent on the higher-affinity interaction with CD40 than other SAgs, such as SEB and SEC. Furthermore, CD40 seems crucial for the disruption of the human vaginal epithelial barrier by pathogens such as *S. aureus* and might be a potential therapeutic target for drug development [227–229].

2.5. Actions of SAgs on the Immune System Cells

Conventional immunity mediated by T cells is given by the interaction of an αβ TCR and a complex MHC II-peptide [230,231]. If the TCR recognizes specifically the foreign antigen, the interactions will trigger signaling pathways that will result in the proliferation and differentiation of several clones of T cells [232]. As TCRs are highly diverse molecules, only ~0.01% of the naïve T cells will be capable of recognizing a particular antigen [233].

T cell activation by SAgs is distinctive from conventional T cell activation both qualitatively and quantitatively [234]. Given that the characteristic feature of SAgs is their ability to activate T cells in a variable *beta* chain-dependent manner [2], between 5 and 20% of the total T cell population is activated as a consequence of SAg exposure. TCR diversity relies on the CD3 loops due to the V (D) J recombination during T cell development; however, there is a limited number of Vβ possible regions of the TCR, around 50 in humans, and therefore, SAgs can activate many more T cells than conventional antigens. In addition, SAgs can act in the order of picograms per mL [235], inducing a CPA-dependent production of cytokines by T cells characterized by a Th1/Th17 profile [16].

Furthermore, T cell anergy, a phenomenon whereby T cells stop responding to stimuli, has been proposed as an *S. aureus* immune evasion tactic. While the first encounter with classical SAgs induces a rapid clonal expansion of CD4+ and CD8+ T cells with a strong production of Th1/Th17 cytokines, repeated SEA or SEB challenges transduce a hyporesponsive state characterized by T cell deletions and anergy in the remaining SEA/SEB-reactive T cells [236,237]. As a consequence of repeated exposure to SEA, SEB and some streptococcal SAgs, there is an induction of CD25+ FOXP3+ CD4+ T cells Vβ specific with a regulatory profile that expresses IL-10 [238,239] as well as a promotion of CD4+ CD25- T cells [240]. This anergy has been attributed at least in part to CD8+ regulatory suppressive T cells in the case of SEA, SEB, TSST-1 and SEC, and may depend on the concentration of SAgs used to stimulate T cells [241–243]. Nonetheless, these regulatory T cells appear to turn ineffective in the massive inflammatory context of TSS [244].

Using SPEA, Sahr and collaborators showed that SAg-stimulated APCs induce pro-inflammatory responses but also promote the initiation of co-inhibitory circuits such as anti-inflammatory cytokines (IL-10), co-inhibitory molecules (PD-L1) and the induction of inhibitory effector programs (IDO) [245]. However, they performed all the assays using only one streptococcal SAg, which seems not enough to generalize for all SAgs.

No reports of T cell anergy using non classical SAgs were found to date; nevertheless, almost every publication using one or two classical staphylococcal SAgs concluded above all SAgs, which may not be true for each of the 30 toxins identified. Although they all are likely to induce anergy, considering that all SAgs stimulate T cells with low specificity, confirmation is required.

Although it is still unclear how exactly anergy occurs, it is clear that not only T cells participate in the reported immunosuppression but also APCs and other immune cells involving several mechanisms, such as anergy, deletion of effector T cells, development of Tregs and induction of tolerogenic profiles in different cell types that should be addressed.

While the superantigenic and enterotoxic effects of SAgs are the most studied mechanisms of the pathogenicity of these toxins, several new works have shown that SAgs have additional capacities.

Dendritic cells (DCs) can uptake SAgs, without inducing maturation, having the ability to maintain themselves as biologically active inside this cell type. Ganem and collaborators reported that DCs effectively incorporated SAgs in a dose-dependent manner by macropinocytosis, retaining unmodified SAgs in endosomes to finally release them to the extracellular medium [214]. Moreover, it was demonstrated that solely the uptake of SEG, SEI, SSA, SEB and SEA did not induce the maturation of DCs [246]. Concerning the retention of biological activity, in this study, SEG conserved its three-dimensional structure since it was recognized by polyclonal antisera during the entire uptake process, inclusion in diverse vesicles and exposure on the membrane, in addition to being able to stimulate T cells in vivo and in vitro. This strongly suggests that SAgs traffic through DCs in intact form. This is not surprising, given that SAgs are known to be very stable and highly resistant to proteases and can resist temperatures of 60 °C or higher, as well as extreme pHs. Earlier studies showed that DCs are more efficient than other APCs, such as B cells or monocytes, to initiate T cell proliferation by presenting TSST-1, SEA, SEB and SEE. Results suggested that picomolar levels of these SAgs are required on DCs to maximally stimulate T cells. This would be related to the high amount of MHC II expressed on DC surface [247].

An ex vivo study in mice showed that DC maturation and migration induced by SEB and TSST-1 require T cell activation. Furthermore, it was evidenced that SEB is also able to upregulate MHC II, CD40, CD205 and CD86 markers [248]. Human DC capability to upregulate MHC II, CD86, CD80, CD83 and CD54 in the presence of SEB was also demonstrated. In contrast with previous data, Coutant et al. found no effect on CD40 expression. An increase in the production of TNF-α by human DCs induced by SEB was shown, compared to other compounds such as LPS. Enhancement of IL-12p270 human DC production by SEB remains controversial because some suggest that it occurs, and others affirm that it does not affect its production [249,250].

As mentioned before, SAgs interact simultaneously with TCR on T cells and MHC II on APC and they have to be in close contact, which occurs in the lymphoid tissues. It was shown in mice DCs that they can internalize SEG, SEI and SSA as biologically active molecules and recycle them into the cell membrane. This fact does not cause DC maturation. SAgs were found intact in the acidic cells compartment and remained active. These results suggest that intracellular trafficking of SAgs in DCs improves their local concentration and promotes their encounter with T cells in lymph tissues [214].

SAgs induce cytokine production in APCs, which promote Th1/Th17 profiles [16]. However, an in vitro study in human DCs suggested that SEB may induce T cell immunoglobulin mucin domain (TIM) 4 expression in these cells when it is processed via TLR2 and NOD1 pathways. The interaction between TIM4 in APCs and TIM1 in CD4+ naïve T cells induces a Th2 profile [250]. These results agree with the idea that SAgs may have diverse effects on immune system cells depending on the interaction pathway.

Plasmacytoid dendritic cells (pDCs) are another innate cell subpopulation that may interact with SAgs. An assay in mice demonstrated that SEA increases the number of pDCs in lymphoid organs and promotes the expression of CD86 and CD40. Furthermore, pDCs would strictly require the presence of IFN-γ, in addition to the interaction with T cells, for maturation. In contrast, DCs do not depend on this cytokine for its activation by SAgs [251].

It has been reported that the interaction of SEB with monocytes enriched from PBMCs induced cell death and IFN-γ production [252]; however, the presence of the remaining T cells, a possible source of IFN-γ, cannot be discarded in those studies. Furthermore, *egc* SAgs inhibited monocytic proliferation in a dose-dependent manner, promoting cell death by apoptosis, and to a lesser extent, by necrosis. The significant production of TNF-α, IL-6 and IL-12 but not of IFN-γ was detected in this scenario [62]. Only SEI induced early apoptosis at the times assayed, similar to SEB that would induce TNF-α dependent apoptosis in THP-1 cells (a cytokine that the classical SAg induces to a greater extent than *egc* SAgs [253]), suggesting that other intracellular pathways might be involved in the death processes promoted by different SAgs. In concordance, Ulett and Anderson [254] proposed that death pathways could vary not only between different cell types but considering the toxin and the molecular conditions of the stimuli. Furthermore, the absence of CD4+ T cells in the cultures of THP-1 may explain the lack of IFN-γ production [255].

Macrophagic cells obtained by PMA-differentiated THP-1 cells inhibited their proliferation and produced pro-inflammatory cytokines in response to *egc* SAgs. In addition, SEI induced significant cell death by apoptosis in this cell type, while SEG, SEM and SEO induced death by apoptosis as well as by necrosis. SEB is the only classical SAg reported to induce apoptosis in THP-1 cells PMA-differentiated to macrophages by caspases 3 and 8 [252], but this does not necessarily explain the death mechanism of *egc* SAgs. All these results not only demonstrate that damage mechanisms are variable according to each toxin, between other factors, but that the response triggered by different SAgs simultaneously may be more diverse than expected.

Despite visible differences, several studies have shown evidence that immune activation generated by classical and new SAgs is similar [256], inducing an intense profile of the Th1/Th17 immune response [235,257–260].

Although Dauwalder (2006) suggested that *egc* SAgs would have a different inflammatory potential than classical SAgs, which may explain the different severity of the symptoms caused by *S. aureus* carriers of one or another type of SAg, this has not been confirmed by other authors. Possibly, their results could be attributed to differences in the production or the purification of SAgs, considering that they used a purchased SEA and a lab-produced SEG, or even to the concentrations used. The most accepted hypothesis is that the differences between the effects of classical and *egc* SAgs lie in a complex network of regulatory pathways that determine the moment and the conditions in which *S. aureus* produces them [256,261], resulting in much lower concentrations of *egc* SAgs compared to classical SAgs [87–89], as was mentioned in Section 2.2.

In contrast with $\alpha\beta$ T cells, $\gamma\delta$ T cells constitute between 1 and 5% of PBMCs and are mainly concentrated in skin and mucosa. Eighty percent of $\gamma\delta$ T cells in peripheral blood co-express the chains Vγ9 (alternatively Vγ2) or Vδ2, and it has been demonstrated that when activated, they can act as antigen-presenting cells [262], in addition to having the cytotoxic capacity and a great source of cytokine production. While some works stated that some classic SAgs, such as SEA, SEB and TSST-1, can stimulate directly or indirectly different subsets of $\gamma\delta$ T cells, differences are referring to each SAg [260,263,264]. These suggest that the activation of this specific cell type could be related to a characteristic affinity of each subset of $\gamma\delta$ TCR and its capacity to interact with one or another SAg, indicating that the residues involved in the binding with the $\alpha\beta$ and $\gamma\delta$ TCR are not shared. The capacity to stimulate this subpopulation of T cells has been extended to other SAgs; however, no evidence supports this fact for non-classical SAgs.

Furthermore, it has been reported that SEB activates NKT cells, and that SEA and SEB can interact with invariant NKT cells (iNKT) [265–267], inducing a pathogenic role of this cell type in SST. Moreover, classical SAgs interact with B cells [268–270], and IgE antibodies against superantigens have been detected [271,272].

It is demonstrated as well that SEA, SEB, SEE and TSST-1 can interact with mast cells [60,273–275], and the novelty that SEB activates mucosal-associated invariant T cells (MAIT cells) to produce high levels of INF-y, TNF-α and IL-2 and then induces anergy in this cell type [276]. MAIT cells are an unconventional T cell subset with a semi-invariant T cell receptor that recognizes antigens presented in the context of classical MHC-like molecules [277] and are involved in microbial immunity displaying protective and pathogenic responses [278,279]. Again, there is no information about the effect of non-classical Sags on these cells.

SSL toxins are another group of exotoxins of *S. aureus* that are unable to induce T cell proliferation but can contribute to immune evasion [48], interfering with complement and neutrophil function [280]. Similarly, Se*l*X can bind both monocytes and neutrophils, impairing neutrophil activation [281,282], and is considered the first line of defense against *S. aureus* [283].

Although the activation of other cell types could have a considerable impact on the immune system response, almost every study published has been developed using the less prevalent classical SAgs, and it is not clear if their effects can be extended to the newer SAgs, which are widely distributed, such as the *egc* operon SAgs.

Due to their ability to interact with several components of the immune system, SAgs have been clinically used as immune modifiers in neoplasm treatment and other pathologies.

2.6. Superantigens as Therapeutic Tools

Considering the complexity of oncological processes, altered superantigens with reducing systemic toxicity that conserves the ability to eliminate tumoral cells have being considered as alternative therapies. Indeed, several cancer cell lines are effectively attacked by PBMCs activated by classical SAgs. Furthermore, various SAgs mutated in the MHC II contact region have been created willing to keep the antitumoral effects of SAgs, thus reducing systemic toxicity. Considering this strategy to diminish side effects, SAgs continue being extensively investigated for oncological applications either alone or in combination with classical anticancer drugs [284]. It was described that PBMCs stimulated with SEA display the ability to promote the death of human lung carcinoma A549 cells [285]. Similar results were observed when SEB-stimulated PBMCs induced apoptosis on transitional cell carcinoma cells [286]. These first studies were carried out with wild-type superantigens, which conserved intact their potent inflammatory activity. With the aim to avoid SAg systemic toxicity, various SAgs altered in hot spot regions have been created. That is the case for SEA D227A, which was conjugated with an antitumor Fab antibody; the fusion protein conserved the capacity to activate T cells but reduced the binding to the MHC-II molecule 500 times, compared to wild-type SEA [287]. Moreover, SEC2 double mutant, T20L/G22E, inhibited the growth of S180 sarcoma with less toxicity in mice [288].

Other alternatives to reduce systemic toxicity in using SAg as therapeutic tools include mutant SAgs that were designed as a part of chimeric single chain antibodies (scFv) specific to tumoral antigens. These SAgs have been mutated in the binding site to the MHC-II molecule with the aim of reducing their superantigenicity. These mutations reduced their affinity to the MHC-II molecule, and the cytotoxic effect on MHC class II- expressing cells [289]. Recently, a new generation of antibody–superantigen fusion proteins was designed, in which the SAg, in this case SEA, was split into two fragments. Individually, each fragment remains inactive, but when the biological formulation reaches the target cell by binding to cell surface antigens, it dimerizes and the SAg regains its biological functionality activating T cells. The effective split SEA design would not affect MHC class II-expressing cells, but when bound to a tumor antigen via a targeting moiety it would activate a T cell response [290].

Non-classical SAgs were also investigated in the immunotherapy field. The *egc* operon superantigens SEG and SEI were linked to the endogenous human MHCII HLA-DQ8 allele in humanized mice inducing a potent antitumor response and extending life in an established melanoma mice model [291].

In addition, the combination of mutant SAgs with other molecules to create chimeras that specifically stimulate the immune system with low toxicity is another strategy not fully studied yet, but with promising results, as in the case of mutant SEG used as a candidate for Chagas vaccination [292]. In that work, a chimeric molecule comprising a *T. cruzi* antigen and a non-toxic form of SEG was used as a novel immunogen to confer protection against *T. cruzi* infection. The mutant SAg used retained its ability to trigger classical activation of macrophages without affecting T cells, because this Th1 profile was adequate to eradicate intracellular protozoa such as *T. cruzi*, proving to be an effective immune modulator against this parasite.

These uses of SAgs as therapeutic agents that permit taking advantage of these toxins allow us to compare them to Dr. Jekyll, but it should not be forgotten that SAgs usually also act as Mr. Hyde, causing several affections and worsening disease conditions.

Unfortunately, everything in nature has its Dr. Jekyll and Mr. Hyde side, and superantigens are one of the most crucial toxin threats in warfare or bioterrorism as they are resistant to heat and can be administrated in contaminated water or as aerosols in the air.

It is our choice to use these toxins either like Dr. Jekyll or Mr. Hyde.

3. Concluding Remarks

Microorganisms evolve to evade the host immune system. All mammal pathogens have to face a strong and complex immune response, designed to fight them back. Manipulating this kind of response could be the most effective mechanism to spread the infection. This situation is the ideal scenario to develop a chronic infection and reach an equilibrium between the effector mechanisms of the immune system and the war for survival of the pathogen.

Staphylococcus aureus causes an acute infection where quick spread and colonization are essential to reach its goal. Considering that inflammation is crucial for the resolution of most bacterial infections, it is not clear how a pro-inflammatory state may help bacteria succeed. However, the inflammation induced by SAgs seems to impair activation and recruitment of important effector cells, promoting a temporal host immune suppression and the survival of *S. aureus*. In view of the numerous diseases in which SAgs participate or are implicated, it can be ascertained that Mr. Hyde's side of SAgs is the easily visible face of these toxins.

For many years, it was considered that SAgs only affected T cell function, driving them to anergy or apoptosis, and promoting host immune suppression. As described throughout this work, SAgs would not only exert their function on T lymphocytes in the presence of antigen-presenting cells, but they would also be capable of eliciting different effects on monocyte/macrophage effector cells, Gδ T lymphocytes, lymphocytes B, neutrophils and other cells not belonging to the immune system, such as adipocytes. In general, this action would lead to nonspecific pro-inflammatory conditions that would deplete effector cell populations, thus being inhibited to eradicate the bacteria that produce them, promoting a state of host immunosuppression from the first to the last stages of the infection, which would favor bacterial propagation. However, this intricate scenario, which is gradually being cleared, can be reversed, by taking advantage of the action of SAgs while restricting their effects for human benefit, thus revealing the protective yet still unclear Dr Jekyll side of SAgs.

Supplementary Materials: The following supporting information can be downloaded at: https://www.mdpi.com/article/10.3390/toxins14110800/s1, Table S1: Rate of sequence similarity between SAgs.

Author Contributions: Conceptualization, writing the original draft preparation, S.N.T.; writing the original draft, D.M.R., writing special section, M.B.S.; editing, E.L.M.; writing, review and supervision, M.M.F. All authors have read and agreed to the published version of the manuscript.

Funding: This research received no external funding.

Institutional Review Board Statement: Not applicable.

Informed Consent Statement: Not applicable.

Data Availability Statement: Not applicable.

Conflicts of Interest: The authors declare no conflict of interest.

References

1. White, J.; Herman, A.; Pullen, A.M.; Kubo, R.; Kappler, J.W.; Marrack, P. The Vβ-specific superantigen staphylococcal enterotoxin B: Stimulation of mature T cells and clonal deletion in neonatal mice. *Cell* **1989**, *56*, 27–35. [CrossRef]
2. Marrack, P.; Kappler, J. The Staphylococcal Enterotoxins and Their Relatives. *Science* **1990**, *248*, 705–711. [CrossRef] [PubMed]
3. Schutzer, S.E.; Fischetti, V.A.; Zabriskie, J.B. Toxic Shock Syndrome and Lysogeny in *Staphylococcus aureus*. *Obstet. Gynecol. Surv.* **1983**, *220*, 316–318.
4. Quimby, F.; Nguyen, H.T.; Bergdoll, M.S. Animal studies of toxic shock syndrome. *Crit. Rev. Microbiol.* **1985**, *12*, 1–44. [CrossRef]
5. Gaventa, S.; Reingold, A.L.; Hightower, A.W.; Broome, C.V.; Schwartz, B.; Hoppe, C.; Harwell, J.; Lefkowitz, L.K.; Makintubee, S.; Cundiff, D.R.; et al. Active surveillance for toxic shock syndrome in the united states, 1986. *Rev. Infect. Dis.* **1989**, *11*, S28–S34. [CrossRef]
6. Blank, C.; Luz, A.; Bendigs, S.; Erdmann, A.; Wagner, H.; Heeg, K. Superantigen and endotoxin synergize in the induction of lethal shock. *Eur. J. Immunol.* **1997**, *27*, 825–833. [CrossRef]
7. Dalpke, A.H.; Heeg, K. Synergistic and antagonistic interactions between LPS and superantigens. *J. Endotoxin Res.* **2003**, *9*, 51–54. [CrossRef]
8. MacIsaac, C.M.; Curtis, N.; Cade, J.; Visvanathan, K. Superantigens in sepsis. *Int. Congr. Ser.* **2006**, *1289*, 121–124. [CrossRef]
9. Cole, B.C.; Atkin, C.L. The Mycoplasma arthritidis T-cell mitogen MAM: A model superantigen. *Immunol. Today* **1991**, *12*, 271–276. [CrossRef]
10. Igwe, E.I.; Shewmaker, P.L.; Facklam, R.R.; Farley, M.M.; Van Beneden, C.; Beall, B. Identification of superantigen genes speM, ssa, and smeZ in invasive strains of beta-hemolytic group C and G streptococci recovered from humans. *FEMS Microbiol. Lett.* **2003**, *229*, 259–264. [CrossRef]
11. Abe, J.; Takeda, T.; Watanabe, Y.; Nakao, H.; Kobayashi, N.; Leung, D.Y.; Kohsaka, T. Evidence for superantigen production by Yersinia pseudotuberculosis. *J. Immunol.* **1993**, *151*, 4183. [PubMed]
12. Wei, B.; Huang, T.; Dalwadi, H.; Sutton, C.L.; Bruckner, D.; Braun, J. Pseudomonas fluorescens encodes the Crohn's disease-associated I2 sequence and T-cell superantigen. *Infect. Immun.* **2002**, *70*, 6567–6575. [CrossRef] [PubMed]
13. Huber, B.T.; Hsu, P.N.; Sutkowski, N. Virus-encoded superantigens. *Microbiol. Rev.* **1996**, *60*, 473–482. [CrossRef] [PubMed]

14. Sutkowski, N.; Conrad, B.; Thorley-Lawson, D.A.; Huber, B.T. Epstein-Barr virus transactivates the human endogenous retrovirus HERV-K18 that encodes a superantigen. *Immunity* **2001**, *15*, 579–589. [CrossRef]
15. Azar, G.A.; Thibodeau, J. Human endogenous retrovirus IDDMK1,222 and mouse mammary tumor virus superantigens differ in their ability to stimulate murine T cell hybridomas. *Immunol. Lett.* **2002**, *81*, 87–91. [CrossRef]
16. Tuffs, S.; Haeryfar, S.; McCormick, J. Manipulation of Innate and Adaptive Immunity by Staphylococcal Superantigens. *Pathogens* **2018**, *7*, 53. [CrossRef]
17. Kluytmans, J.; Van Belkum, A.; Verbrugh, H. Nasal Carriage of *Staphylococcus aureus*: Epidemiology, Underlying Mechanisms, and Associated Risks. *Clin. Microbiol. Rev.* **1997**, *10*, 505–520. [CrossRef]
18. Laux, C.; Peschel, A.; Krismer, B. *Staphylococcus aureus* Colonization of the Human Nose and Interaction with Other Microbiome Members. *Microbiol. Spectr.* **2019**, *7*, GPP3-0029-2018. [CrossRef]
19. Verhoeven, P.O.; Gagnaire, J.; Botelho-Nevers, E.; Grattard, F.; Carricajo, A.; Lucht, F.; Pozzetto, B.; Berthelot, P. Detection and clinical relevance of *Staphylococcus aureus* nasal carriage: An update. *Expert Rev. Anti. Infect. Ther.* **2014**, *12*, 75–89. [CrossRef]
20. Williams, R.E. Healthy carriage of *Staphylococcus aureus*: Its prevalence and importance. *Bacteriol. Rev.* **1963**, *27*, 56–71. [CrossRef]
21. Armstrong-Esther, C.A.; Smith, J.E. Carriage patterns of staphylococcus aureus in a healthy non-hospital population of adults and children. *Ann. Hum. Biol.* **1976**, *3*, 221–227. [CrossRef] [PubMed]
22. Gagnaire, J.; Verhoeven, P.O.; Grattard, F.; Rigaill, J.; Lucht, F.; Pozzetto, B.; Berthelot, P.; Botelho-Nevers, E. Epidemiology and clinical relevance of *Staphylococcus aureus* intestinal carriage: A systematic review and meta-analysis. *Expert Rev. Anti. Infect. Ther.* **2017**, *15*, 767–785. [CrossRef] [PubMed]
23. Ridley, M. Perineal carriage of staph. aureus. *Br. Med. J.* **1959**, *1*, 270–273. [CrossRef] [PubMed]
24. Wertheim, H.F.L.; Verveer, J.; Boelens, H.A.M.; Van Belkum, A.; Verbrugh, H.A.; Vos, M.C. Effect of mupirocin treatment on nasal, pharyngeal, and perineal carriage of *Staphylococcus aureus* in healthy adults. *Antimicrob. Agents Chemother.* **2005**, *49*, 1465–1467. [CrossRef] [PubMed]
25. von Eiff, C.; Becker, K.; Machka, K.; Stammer, H.; Peters, G. Nasal Carriage as a Source of *Staphylococcus aureus* Bacteremia. *N. Engl. J. Med.* **2001**, *344*, 11–16. [CrossRef]
26. Bergdoll, M.S.; Borja, C.R.; Avena, R.M. Identification of a new enterotoxin as enterotoxin C. *J. Bacteriol.* **1965**, *90*, 1481–1485. [CrossRef]
27. Bergdoll, M.S.; Borja, C.R.; Robbins, R.N.; Weiss, K.F. Identification of enterotoxin E. *Infect. Immun.* **1971**, *4*, 593–595. [CrossRef]
28. Casman, E.P.; Bennett, R.W.; Dorsey, A.E.; Issa, J.A. Identification of a fourth staphylococcal enterotoxin, enterotoxin D. *J. Bacteriol.* **1967**, *94*, 1875–1882. [CrossRef]
29. Casman, E.P.; Bennett, R.W. Culture medium for the production of staphylococcal enterotoxin A. *J. Bacteriol.* **1963**, *86*, 18–23. [CrossRef]
30. Casman, E.P.; Bergdoll, M.S.; Robinson, J. Designation of staphylococcal enterotoxins. *J. Bacteriol.* **1963**, *85*, 715–716. [CrossRef]
31. Jones, C.L.; Khan, S.A. Nucleotide Sequence of the Enterotoxin B Gene from *Staphylococcus aureus*. *J. Bacteriol.* **1986**, *166*, 29–33. [CrossRef] [PubMed]
32. Peavy, D.L.; Adler, W.H.; Smith, R.T. The mitogenic effects of endotoxin and staphylococcal enterotoxin B on mouse spleen cells and human peripheral lymphocytes. *J. Immunol.* **1970**, *105*, 1453–1458. [PubMed]
33. Spaulding, A.R.; Salgado-Pabón, W.; Kohler, P.L.; Horswill, A.R.; Leung, D.Y.M.; Schlievert, P.M. Staphylococcal and streptococcal superantigen exotoxins. *Clin. Microbiol. Rev.* **2013**, *26*, 422–447. [CrossRef] [PubMed]
34. Ren, K.; Bannan, J.D.; Pancholi, V.; Cheung, A.L.; Robbins, J.C.; Fischetti, V.A.; Zabriskie, J.B. Characterization and Biological Properties of a New Staphylococcal Exotoxin. *J. Exp. Med.* **1994**, *180*, 1675–1683. [CrossRef]
35. Zhang, S.; Iandolo, J.J.; Stewart, G.C. The enterotoxin D plasmid of *Staphylococcus aureus* encodes a second enterotoxin determinant (sej). *FEMS Microbiol. Lett.* **1998**, *168*, 227–233. [CrossRef]
36. Munson, S.H.; Tremaine, M.T.; Betley, M.J.; Welch, R.A. Identification and Characterization of Staphylococcal Enterotoxin Types G and I from *Staphylococcus aureus* Identification and Characterization of Staphylococcal Enterotoxin Types G and I from *Staphylococcus aureus*. *Infect. Immun.* **1998**, *66*, 3337–3348. [CrossRef]
37. Jarraud, S.; Peyrat, M.A.; Lim, A.; Tristan, A.; Bes, M.; Mougel, C.; Etienne, J.; Vandenesch, F.; Bonneville, M.; Lina, G. egc, a highly prevalent operon of enterotoxin gene, forms a putative nursery of superantigens in *Staphylococcus aureus*. *J. Immunol.* **2001**, *166*, 669–677. [CrossRef]
38. Fitzgerald, J.R.; Monday, S.R.; Foster, T.J.; Bohach, G.A.; Hartigan, P.J.; Meaney, W.J.; Smyth, C.J. Characterization of a Putative Pathogenicity Island from Bovine *Staphylococcus aureus* Encoding Multiple Superantigens. *J. Bacteriol.* **2001**, *183*, 63–70. [CrossRef]
39. Orwin, P.M.; Leung, D.Y.M.; Donahue, H.L.; Novick, R.P.; Schlievert, P.M. Biochemical and Biological Properties of Staphylococcal Enterotoxin K. *Infect. Immun.* **2001**, *69*, 360–366. [CrossRef]
40. Correction. *J. Immunol.* **2001**, *166*, 4260. [CrossRef]
41. Letertre, C.; Perelle, S.; Dilasser, F.; Fach, P. Identification of a new putative enterotoxin SEU encoded by the egc cluster of *Staphylococcus aureus*. *J. Appl. Microbiol.* **2003**, *95*, 38–43. [CrossRef] [PubMed]
42. Thomas, D.Y.; Jarraud, S.; Lemercier, B.; Cozon, G.; Echasserieau, K.; Etienne, J.; Gougeon, M.L.; Lina, G.; Vandenesch, F. Staphylococcal enterotoxin-like toxins U2 and V, two new staphylococcal superantigens arising from recombination within the enterotoxin gene cluster. *Infect. Immun.* **2006**, *74*, 4724–4734. [CrossRef] [PubMed]

43. Omoe, K.; Ichi Imanishi, K.; Hu, D.L.; Kato, H.; Fugane, Y.; Abe, Y.; Hamaoka, S.; Watanabe, Y.; Nakane, A.; Uchiyama, T.; et al. Characterization of Novel Staphylococcal Enterotoxin-Like Toxin Type P. *Infect. Immun.* **2005**, *73*, 5540–5546. [CrossRef] [PubMed]

44. Ono, H.K.; Omoe, K.; Imanishi, K.; Iwakabe, Y.; Hu, D.L.; Kato, H.; Saito, N.; Nakane, A.; Uchiyama, T.; Shinagawa, K. Identification and Characterization of Two Novel Staphylococcal Enterotoxins, Types S and T. *Infect. Immun.* **2008**, *76*, 4999–5005. [CrossRef] [PubMed]

45. Orwin, P.M.; Leung, D.Y.M.; Tripp, T.J.; Bohach, G.A.; Earhart, C.A.; Ohlendorf, D.H.; Schlievert, P.M. Characterization of a novel staphylococcal enterotoxin-like superantigen, a member of the group V subfamily of pyrogenic toxins. *Biochemistry* **2002**, *41*, 14033–14040. [CrossRef] [PubMed]

46. Omoe, K.; Ichi Imanishi, K.; Hu, D.L.; Kato, H.; Takahashi-Omoe, H.; Nakane, A.; Uchiyama, T.; Shinagawa, K. Biological Properties of Staphylococcal Enterotoxin-Like Toxin Type R. *Infect. Immun.* **2004**, *72*, 3664–3667. [CrossRef]

47. Okumura, K.; Shimomura, Y.; Yamagata Murayama, S.; Yagi, J.; Ubukata, K.; Kirikae, T.; Miyoshi-Akiyama, T. Evolutionary paths of streptococcal and staphylococcal superantigens. *BMC Genom.* **2012**, *13*, 404. [CrossRef]

48. Wilson, G.J.; Seo, K.S.; Cartwright, R.A.; Connelley, T.; Chuang-Smith, O.N.; Merriman, J.A.; Guinane, C.M.; Park, J.Y.; Bohach, G.A.; Schlievert, P.M.; et al. A Novel Core Genome-Encoded Superantigen Contributes to Lethality of Community-Associated MRSA Necrotizing Pneumonia. *PLoS Pathog.* **2011**, *7*, e1002271. [CrossRef]

49. Ono, H.K.; Sato'o, Y.; Narita, K.; Naito, I.; Hirose, S.; Hisatsune, J.; Asano, K.; Hu, D.L.; Omoe, K.; Sugai, M.; et al. Identification and characterization of a novel staphylococcal emetic toxin. *Appl. Environ. Microbiol.* **2015**, *81*, 7034–7040. [CrossRef]

50. Spoor, L.E.; Richardson, E.; Richards, A.C.; Wilson, G.J.; Mendonca, C.; Gupta, R.K.; Mcadam, P.R.; Nutbeam-Tuffs, S.; Black, N.S.; O'gara, J.P.; et al. Recombination-mediated remodelling of host-pathogen interactions during *Staphylococcus aureus* niche adaptation. *Microb. Genom.* **2015**, *1*, e000036. [CrossRef]

51. Hisatsune, J.; Hagiya, H.; Shiota, S.; Sugai, M. Complete Genome Sequence of Systemically Disseminated Sequence Type 8 Staphylococcal Cassette Chromosome mec Type IVl Community-Acquired Methicillin-Resistant *Staphylococcus aureus*. *Genome Announc.* **2017**, *5*, e00852-17. [CrossRef] [PubMed]

52. Suzuki, Y.; Ono, H.K.; Shimojima, Y.; Kubota, H.; Kato, R.; Kakuda, T.; Hirose, S.; Hu, D.L.; Nakane, A.; Takai, S.; et al. A novel staphylococcal enterotoxin SE02 involved in a staphylococcal food poisoning outbreak that occurred in Tokyo in 2004. *Food Microbiol.* **2020**, *92*, 103588. [CrossRef] [PubMed]

53. Zhang, D.F.; Yang, X.Y.; Zhang, J.; Qin, X.; Huang, X.; Cui, Y.; Zhou, M.; Shi, C.; French, N.P.; Shi, X. Identification and characterization of two novel superantigens among *Staphylococcus aureus* complex. *Int. J. Med. Microbiol.* **2018**, *308*, 438–446. [CrossRef] [PubMed]

54. Lina, G.; Bohach, G.A.; Nair, S.P.; Hiramatsu, K.; Jouvin-Marche, E.; Mariuzza, R. Standard Nomenclature for the Superantigens Expressed by Staphylococcus. *J. Infect. Dis.* **2004**, *189*, 2334–2336. [CrossRef] [PubMed]

55. Wang, T.; Tao, X.X.; Meng, F.L.; Li, X.P.; Ono-Hisaya, O.H.; Wang, D.; Hu, D.L.; Zhang, J.Z.; Wang, G.Q.; Yan, X.M. Cloning and expression of recombinant truncated SElX protein and evaluation on the related emetic activities. *Zhonghua Liu Xing Bing Xue Za Zhi* **2020**, *41*, 567–570. [CrossRef] [PubMed]

56. Hu, D.L.; Omoe, K.; Shimoda, Y.; Nakane, A.; Shinagawa, K. Induction of Emetic Response to Staphylococcal Enterotoxins in the House Musk Shrew (*Suncus murinus*). *Infect. Immun.* **2003**, *71*, 567–570. [CrossRef] [PubMed]

57. Ueno, S.; Matsuki, N.; Saito, H. *Suncus murinus* as a new experimental model for motion sickness. *Life Sci.* **1988**, *43*, 413–420. [CrossRef]

58. Omoe, K.; Hu, D.L.; Takahashi-Omoe, H.; Nakane, A.; Shinagawa, K. Comprehensive analysis of classical and newly described staphylococcal superantigenic toxin genes in *Staphylococcus aureus* isolates. *FEMS Microbiol. Lett.* **2005**, *246*, 191–198. [CrossRef]

59. Hu, D.L.; Ono, H.K.; Isayama, S.; Okada, R.; Okamura, M.; Lei, L.C.; Liu, Z.S.; Zhang, X.C.; Liu, M.Y.; Cui, J.C.; et al. Biological characteristics of staphylococcal enterotoxin Q and its potential risk for food poisoning. *J. Appl. Microbiol.* **2017**, *122*, 1672–1679. [CrossRef]

60. Ono, H.K.; Hirose, S.; Narita, K.; Sugiyama, M.; Asano, K.; Hu, D.L.; Nakane, A. Histamine release from intestinal mast cells induced by staphylococcal enterotoxin A (SEA) evokes vomiting reflex in common marmoset. *PLoS Pathog.* **2019**, *15*, e1007803. [CrossRef]

61. Ferry, T.; Thomas, D.; Genestier, A.L.; Bes, M.; Lina, G.; Vandenesch, F.; Etienne, J. Comparative Prevalence of Superantigen Genes in *Staphylococcus aureus* Isolates Causing Sepsis with and without Septic Shock. *Clin. Infect. Dis.* **2005**, *41*, 771–777. [CrossRef] [PubMed]

62. Noli Truant, S.; De Marzi, M.C.; Sarratea, M.B.; Antonoglou, M.B.; Meo, A.P.; Iannantuono López, L.V.; Fernández Lynch, M.J.; Todone, M.; Malchiodi, E.L.; Fernández, M.M. egc Superantigens Impair Monocytes/Macrophages Inducing Cell Death and Inefficient Activation. *Front. Immunol.* **2020**, *10*, 3008. [CrossRef] [PubMed]

63. Zhang, P.; Liu, X.; Zhang, M.; Kou, M.; Chang, G.; Wan, Y.; Xu, X.; Ruan, F.; Wang, Y.; Wang, X. Prevalence, Antimicrobial Resistance, and Molecular Characteristics of *Staphylococcus aureus* and Methicillin-Resistant *Staphylococcus aureus* from Retail Ice Cream in Shaanxi Province, China. *Foodborne Pathog. Dis.* **2022**, *19*, 217–225. [CrossRef] [PubMed]

64. Aung, M.S.; Urushibara, N.; Kawaguchiya, M.; Ito, M.; Habadera, S.; Kobayashi, N. Prevalence and Genetic Diversity of Staphylococcal Enterotoxin (-Like) Genes sey, selw, selx, selz, sel26 and sel27 in Community-Acquired Methicillin-Resistant *Staphylococcus aureus*. *Toxins* **2020**, *12*, 347. [CrossRef] [PubMed]

65. Becker, K.; Friedrich, A.W.; Lubritz, G.; Weilert, M.; Peters, G.; Von Eiff, C. Prevalence of Genes Encoding Pyrogenic Toxin Superantigens and Exfoliative Toxins among Strains of *Staphylococcus aureus* Isolated from Blood and Nasal Specimens. *J. Clin. Microbiol.* **2003**, *41*, 1434–1439. [CrossRef] [PubMed]
66. Fischer, A.J.; Kilgore, S.H.; Singh, S.B.; Allen, P.D.; Hansen, A.R.; Limoli, D.H.; Schlievert, P.M. High Prevalence of *Staphylococcus aureus* Enterotoxin Gene Cluster Superantigens in Cystic Fibrosis Clinical Isolates. *Genes* **2019**, *10*, 1036. [CrossRef]
67. Park, K.H.; Greenwood-Quaintance, K.E.; Cunningham, S.A.; Chia, N.; Jeraldo, P.R.; Sampathkumar, P.; Nelson, H.; Patel, R. Lack of correlation of virulence gene profiles of *Staphylococcus aureus* bacteremia isolates with mortality. *Microb. Pathog.* **2019**, *133*, 103543. [CrossRef]
68. Garbacz, K.; Piechowicz, L.; Podkowik, M.; Mroczkowska, A.; Empel, J.; Bania, J. Infection and Drug Resistance Dovepress emergence and spread of worldwide *Staphylococcus aureus* clones among cystic fibrosis patients. *Infect. Drug Resist.* **2018**, *11*, 247. [CrossRef]
69. Mempel, M.; Lina, G.; Hojka, M.; Schnopp, C.; Seidl, H.P.; Schäfer, T.; Ring, J.; Vandenesch, F.; Abeck, D. High prevalence of superantigens associated with the egc locus in *Staphylococcus aureus* isolates from patients with atopic eczema. *Eur. J. Clin. Microbiol. Infect. Dis.* **2003**, *22*, 306–309. [CrossRef]
70. Merriman, J.A.; Mueller, E.A.; Cahill, M.P.; Beck, L.A.; Paller, A.S.; Hanifin, J.M.; Ong, P.Y.; Schneider, L.; Babineau, D.C.; David, G.; et al. Temporal and Racial Differences Associated with Atopic Dermatitis *Staphylococcus aureus* and Encoded Virulence Factors. *Msphere* **2016**, *1*, e00295-16. [CrossRef]
71. Van Belkum, A.; Melles, D.C.; Snijders, S.V.; Van Leeuwen, W.B.; Wertheim, H.F.L.; Nouwen, J.L.; Verbrugh, H.A.; Etienne, J. Clonal Distribution and Differential Occurrence of the Enterotoxin Gene Cluster, egc, in Carriage-versus Bacteremia-Associated Isolates of *Staphylococcus aureus*. *J. Clin. Microbiol.* **2006**, *44*, 1555–1557. [CrossRef] [PubMed]
72. Bronner, S.; Monteil, H.; Prévost, G. Regulation of virulence determinants in *Staphylococcus aureus*: Complexity and applications. *FEMS Microbiol. Rev.* **2006**, *28*, 183–200. [CrossRef] [PubMed]
73. Cheung, A.L.; Bayer, A.S.; Zhang, G.; Gresham, H.; Xiong, Y.Q. Regulation of virulence determinants in vitro and in vivo in *Staphylococcus aureus*. *FEMS Immunol. Med. Microbiol.* **2006**, *40*, 1–9. [CrossRef]
74. Jenul, C.; Horswill, A.R. Regulation of *Staphylococcus aureus* virulence. *Microbiol. Spectr.* **2019**, *7*, 7-2. [CrossRef]
75. Bergdoll, M.S.; Czop, J.K.; Gould, S.S. Enterotoxin Synthesis by the Staphylococci. *Ann. N. Y. Acad. Sci.* **1974**, *236*, 307–316. [CrossRef]
76. Bernardo, K.; Pakulat, N.; Fleer, S.; Schnaith, A.; Utermöhlen, O.; Krut, O.; Müller, S.; Krönke, M. Subinhibitory Concentrations of Linezolid Reduce *Staphylococcus aureus* Virulence Factor Expression. *Antimicrob. Agents Chemother.* **2004**, *48*, 546–555. [CrossRef]
77. Czop, J.K.; Bergdoll, M.S. Staphylococcal Enterotoxin Synthesis during the Exponential, Transitional, and Stationary Growth Phases. *Infect. Immun.* **1974**, *9*, 229–235. [CrossRef]
78. Gaskill, M.E.; Khan, S.A. Regulation of the enterotoxin B gene in *Staphylococcus aureus*. *J. Biol. Chem.* **1988**, *263*, 6276–6280. [CrossRef]
79. Otero, A.; Garcia, M.L.; Garcia, M.C.; Moreno, B.; Bergdoll, M.S. Production of Staphylococcal Enterotoxins C1 and C2 and Thermonuclease throughout the Growth Cycle. *Appl. Environ. Microbiol.* **1990**, *56*, 555–559. [CrossRef]
80. Regassa, L.B.; Couch, J.L.; Betley, M.J. Steady-State Staphylococcal Enterotoxin Type C mRNA Is Affected by a Product of the Accessory Gene Regulator (agr) and by Glucose. *Infect. Immun.* **1991**, *59*, 955–962. [CrossRef]
81. Sihto, H.M.; Tasara, T.; Stephan, R.; Johler, S. Temporal expression of the staphylococcal enterotoxin D gene under NaCl stress conditions. *FEMS Microbiol. Lett.* **2015**, *362*, 24. [CrossRef] [PubMed]
82. Macori, G.; Bellio, A.; Bianchi, D.M.; Chiesa, F.; Gallina, S.; Romano, A.; Zuccon, F.; Cabrera-Rubio, R.; Cauquil, A.; Merda, D.; et al. Genome-wide profiling of enterotoxigenic *Staphylococcus aureus* strains used for the production of naturally contaminated cheeses. *Genes* **2020**, *11*, 33. [CrossRef] [PubMed]
83. Andrey, D.O.; Jousselin, A.; Villanueva, M.; Renzoni, A.; Monod, A.; Barras, C.; Rodriguez, N.; Kelley, W.L. Impact of the regulators SigB, rot, SarA and sarS on the toxic shock tst promoter and TSST-1 expression in *Staphylococcus aureus*. *PLoS ONE* **2015**, *10*, e0135579. [CrossRef] [PubMed]
84. Baroja, M.L.; Herfst, C.A.; Kasper, K.J.; Xu, S.X.; Gillett, D.A.; Li, J.; Reid, G.; McCormick, J.K. The SaeRS two-component system is a direct and dominant transcriptional activator of toxic shock syndrome toxin 1 in *Staphylococcus aureus*. *J. Bacteriol.* **2016**, *198*, 2732–2742. [CrossRef]
85. Nagao, M.; Okamoto, A.; Yamada, K.; Hasegawa, T.; Hasegawa, Y.; Ohta, M. Variations in amount of TSST-1 produced by clinical methicillin resistant *Staphylococcus aureus* (MRSA) isolates and allelic variation in accessory gene regulator (agr) locus. *BMC Microbiol.* **2009**, *9*, 52. [CrossRef]
86. Derzelle, S.; Dilasser, F.; Duquenne, M.; Deperrois, V. Differential temporal expression of the staphylococcal enterotoxins genes during cell growth. *Food Microbiol.* **2009**, *26*, 896–904. [CrossRef]
87. Stach, C.S.; Vu, B.G.; Merriman, J.A.; Herrera, A.; Cahill, M.P.; Schlievert, P.M.; Salgado-Pabón, W. Novel Tissue Level Effects of the *Staphylococcus aureus* Enterotoxin Gene Cluster Are Essential for Infective Endocarditis. *PLoS ONE* **2016**, *11*, e0154762. [CrossRef]
88. Strandberg, K.L.; Rotschafer, J.H.; Vetter, S.M.; Buonpane, R.A.; Kranz, D.M.; Schlievert, P.M. Staphylococcal Superantigens Cause Lethal Pulmonary Disease in Rabbits. *J. Infect. Dis.* **2010**, *202*, 1690–1697. [CrossRef]

89. Schlievert, P.M.; Peterson, M.L. Glycerol Monolaurate Antibacterial Activity in Broth and Biofilm Cultures. *PLoS ONE* **2012**, *7*, e40350. [CrossRef]

90. Goerke, C.; Campana, S.; Bayer, M.G.; Döring, G.; Botzenhart, K.; Wolz, C. Direct quantitative transcript analysis of the agr regulon of *Staphylococcus aureus* during human infection in comparison to the expression profile in vitro. *Infect. Immun.* **2000**, *68*, 1304–1311. [CrossRef]

91. Goerke, C.; Wolz, C. Regulatory and genomic plasticity of *Staphylococcus aureus* during persistent colonization and infection. *Int. J. Med. Microbiol.* **2004**, *294*, 195–202. [CrossRef] [PubMed]

92. Banks, M.C.; Kamel, N.S.; Zabriskie, J.B.; Larone, D.H.; Ursea, D.; Posnett, D.N. *Staphylococcus aureus* Express Unique Superantigens Depending on the Tissue Source. *J. Infect. Dis.* **2003**, *187*, 77–86. [CrossRef] [PubMed]

93. Omoe, K.; Ishikawa, M.; Shimoda, Y.; Hu, D.L.; Ueda, S.; Shinagawa, K. Detection of seg, seh, and sei genes in *Staphylococcus aureus* Isolates and Determination of the Enterotoxin Productivities of *S. aureus* Isolates Harboring seg, seh, or sei Genes. *J. Clin. Microbiol.* **2002**, *40*, 857–862. [CrossRef] [PubMed]

94. Burlak, C.; Hammer, C.H.; Robinson, M.A.; Whitney, A.R.; Mcgavin, M.J.; Kreiswirth, B.N.; Deleo, F.R. Global analysis of community-associated methicillin-resistant *Staphylococcus aureus* exoproteins reveals molecules produced in vitro and during infection. *Cell. Microbiol.* **2007**, *9*, 1172–1190. [CrossRef]

95. Enany, S.; Yoshida, Y.; Magdeldin, S.; Zhang, Y.; Bo, X.; Yamamoto, T. Extensive proteomic profiling of the secretome of European community acquired methicillin resistant *Staphylococcus aureus* clone. *Peptides* **2012**, *37*, 128–137. [CrossRef]

96. Pocsfalvi, G.; Cacace, G.; Cuccurullo, M.; Serluca, G.; Sorrentino, A.; Schlosser, G.; Blaiotta, G.; Malorni, A. Proteomic analysis of exoproteins expressed by enterotoxigenic *Staphylococcus aureus* strains. *Proteomics* **2008**, *8*, 2462–2476. [CrossRef]

97. Enany, S.; Yoshida, Y.; Magdeldin, S.; Bo, X.; Zhang, Y.; Enany, M.; Yamamoto, T. Two dimensional electrophoresis of the exo-proteome produced from community acquired methicillin resistant *Staphylococcus aureus* belonging to clonal complex 80. *Microbiol. Res.* **2013**, *168*, 504–511. [CrossRef] [PubMed]

98. Smith, D.S.; Siggins, M.K.; Gierula, M.; Pichon, B.; Turner, C.E.; Lynskey, N.N.; Mosavie, M.; Kearns, A.M.; Edwards, R.J.; Sriskandan, S. Identification of commonly expressed exoproteins and proteolytic cleavage events by proteomic mining of clinically relevant UK isolates of *Staphylococcus aureus*. *Microb. Genom.* **2016**, *2*, e000049. [CrossRef]

99. Ziebandt, A.K.; Kusch, H.; Degner, M.; Jaglitz, S.; Sibbald, M.J.J.B.; Arends, J.P.; Chlebowicz, M.A.; Albrecht, D.; Pantuček, R.; Doškar, J.; et al. Proteomics uncovers extreme heterogeneity in the *Staphylococcus aureus* exoproteome due to genomic plasticity and variant gene regulation. *Proteomics* **2010**, *10*, 1634–1644. [CrossRef]

100. Brizzio, A.A.; Tedeschi, F.A.; Zalazar, F.E. Multiplex PCR strategy for the simultaneous identification of *Staphylococcus aureus* and detection of staphylococcal enterotoxins in isolates from food poisoning outbreaks. *Biomed. Rev. Inst. Nac. Salud* **2013**, *33*, 122–127. [CrossRef]

101. Chiang, Y.C.; Liao, W.W.; Fan, C.M.; Pai, W.Y.; Chiou, C.S.; Tsen, H.Y. PCR detection of Staphylococcal enterotoxins (SEs) N, O, P, Q, R, U, and survey of SE types in *Staphylococcus aureus* isolates from food-poisoning cases in Taiwan. *Int. J. Food Microbiol.* **2008**, *121*, 66–73. [CrossRef] [PubMed]

102. Hait, J.M.; Tallent, S.M.; Bennett, R.W. Screening, Detection, and Serotyping Methods for Toxin Genes and Enterotoxins in Staphylococcus Strains. *J. AOAC Int.* **2014**, *97*, 1078–1083. [CrossRef] [PubMed]

103. Lefebvre, D.; Blanco-Valle, K.; Hennekinne, J.A.; Simon, S.; Fenaille, F.; Becher, F.; Nia, Y.; Lefebvre, D.; Blanco-Valle, K.; Hennekinne, J.-A.; et al. Multiplex Detection of 24 Staphylococcal Enterotoxins in Culture Supernatant Using Liquid Chromatography Coupled to High-Resolution Mass Spectrometry. *Toxins* **2022**, *14*, 249. [CrossRef] [PubMed]

104. Sarratea, M.B.; Noli, T.S.; Mitarotonda, R.; Antonoglou, M.B.; Chiappini, S.; Fernández Lynch, M.J.; Romasanta, P.; Vescina, C.; Desimone, M.; De Marzi, M.; et al. Optimized surface plasmon resonance immunoassay for staphylococcal enterotoxin G detection using silica nanoparticles. *Biochem. Biophys. Res. Commun.* **2021**, *558*, 168–174. [CrossRef]

105. Gupta, G.; Singh, P.K.; Boopathi, M.; Kamboj, D.V.; Singh, B.; Vijayaraghavan, R. Surface plasmon resonance detection of biological warfare agent Staphylococcal enterotoxin B using high affinity monoclonal antibody. *Thin Solid Film* **2010**, *519*, 1171–1177. [CrossRef]

106. Hait, J.M.; Nguyen, A.T.; Tallent, S.M. Analysis of the VIDAS® Staph Enterotoxin III (SET3) for detection of staphylococcal enterotoxins G, H, and I in foods. *J. AOAC Int.* **2018**, *101*, 1482–1489. [CrossRef]

107. Nagaraj, S.; Ramlal, S.; Kingston, J.; Batra, H.V. Development of IgY based sandwich ELISA for the detection of staphylococcal enterotoxin G (SEG), an egc toxin. *Int. J. Food Microbiol.* **2016**, *237*, 136–141. [CrossRef]

108. Wu, S.; Duan, N.; Gu, H.; Hao, L.; Ye, H.; Gong, W.; Wang, Z. A Review of the Methods for Detection of *Staphylococcus aureus* Enterotoxins. *Toxins* **2016**, *8*, 176. [CrossRef]

109. Rasooly, L.; Rasooly, A. Real time biosensor analysis of Staphylococcal enterotoxin A in food. *Int. J. Food Microbiol.* **1999**, *49*, 119–127. [CrossRef]

110. Schlievert, P.M.; Blomster, D.A. Production of Staphylococcal Pyrogenic Exotoxin Type C: Influence of Physical and Chemical Factors. *J. Infect. Dis.* **1983**, *147*, 236–242. [CrossRef]

111. Hajjeh, R.A.; Reingold, A.; Weil, A.; Shutt, K.; Schuchat, A.; Perkins, B.A. Hajjeh Toxic Shock Syndrome in the United States: Surveillance Update, 1979–1996. *Emerg. Infect. Dis.* **1996**, *5*, 807. [CrossRef] [PubMed]

112. Davis, J.P. Toxic-Shock Syndrome Epidemiologic Features, Recurrence, Risk Factors, and Prevention. *N. Engl. J. Med.* **1986**, *314*, 144–149. [CrossRef]

113. Lee, V.T.P.; Chang, A.H.; Chow, A.W. Detection of Staphylococcal Enterotoxin B among Toxic Shock Syndrome (TSS)-and Non-TSS-Associated *Staphylococcus aureus* Isolates. *J. Infect. Dis.* **1992**, *166*, 911–915. [CrossRef] [PubMed]
114. De Vries, A.S.; Lesher, L.; Schlievert, P.M.; Rogers, T.; Villaume, L.G.; Danila, R.; Lynfield, R. Staphylococcal toxic shock syndrome 2000–2006: Epidemiology, clinical features, and molecular characteristics. *PLoS ONE* **2011**, *6*, e22997. [CrossRef]
115. Bohach, G.A.; Fast, D.J.; Nelson, R.D.; Schlievert, P.M. Staphylococcal and Streptococcal Pyrogenic Toxins Involved in Toxic Shock Syndrome and Related Illnesses. *Crit. Rev. Microbiol.* **1990**, *17*, 251–272. [CrossRef]
116. Chang, A.; Musser, J.; Chow, A.W. A single clone of S. aureus which produces both TSST-1 and SEA causes the majority of menstrual toxic shock syndrome. In *Clinical Research*; Slack Inc.: Thorofare, NJ, USA, 1991.
117. Jarraud, S.; Cozon, G.; Vandenesch, O.I.S.; Etienne, J.; Lina, G. Involvement of Enterotoxins G and I in Staphylococcal Toxic Shock Syndrome and Staphylococcal Scarlet Fever. *J. Clin. Microbiol.* **1999**, *37*, 2446–2449. [CrossRef]
118. Peterson, M.L.; Ault, K.; Kremer, M.J.; Klingelhutz, A.J.; Davis, C.C.; Squier, C.A.; Schlievert, P.M. The innate immune system is activated by stimulation of vaginal epithelial cells with *Staphylococcus aureus* and toxic shock syndrome toxin 1. *Infect. Immun.* **2005**, *73*, 2164–2174. [CrossRef]
119. Asano, K.; Narita, K.; Hirose, S.; Nakane, A. Contribution of toxic shock syndrome toxin-1 to systemic inflammation investigated by a mouse model of cervicovaginal infection with *Staphylococcus aureus*. *Med. Microbiol. Immunol.* **2018**, *207*, 297–306. [CrossRef]
120. Asano, K.; Asano, Y.; Ono, H.K.; Nakane, A. Suppression of starvation-induced autophagy by recombinant toxic shock syndrome toxin-1 in epithelial cells. *PLoS ONE* **2014**, *9*, e113018. [CrossRef]
121. Schlievert, P.M.; Davis, C.C. Device-Associated Menstrual Toxic Shock Syndrome. *Clin. Microbiol. Rev.* **2020**, *33*, e00032-19. [CrossRef]
122. Billon, A.; Gustin, M.-P.; Tristan, A.; Bénet, T.; Berthiller, J.; Gustave, C.A.; Vanhems, P.; Lina, G. Association of characteristics of tampon use with menstrual toxic shock syndrome in France. *EClinicalMedicine* **2020**, *21*, 100308. [CrossRef] [PubMed]
123. Davis, C.C.; Baccam, M.; Mantz, M.J.; Osborn, T.W.; Hill, D.R.; Squier, C.A. Use of porcine vaginal tissue ex-vivo to model environmental effects on vaginal mucosa to toxic shock syndrome toxin-1. *Toxicol. Appl. Pharmacol.* **2014**, *274*, 240–248. [CrossRef] [PubMed]
124. Nonfoux, L.; Chiaruzzi, M.; Badiou, C.; Baude, J.; Tristan, A.; Thioulouse, J.; Muller, D.; Prigent-Combaret, C.; Lina, G. Impact of currently marketed tampons and menstrual cups on *Staphylococcus aureus* growth and toxic shock syndrome toxin 1 production in vitro. *Appl. Environ. Microbiol.* **2018**, *84*, e00351-18. [CrossRef]
125. Schlievert, P.M. Effect of non-absorbent intravaginal menstrual/contraceptive products on *Staphylococcus aureus* and production of the superantigen TSST-1. *Eur. J. Clin. Microbiol. Infect. Dis.* **2020**, *39*, 31–38. [CrossRef] [PubMed]
126. Tierno, P.M.; Hanna, B.A. Propensity of Tampons and Barrier Contraceptives to Amplify *Staphylococcus aureus* Toxic Shock Syndrome Toxin-1. *Infect. Dis. Obstet. Gynecol.* **1994**, *2*, 140–145. [CrossRef] [PubMed]
127. Herrmann, T.; Accolla, R.S.; MacDonald, H.R. Different staphylococcal enterotoxins bind preferentially to distinct major histocompatibility complex class ii isotypes. *Eur. J. Immunol.* **1989**, *19*, 2171–2174. [CrossRef] [PubMed]
128. Reingold, A.L.; Hargrett, N.T.; Dan, B.B.; Shands, K.N.; Strickland, B.Y.; Broome, C.V.; Atlanta, G. Nonmenstrual Toxic Shock Syndrome A Review of 130 Cases. *Ann. Intern. Med.* **1982**, *96*, 871–874. [CrossRef]
129. Whiting, J.L.; Rosten, P.M.; Chow, A.W. Determination by Western blot (immunoblot) of seroconversions to toxic shock syndrome (TSS) toxin 1 and enterotoxin A, B, or C during infection with TSS- and Non-TSS-associated *Staphylococcus aureus*. *Infect. Immun.* **1989**, *57*, 231–234. [CrossRef]
130. Ferro, J.M.; Fonseca, A.C. Infective endocarditis. *Handb. Clin. Neurol.* **2014**, *119*, 75–91. [CrossRef]
131. Fowler, V.G.; Olsen, M.K.; Corey, G.R.; Cheng, A.C.; Dudley, T.; Oddone, E.Z. Clinical Identifiers of Complicated *S. aureus* bacteremia. *Arch. Intern. Med.* **2003**, *163*, 2066–2072. [CrossRef]
132. Chung, J.W.; Karau, M.J.; Greenwood-Quaintance, K.E.; Ballard, A.D.; Tilahun, A.; Khaleghi, S.R.; David, C.S.; Patel, R.; Rajagopalan, G. Superantigen profiling of *Staphylococcus aureus* infective endocarditis isolates. *Diagn. Microbiol. Infect. Dis.* **2014**, *79*, 119–124. [CrossRef] [PubMed]
133. Kinney, K.J.; Tran, P.M.; Gibson-Corley, K.N.; Forsythe, A.N.; Kulhankova, K.; Salgado-Pabón, W. Staphylococcal Enterotoxin C promotes *Staphylococcus aureus* Infective Endocarditis Independent of Superantigen Activity. *bioRxiv* **2019**, *14*. [CrossRef]
134. Salgado-Pabón, W.; Injury, A.K.; Peterson, M.L.; Schlievert, M. Superantigens Are Critical for *Staphylococcus aureus* Infective Endocarditis, Sepsis, and Acute Kidney Injury. *MBio* **2013**, *4*, e00494-13. [CrossRef]
135. Kulhankova, K. The Superantigen Toxic Shock Syndrome Toxin 1 Alters Human Aortic Endothelial Cell Function. *Infect. Immun.* **2018**, *86*, e00848-17. [CrossRef] [PubMed]
136. Becker, A.; Kerr, E. *Staphylococcus aureus* pneumonia: A "superbug" infection in community and hospital settings. *Chest* **2005**, *128*, 99–111. [CrossRef]
137. Torres, A.; Menéndez, R.; Wundernik, R.G. Bacterial pneumonia and lung abscess. *Murray Nadel's Textb. Respir. Med.* **2020**, 557–582. [CrossRef]
138. Hyvernat, H.; Pulcini, C.; Carles, D.; Roques, A.; Lucas, P.; Hofman, V.; Hofman, P.; Bernardin, G. Fatal *Staphylococcus aureus* haemorrhagic pneumonia producing Panton-Valentine leucocidin. *Scand. J. Infect. Dis.* **2007**, *39*, 183–185. [CrossRef]
139. Ippolito, G.; Leone, S.; Lauria, F.N.; Nicastri, E.; Wenzel, R.P. Methicillin-resistant *Staphylococcus aureus*: The superbug. *Int. J. Infect. Dis.* **2010**, *14*, S7–S11. [CrossRef]

140. Aliberti, S.; Reyes, L.F.; Faverio, P.; Sotgiu, G.; Dore, S.; Rodriguez, A.H.; Soni, N.J.; Restrepo, M.I.; Aruj, P.K.; Attorri, S.; et al. Global initiative for meticillin-resistant *Staphylococcus aureus* pneumonia (GLIMP): An international, observational cohort study. *Lancet Infect. Dis.* **2016**, *16*, 1364–1376. [CrossRef]

141. Centers for Disease Control and Prevention. Four pediatric deaths from community-acquired methicillin-resistant *Staphylococcus aureus*—Minnesota and North Dakota, 1997–1999. *JAMA* **1999**, *282*, 1123–1125.

142. Strandberg, K.L.; Rotschafer, J.H.; Schlievert, P.M. Extreme Pyrexia and Rapid Death Due to *Staphylococcus aureus* Infection: Analysis of 2 Cases. *Clin. Infect. Dis.* **2009**, *48*, 612–614. [CrossRef]

143. Kulhankova, K.; King, J.; Salgado-Pabón, W. Staphylococcal toxic shock syndrome: Superantigen-mediated enhancement of endotoxin shock and adaptive immune suppression. *Immunol. Res.* **2014**, *59*, 182–187. [CrossRef] [PubMed]

144. Parker, D.; Ryan, C.L.; Alonzo, F.; Torres, V.J.; Planet, P.J.; Prince, A.S. CD4+ T cells promote the pathogenesis of *Staphylococcus aureus* pneumonia. *J. Infect. Dis.* **2015**, *211*, 835–845. [CrossRef] [PubMed]

145. Chao, G.; Bao, G.; Cao, Y.; Yan, W.; Wang, Y.; Zhang, X.; Zhou, L.; Wu, Y. Prevalence and diversity of enterotoxin genes with genetic background of *Staphylococcus aureus* isolates from different origins in China. *Int. J. Food Microbiol.* **2015**, *211*, 142–147. [CrossRef] [PubMed]

146. Johler, S.; Giannini, P.; Jermini, M.; Hummerjohann, J.; Baumgartner, A.; Stephan, R. Further evidence for staphylococcal food poisoning outbreaks caused by egc-Encoded enterotoxins. *Toxins* **2015**, *7*, 997–1004. [CrossRef]

147. Lv, G.P.; Xu, B.H.; Wei, P.N.; Song, J.; Zhang, H.Y.; Zhao, C.; Qin, L.Y.; Zhao, B.H. Molecular characterization of foodborne-associated *Staphylococcus aureus* strains isolated in Shijiazhuang, China, from 2010 to 2012. *Diagn. Microbiol. Infect. Dis.* **2014**, *78*, 462–468. [CrossRef]

148. Shen, M.; Li, Y.; Zhang, L.; Dai, S.; Wang, J.; Li, Y.; Zhang, L.; Huang, J. Staphylococcus enterotoxin profile of China isolates and the superantigenicity of some novel enterotoxins. *Arch. Microbiol.* **2017**, *199*, 723–736. [CrossRef]

149. Wattinger, L.; Stephan, R.; Layer, F.; Johler, S. Comparison of staphylococcus aureus isolates associated with food intoxication with isolates from human nasal carriers and human infections. *Eur. J. Clin. Microbiol. Infect. Dis.* **2012**, *31*, 455–464. [CrossRef]

150. Yan, X.; Wang, B.; Tao, X.; Hu, Q.; Cui, Z.; Zhang, J.; Lin, Y.; You, Y.; Shi, X.; Grundmann, H. Characterization of *Staphylococcus aureus* strains associated with food poisoning in Shenzhen, China. *Appl. Environ. Microbiol.* **2012**, *78*, 6637–6642. [CrossRef]

151. Blaiotta, G.; Ercolini, D.; Pennacchia, C.; Fusco, V.; Casaburi, A.; Pepe, O.; Villani, F. PCR detection of staphylococcal enterotoxin genes in Staphylococcus spp. strains isolated from meat and dairy products. Evidence for new variants of seG and seI in *S. aureus* AB-8802. *J. Appl. Microbiol.* **2004**, *97*, 719–730. [CrossRef]

152. McLauchlin, J.; Narayanan, G.L.; Mithani, V.; O'Neill, G. The detection of enterotoxins and toxic shock syndrome toxin genes in *Staphylococcus aureus* by polymerase chain reaction. *J. Food Prot.* **2000**, *63*, 479–488. [CrossRef] [PubMed]

153. Viçosa, G.N.; Le Loir, A.; Le Loir, Y.; de Carvalho, A.F.; Nero, L.A. Egc characterization of enterotoxigenic *Staphylococcus aureus* isolates obtained from raw milk and cheese. *Int. J. Food Microbiol.* **2013**, *165*, 227–230. [CrossRef] [PubMed]

154. Ikeda, T.; Tamate, N.; Yamaguchi, K.; Makino, S.I. Mass outbreak of food poisoning disease caused by small amounts of staphylococcal enterotoxins A and H. *Appl. Environ. Microbiol.* **2005**, *71*, 2793–2795. [CrossRef] [PubMed]

155. Jørgensen, H.J.; Mathisen, T.; Løvseth, A.; Omoe, K.; Qvale, K.S.; Loncarevic, S. An outbreak of staphylococcal food poisoning caused by enterotoxin H in mashed potato made with raw milk. *FEMS Microbiol. Lett.* **2005**, *252*, 267–272. [CrossRef] [PubMed]

156. de Freitas, M.F.; Luz, I.D.S.; Silveira-Filho, V.D.M.; Júnior, J.W.P.; Stamford, T.L.; Mota, R.A.; de Sena, M.J.; de Almeida, A.M.P.; de Q. Balbino, V.; Leal-Balbino, T.C. Staphylococcal toxin genes in strains isolated from cows with subclinical mastitis. *Pesqui. Veterinária Bras.* **2008**, *28*, 617–621. [CrossRef]

157. Bianchi, D.M.; Gallina, S.; Bellio, A.; Chiesa, F.; Civera, T.; Decastelli, L. Enterotoxin gene profiles of *Staphylococcus aureus* isolated from milk and dairy products in Italy. *Lett. Appl. Microbiol.* **2014**, *58*, 190–196. [CrossRef]

158. Carfora, V.; Caprioli, A.; Marri, N.; Sagrafoli, D.; Boselli, C.; Giacinti, G.; Giangolini, G.; Sorbara, L.; Dottarelli, S.; Battisti, A.; et al. Enterotoxin genes, enterotoxin production, and methicillin resistance in *Staphylococcus aureus* isolated from milk and dairy products in Central Italy. *Int. Dairy J.* **2015**, *42*, 12–15. [CrossRef]

159. Rosec, J.P.; Guiraud, J.P.; Dalet, C.; Richard, N. Enterotoxin production by staphylococci isolated from foods in France. *Int. J. Food Microbiol.* **1997**, *35*, 213–221. [CrossRef]

160. Zhao, Y.; Zhu, A.; Tang, J.; Tang, C.; Chen, J.; Liu, J. Identification and measurement of staphylococcal enterotoxin-like protein I (SEll) secretion from *Staphylococcus aureus* clinical isolate. *J. Appl. Microbiol.* **2016**, *121*, 539–546. [CrossRef]

161. Cheng, F.H.M.; Andrews, P.L.R.; Moreaux, B.; Ngan, M.P.; Rudd, J.A.; Sam, T.S.W.; Wai, M.K.; Wan, C. Evaluation of the anti-emetic potential of anti-migraine drugs to prevent resiniferatoxin-induced emesis in *Suncus murinus* (house musk shrew). *Eur. J. Pharmacol.* **2005**, *508*, 231–238. [CrossRef]

162. Hu, D.L.; Zhu, G.; Mori, F.; Omoe, K.; Okada, M.; Wakabayashi, K.; Kaneko, S.; Shinagawa, K.; Nakane, A. Staphylococcal enterotoxin induces emesis through increasing serotonin release in intestine and it is downregulated by cannabinoid receptor 1. *Cell. Microbiol.* **2007**, *9*, 2267–2277. [CrossRef] [PubMed]

163. Sugiyama, H.; Hayama, T. Abdominal viscera as site of emetic action for staphylococcal enterotoxin in the monkey. *J. Infect. Dis.* **1965**, *115*, 330–336. [CrossRef] [PubMed]

164. Shupp, J.W.; Jett, M.; Pontzer, C.H. Identification of a transcytosis epitope on staphylococcal enterotoxins. *Infect. Immun.* **2002**, *70*, 2178–2186. [CrossRef] [PubMed]

165. Spiekermann, G.M.; Anderson, C.N. Oral Administration of the Bacterial Superantigen Staphylococcal Enterotoxin B Induces Activation and Cytokine Production by T Cells in Murine Gut-Associated Lymphoid Tissue. *J. Immunol.* **1998**, *161*, 5825–5831. [CrossRef]

166. McKay, D.M.; Benjamin, M.A.; Lu, J. CD4+ T cells mediate superantigen-induced abnormalities in murine jejunal ion transport. *Am. J. Physiol.-Gastrointest. Liver Physiol.* **1998**, *275*, 29–38. [CrossRef]

167. Pinchuk, I.V.; Beswick, E.J.; Reyes, V.E. Staphylococcal enterotoxins. *Toxins* **2010**, *2*, 2177–2197. [CrossRef]

168. Friedman, S.M.; Posnett, D.N.; Tumang, J.R.; Crow, M.K.; Cole, B.C. A potential role for microbial superantigens in the pathogenesis of systemic autoimmune disease. *Arthritis Rheum.* **1991**, *34*, 468–480. [CrossRef]

169. Wucherpfennig, K.W. Mechanisms for the induction of autoimmunity by infectious agents. *J. Clin. Investig.* **2001**, *108*, 1097–1104. [CrossRef]

170. Schiffenbauer, J.; Soos, J.; Johnson, H. The possible role of bacterial superantigens in the pathogenesis of autoimmune disorders. *Immunol. Today* **1998**, *19*, 117–120. [CrossRef]

171. Dietz, S.M.; van Stijn, D.; Burgner, D.; Levin, M.; Kuipers, I.M.; Hutten, B.A.; Kuijpers, T.W. Dissecting Kawasaki disease: A state-of-the-art review. *Eur. J. Pediatr.* **2017**, *176*, 995–1009. [CrossRef]

172. Matsubara, K.; Fukaya, T. The role of superantigens of group A Streptococcus and *Staphylococcus aureus* in Kawasaki disease. *Curr. Opin. Infect. Dis.* **2007**, *20*, 298–303. [CrossRef] [PubMed]

173. Nagata, S. Causes of Kawasaki disease—From past to present. *Front. Pediatr.* **2019**, *7*, 1–7. [CrossRef] [PubMed]

174. Yeung, R.S.M. The etiology of Kawasaki disease: A superantigen-mediated process. *Prog. Pediatr. Cardiol.* **2004**, *19*, 115–122. [CrossRef]

175. Nagata, S.; Yamashiro, Y.; Ohtsuka, Y.; Shimizu, T.; Sakurai, Y.; Misawa, S.; Ito, T. Heat shock proteins and superantigenic properties of bacteria from the gastrointestinal tract of patients with Kawasaki disease. *Immunology* **2009**, *128*, 511–520. [CrossRef]

176. Hall, M.; Schlievert, P.M. Kawasaki Syndrome-Like Illness Associated with Infection Caused by Enterotoxin B-Secreting *Staphylococcus aureus*. *Clin. Infect. Dis.* **1999**, *29*, 586–589. [CrossRef]

177. Leung, D.Y.M.; Meissner, H.C.; Fulton, D.; Murray, D.L.; Kotzin, B.; Schlievert, P.M. Toxic shock syndrome toxin-secreting *Staphylococcus aureus* in Kawasaki syndrome. *Lancet* **1994**, *343*, 300. [CrossRef]

178. Leung, D.Y.M.; Meissner, H.C.; Shulman, S.T.; Mason, W.H.; Gerber, M.A.; Glode, M.P.; Myones, B.L.; Wheeler, J.G.; Ruthazer, R.; Schlievert, P.M. Prevalence of superantigen-secreting bacteria in patients with Kawasaki disease. *J. Pediatr.* **2002**, *140*, 742–746. [CrossRef]

179. Matsubara, K.; Fukaya, T.; Miwa, K.; Shibayama, N.; Nigami, H.; Harigaya, H.; Nozaki, H.; Hirata, T.; Baba, K.; Suzuki, T.; et al. Development of serum IgM antibodies against superantigens of *Staphylococcus aureus* and Streptococcus pyogenes in Kawasaki disease. *Clin. Exp. Immunol.* **2006**, *143*, 427–434. [CrossRef]

180. Nomura, Y.; Yoshinaga, M.; Masuda, K.; Takei, S.; Miyata, K. Maternal Antibody against Toxic Shock Syndrome Toxin–1 May Protect Infants Younger than 6 Months of Age from Developing Kawasaki Syndrome. *J. Infect. Dis.* **2002**, *185*, 1677–1680. [CrossRef]

181. Terai, M.; Miwa, K.; Williams, T.; Kabat, W.; Fukuyama, M.; Okajima, Y.; Igarashi, H.; Shulman, S.T. The absence of evidence of staphylococcal toxin involvement in the pathogenesis of kawasaki disease. *J. Infect. Dis.* **1995**, *172*, 558–561. [CrossRef]

182. Yim, D.; Ramsay, J.; Kothari, D.; Burgner, D. Coronary artery dilatation in toxic shock-like syndrome: The kawasaki disease shock syndrome. *Pediatr. Cardiol.* **2010**, *31*, 1232–1235. [CrossRef] [PubMed]

183. Vu, B.G.; Stach, C.S.; Kulhankova, K.; Salgado-Pabón, W.; Klingelhutz, A.J.; Schlievert, P.M. Chronic superantigen exposure induces systemic inflammation, elevated bloodstream endotoxin, and abnormal glucose tolerance in rabbits: Possible role in diabetes. *MBio* **2015**, *6*, e02554-14. [CrossRef] [PubMed]

184. Banke, E.; Rödström, K.; Ekelund, M.; Dalla-Riva, J.; Lagerstedt, J.O.; Nilsson, S.; Degerman, E.; Lindkvist-Petersson, K.; Nilson, B. Superantigen activates the gp130 receptor on adipocytes resulting in altered adipocyte metabolism. *Metabolism* **2014**, *63*, 831–840. [CrossRef] [PubMed]

185. Vu, B.G.; Gourronc, F.A.; Bernlohr, D.A.; Schlievert, P.M.; Klingelhutz, A.J. Staphylococcal superantigens stimulate immortalized human adipocytes to produce chemokines. *PLoS ONE* **2013**, *8*, e77988. [CrossRef]

186. Paliard, X.; West, S.G.; Lafferty, J.A.; Clements, J.R.; Kappler, J.W.; Marrack, P.; Kotzin, B.L. Evidence for the effects of a superantigen in rheumatoid arthritis. *Science* **1991**, *253*, 325–329. [CrossRef]

187. Goodacre, J.A.; Brownlie, C.E.D.; Ross, D.A. Bacterial superantigens in autoimmune arthritis. *Rheumatology* **1994**, *33*, 413–419. [CrossRef]

188. Ataee, R.A.; Ataee, M.H.; Alishiri, G.H.; Esmaeili, D. Staphylococcal enterotoxin C in synovial fluid of patients with rheumatoid arthritis. *Iran. Red Crescent Med. J.* **2014**, *16*, e16075. [CrossRef]

189. Ataee, R.A.; Kahani, M.S.; Alishiri, G.H.; Ahamadi, Z. Staphylococcal Enterotoxin A Detection from Rheumatoid Arthritis Patients' Blood and Synovial Fluid Ramezan. *Electron. Physician* **2016**, *8*, 1850. [CrossRef]

190. Ataee, R.A.; Kashefi, R.; Alishiri, G.H.; Esmaieli, D. *Staphylococcus aureus* enterotoxin D: Absence of bacteria but presence of its toxin. *Jundishapur J. Microbiol.* **2015**, *8*, e28395. [CrossRef]

191. Rashki, M.; Ataee, R.A.; Alishiri, G.H.; Esmaeili, D. Molecular Assay of Staphylococcal Enterotoxin E in Synovial Fluid of Patients with Rheumatoid Arthritis. *Int. J. Infect.* **2018**, *5*, 1–4. [CrossRef]

192. Shokrollahi, M.R.; Noorbakhsh, S.; Aliakbari, M.; Tabatabaei, A. Searching the staphylococcal superantigens: Enterotoxins A, B, C, and TSST1 in synovial fluid of cases with negative culture inflammatory arthritis. *Jundishapur J. Microbiol.* **2014**, *7*, e11647. [CrossRef] [PubMed]

193. Tabarya, D.; Hoffman, W.L. *Staphylococcus aureus* nasal carriage in rheumatoid arthritis: Antibody response to toxic shock syndrome toxin-1. *Ann. Rheum. Dis.* **1996**, *55*, 823–828. [CrossRef] [PubMed]

194. Origuchi, T.; Eguchi, K.; Kawabe, Y.; Yamashita, I.; Mizokami, A.; Ida, H.; Nagataki, S. Increased levels of serum IgM antibody to staphylococcal enterotoxin B in patients with rheumatoid arthritis. *Ann. Rheum. Dis.* **1995**, *54*, 713–720. [CrossRef] [PubMed]

195. Gerlach, K.; Tomuschat, C.; Finke, R.; Staege, M.S.; Brütting, C.; Brandt, J.; Jordan, B.; Schwesig, R.; Rosemeier, A.; Delank, K.S.; et al. Experimental Arthritis in the Rat Induced by the Superantigen Staphylococcal Enterotoxin A. *Scand. J. Immunol.* **2017**, *85*, 191–196. [CrossRef]

196. Banaei, M.; Alishiri, G.H.; Ataee, R.A.; Mahalati, A.H. Rheumatoid arthritis mediator CD18 expression by *Staphylococcus aureus* superantigen C in rats. *Iran. J. Microbiol.* **2019**, *11*, 337–344. [CrossRef]

197. Abdelnour, A.; Bremell, T.; Holmdahl, R.; Tarkowski, A. Clonal expansion of T lymphocytes causes arthritis and mortality in mice infected with toxic shock syndrome toxin-1-producing staphylococci. *Eur. J. Immunol.* **1994**, *24*, 1161–1166. [CrossRef] [PubMed]

198. Wooley, P.H.; Cingel, B. Staphylococcal enterotoxin B increases the severity of type II collagen induced arthritis in mice. *Ann. Rheum. Dis.* **1995**, *54*, 298–304. [CrossRef]

199. Schwab, J.H.; Brown, R.R.; Anderle, S.K.; Schlievert, P.M. Superantigen can reactivate bacterial cell wall-induced arthritis. *J. Immunol.* **1993**, *150*, 4151–4159.

200. Tsai, C.; Diaz, L.A.; Singer, N.G.; Li, L.L.; Kirsch, A.H.; Mitra, R.; Nickoloff, B.J.; Crofford, L.J.; Fox, D.A. Responsiveness of human T lymphocytes to bacterial superantigens presented by cultured rheumatoid arthritis synoviocytes. *Arthritis Rheum.* **1996**, *39*, 125–136. [CrossRef]

201. Leung, D.Y.M.; Harbeck, R.; Bina, P.; Reiser, R.F.; Yang, E.; Norris, D.A.; Hanifin, J.M.; Sampson, H.A. Presence of IgE antibodies to staphylococcal exotoxins on the skin of patients with atopic dermatitis evidence for a new group of allergens. *J. Clin. Investig.* **1993**, *92*, 1374–1380. [CrossRef]

202. Leyden, J.J.; Marples, R.R.; Kligman, A.M. *Staphylococcus aureus* in the lesions of atopic dermatitis. *Br. J. Dermatol.* **1974**, *90*, 525. [CrossRef] [PubMed]

203. Skov, L.; Baadsgaard, O. Bacterial superantigens and inflammatory skin diseases. *Clin. Exp. Dermatol.* **2000**, *25*, 57–61. [CrossRef] [PubMed]

204. Jappe, U. Superantigens and their association with dermatological inflammatory diseases: Facts and hypotheses. *Acta Derm. Venereol.* **2000**, *80*, 321–328. [CrossRef] [PubMed]

205. Bunikowski, R.; Mielke, M.; Skarabis, H.; Herz, U.; Bergmann, R.L.; Wahn, U.; Renz, H. Prevalence and role of serum IgE antibodies to the *Staphylococcus aureus*-derived superantigens SEA and SEB in children with atopic dermatitis. *J. Allergy Clin. Immunol.* **1999**, *103*, 119–124. [CrossRef]

206. Lin, Y.T.; Yang, Y.H.; Hwang, Y.W.; Tsai, M.J.; Tsao, P.N.; Chiang, B.L.; Shau, W.Y.; Wang, L.F. Comparison of serum specific IgE antibodies to staphylococcal enterotoxins between atopic children with and without atopic dermatitis. *Allergy Eur. J. Allergy Clin. Immunol.* **2000**, *55*, 641–646. [CrossRef]

207. Tomczak, H.; Wróbel, J.; Jenerowicz, D.; Sadowska-Przytocka, A.; Wachal, M.; Adamski, Z.; Czarnecka-Operacz, M.M. The role of *Staphylococcus aureus* in atopic dermatitis: Microbiological and immunological implications. *Postep. Dermatol. Alergol.* **2019**, *36*, 485–491. [CrossRef]

208. Schlievert, P.M.; Case, L.C.; Strandberg, K.L.; Abrams, B.B.; Leung, D.Y.M. Superantigen profile of *Staphylococcus aureus* isolates from patients with steroid-resistant atopic dermatitis. *Clin. Infect. Dis.* **2008**, *46*, 1562–1567. [CrossRef]

209. Strickland, I.; Hauk, P.J.; Trumble, A.E.; Picker, L.J.; Leung, D.Y.M. Evidence for superantigen involvement in skin homing of T cells in atopic dermatitis. *J. Investig. Dermatol.* **1999**, *112*, 249–253. [CrossRef]

210. McFadden, J.P.; Noble, W.C.; Camp, R.D.R. Superantigenic exotoxin-secreting potential of staphylococci isolated from atopic eczematous skin. *Br. J. Dermatol.* **1993**, *128*, 631–632. [CrossRef]

211. Arad, G.; Levy, R.; Nasie, I.; Hillman, D.; Rotfogel, Z.; Barash, U.; Supper, E.; Shpilka, T.; Minis, A.; Kaempfer, R. Binding of superantigen toxins into the CD28 homodimer interface is essential for induction of cytokine genes that mediate lethal shock. *PLoS Biol.* **2011**, *9*, e1001149. [CrossRef]

212. Ramachandran, G.; Tulapurkar, M.E.; Harris, K.M.; Arad, G.; Shirvan, A.; Shemesh, R.; Detolla, L.J.; Benazzi, C.; Opal, S.M.; Kaempfer, R.; et al. A peptide antagonist of CD28 signaling attenuates toxic shock and necrotizing soft-tissue infection induced by streptococcus pyogenes. *J. Infect. Dis.* **2013**, *207*, 1869–1877. [CrossRef] [PubMed]

213. Levy, R.; Rotfogel, Z.; Hillman, D.; Popugailo, A.; Arad, G.; Supper, E.; Osman, F.; Kaempfer, R. Superantigens hyperinduce inflammatory cytokines by enhancing the B7-2/CD28 costimulatory receptor interaction. *Proc. Natl. Acad. Sci. USA* **2016**, *113*, E6437–E6446. [CrossRef] [PubMed]

214. Ganem, M.B.; De Marzi, M.C.; Fernández-Lynch, M.J.; Jancic, C.; Vermeulen, M.; Geffner, J.; Mariuzza, R.A.; Fernández, M.M.; Malchiodi, E.L. Uptake and Intracellular Trafficking of Superantigens in Dendritic Cells. *PLoS ONE* **2013**, *8*, e66244. [CrossRef] [PubMed]

215. Hovde, C.J.; Marr, J.C.; Hoffmann, M.L.; Hackett, S.P.; Chi, Y.I.; Crum, K.K.; Stevens, D.L.; Stauffacher, C.V.; Bohach, G.A. Investigation of the role of the disulphide bond in the activity and structure of staphylococcal enterotoxin C1. *Mol. Microbiol.* **1994**, *13*, 897–909. [CrossRef]

216. Baker, M.D.; Acharya, K.R. Superantigens: Structure-function relationships. *Int. J. Med. Microbiol.* **2004**, *293*, 529–537. [CrossRef]

217. Petersson, K.; Håkansson, M.; Nilsson, H.; Forsberg, G.; Svensson, L.A.; Liljas, A.; Walse, B. Crystal structure of a superantigen bound to MHC class II displays zinc and peptide dependence. *EMBO J.* **2001**, *20*, 3306–3312. [CrossRef]

218. Jardetzky, T.S.; Brown, J.H.; Gorga, J.C.; Stern, L.J.; Urban, R.G.; Chi, Y.I.; Stauffacher, C.; Strominger, J.L.; Wiley, D.C. Three-dimensional structure of a human class II histocompatibility molecule complexed with superantigen. *Nature* **1994**, *368*, 711–718. [CrossRef]

219. Fernández, M.M.; Guan, R.; Swaminathan, C.P.; Malchiodi, E.L.; Mariuzza, R.A. Crystal structure of staphylococcal enterotoxin I (SEI) in complex with a human major histocompatibility complex class II molecule. *J. Biol. Chem.* **2006**, *281*, 25356–25364. [CrossRef]

220. Leder, L.; Llera, A.; Lavoie, P.M.; Lebedeva, M.I.; Li, H.; Sékaly, R.P.; Bohach, G.A.; Gahr, P.J.; Schlievert, P.M.; Karjalainen, K.; et al. A Mutational Analysis of the Binding of Staphylococcal Enterotoxins B and C3 to the T Cell Receptor Beta Chain and Major Histocompatibility Complex Class II. *J. Exp. Med.* **1998**, *187*, 823–833. [CrossRef]

221. Fernández, M.M.; Cho, S.; De Marzi, M.C.; Kerzic, M.C.; Robinson, H.; Mariuzza, R.A.; Malchiodi, E.L. Crystal structure of Staphylococcal Enterotoxin G (SEG) in complex with a mouse T-cell receptor β chain. *J. Biol. Chem.* **2011**, *286*, 1189–1195. [CrossRef]

222. Fernández, M.M.; De Marzi, M.C.; Berguer, P.; Burzyn, D.; Langley, R.J.; Piazzon, I.; Mariuzza, R.A.; Malchiodi, E.L. Binding of natural variants of staphylococcal superantigens SEG and SEI to TCR and MHC class II molecule. *Mol. Immunol.* **2006**, *43*, 927–938. [CrossRef] [PubMed]

223. Saline, M.; Rödström, K.E.J.; Fischer, G.; Orekhov, V.Y.; Karlsson, B.G.; Lindkvist-Petersson, K. The structure of superantigen complexed with TCR and MHC reveals novel insights into superantigenic T cell activation. *Nat. Commun.* **2010**, *1*, 119. [CrossRef] [PubMed]

224. Popugailo, A.; Rotfogel, Z.; Supper, E.; Hillman, D.; Kaempfer, R. Staphylococcal and streptococcal superantigens trigger B7/CD28 costimulatory receptor engagement to hyperinduce inflammatory cytokines. *Front. Immunol.* **2019**, *10*, 942. [CrossRef]

225. Gregory, S.; Zilber, M.T.; Charron, D.; Gelin, C. Human CD1a Molecule Expressed on Monocytes Plays an Accessory Role in the Superantigen-induced Activation of T Lymphocytes. *Hum. Immunol.* **2000**, *61*, 193–201. [CrossRef]

226. Uzunçayır, S.; Vera-Rodriguez, A.; Regenthal, P.; Åbacka, H.; Emanuelsson, C.; Bahl, C.D.; Lindkvist-Petersson, K. Analyses of the complex formation of staphylococcal enterotoxin A and the human gp130 cytokine receptor. *FEBS Lett.* **2022**, *596*, 910–923. [CrossRef] [PubMed]

227. Schlievert, P.M.; Cahill, M.P.; Hostager, B.S.; Brosnahan, A.J.; Klingelhutz, A.J.; Gourronc, F.A.; Bishop, G.A.; Leung, D.Y.M. Staphylococcal superantigens stimulate epithelial cells through CD40 to produce chemokines. *MBio* **2019**, *10*, e00214-19. [CrossRef]

228. Schlievert, P.M.; Gourronc, F.A.; Leung, D.Y.M.; Klingelhutz, A.J. Human Keratinocyte Response to Superantigens. *mSphere* **2020**, *5*, e00803-20. [CrossRef]

229. Schlievert, P.M.; Kilgore, S.H.; Benavides, A.; Klingelhutz, A.J. Pathogen Stimulation of Interleukin-8 from Human Vaginal Epithelial Cells through CD40. *Microbiol. Spectr.* **2022**, *10*, e00106-22. [CrossRef]

230. Garcia, K.C.; Teyton, L.; Wilson, I.A. Structural basis of T cell recognition. *Annu. Rev. Immunol.* **1999**, *17*, 369–397. [CrossRef] [PubMed]

231. Garcia, K.C.; Adams, E.J. How the T cell receptor sees antigen—A structural view. *Cell* **2005**, *122*, 333–336. [CrossRef]

232. Smith-Garvin, J.E.; Koretzky, G.A.; Jordan, M.S. T cell activation. *Annu. Rev. Immunol.* **2009**, *27*, 592–619. [CrossRef] [PubMed]

233. Givan, A.L.; Fisher, J.L.; Waugh, M.; Ernstoff, M.S.; Wallace, P.K. A flow cytometric method to estimate the precursor frequencies of cells proliferating in response to specific antigens. *J. Immunol. Methods* **1999**, *230*, 99–112. [CrossRef]

234. Bueno, C.; Criado, G.; McCormick, J.K.; Madrenas, J. T Cell Signalling Induced by Bacterial Superantigens. In *Superantigens and Superallergens*; KARGER: Basel, Switzerland, 2007; pp. 161–180.

235. Mccormick, J.K.; Yarwood, J.M.; Schlievert, P.M. Toxic shock syndrome and bacterial superantigens: An update. *Annu. Rev. Microbiol.* **2001**, *55*, 77. [CrossRef] [PubMed]

236. Kawabe, Y.; Ochi, A. Programmed cell death and extrathymic reduction of Vβ8+ CD4+ T cells in mice tolerant to *Staphylococcus aureus* enterotoxin B. *Nature* **1991**, *349*, 245–248. [CrossRef]

237. MacDonald, H.R.; Baschieri, S.; Lees, R.K. Clonal expansion precedes anergy and death of Vβ8+ peripheral T cells responding to staphylococcal enterotoxin B in vivo. *Eur. J. Immunol.* **1991**, *21*, 1963–1966. [CrossRef] [PubMed]

238. Sundstedt, A.; Höiden, I.; Rosendahl, A.; Kalland, T.; van Rooijen, N.; Dohlsten, M. Immunoregulatory role of IL-10 during superantigen-induced hyporesponsiveness in vivo. *J. Immunol.* **1997**, *158*, 180–186.

239. Taylor, A.L.; Llewelyn, M.J. Superantigen-Induced Proliferation of Human CD4 + CD25− T Cells Is Followed by a Switch to a Functional Regulatory Phenotype. *J. Immunol.* **2010**, *185*, 6591–6598. [CrossRef]

240. Feunou, P.; Poulin, L.; Habran, C.; Le Moine, A.; Goldman, M.; Braun, M.Y. CD4+ CD25+ and CD4+ CD25− T Cells Act Respectively as Inducer and Effector T Suppressor Cells in Superantigen-Induced Tolerance. *J. Immunol.* **2003**, *171*, 3475–3484. [CrossRef]

241. Lee, J.; Park, N.; Park, J.Y.; Kaplan, B.L.F.; Pruett, S.B.; Park, J.W.; Ho Park, Y.; Seo, K.S. Induction of Immunosuppressive CD8+ CD25+ FOXP3+ Regulatory T Cells by Suboptimal Stimulation with Staphylococcal Enterotoxin C1. *J. Immunol.* **2018**, *200*, 669–680. [CrossRef]
242. Taylor, A.L.; Cross, E.L.A.; Llewelyn, M.J. Induction of contact-dependent CD8+ regulatory T cells through stimulation with staphylococcal and streptococcal superantigens*. *Immunology* **2012**, *135*, 158–167. [CrossRef]
243. Wang, Z.Q.; Orlikowsky, T.; Dudhane, A.; Trejo, V.; Dannecker, G.E.; Pernist, B.; Hoffmann, M.K. Staphylococcal enterotoxin B-induced T-cell anergy is mediated by regulatory T cells. *Immunology* **1998**, *94*, 331–339. [CrossRef] [PubMed]
244. Tilahun, A.Y.; Chowdhary, V.R.; David, C.S.; Rajagopalan, G. Systemic Inflammatory Response Elicited by Superantigen Destabilizes T Regulatory Cells Rendering them Ineffective during Toxic Shock Syndrome NIH Public Access. *J. Immunol.* **2014**, *193*, 2919–2930. [CrossRef] [PubMed]
245. Parker, D.; Karauzum, H.; Sähr, A.; Förmer, S.; Hildebrand, D.; Heeg, K. T-cell activation or tolerization: The Yin and Yang of bacterial superantigens. *Front. Microbiol.* **2015**, *6*, 1153. [CrossRef]
246. Kato, M.; Nakamura, Y.; Suda, T.; Ozawa, Y.; Inui, N.; Seo, N.; Nagata, T.; Koide, Y.; Kalinski, P.; Nakamura, H.; et al. Enhanced anti-tumor immunity by superantigen-pulsed dendritic cells. *Cancer Immunol. Immunother.* **2011**, *60*, 1029–1038. [CrossRef] [PubMed]
247. Bhardwaj, N.; Friedman, S.M.; Cole, B.C.; Nisanian, A.J. Dendritic Cells Are Potent Antigen-presenting Cells for Microbial Superantigens. *J. Exp. Med.* **1992**, *175*, 267–273. [CrossRef] [PubMed]
248. Muraille, E.; De Trez, C.; Pajak, B.; Brait, M.; Urbain, J.; Leo, O. T Cell-Dependent Maturation of Dendritic Cells in Response to Bacterial Superantigens. *J. Immunol.* **2002**, *168*, 4352–4360. [CrossRef] [PubMed]
249. Coutant, K.D.; De Fraissinette, A.D.B.; Cordier, A.; Ulrich, P. Modulation of the activity of human monocyte-derived dendritic cells by chemical haptens, a metal allergen, and a staphylococcal superantigen. *Toxicol. Sci.* **1999**, *52*, 189–198. [CrossRef]
250. Liu, T.; He, S.H.; Zheng, P.Y.; Zhang, T.Y.; Wang, B.Q.; Yang, P.C. Staphylococcal enterotoxin B increases TIM4 expression in human dendritic cells that drives naïve CD4 T cells to differentiate into Th2 cells. *Mol. Immunol.* **2007**, *44*, 3580–3587. [CrossRef]
251. Muralimohan, G.; Vella, A.T. A role for IFNγ in differential superantigen stimulation of conventional versus plasmacytoid DCs. *Cell. Immunol.* **2006**, *23*, 9–22. [CrossRef]
252. Takahashi, M.; Takahashi, M.; Shinohara, F.; Takada, H.; Rikiishi, H. Effects of superantigen and lipopolysaccharide on induction of CD80 through apoptosis of human monocytes. *Infect. Immun.* **2001**, *69*, 3652–3657. [CrossRef]
253. Zhang, X.; Shang, W.; Yuan, J.; Hu, Z.; Peng, H.; Zhu, J.; Hu, Q.; Yang, Y.; Liu, H.; Jiang, B.; et al. Positive Feedback Cycle of TNFα Promotes Staphylococcal Enterotoxin B-Induced THP-1 Cell Apoptosis. *Front. Cell. Infect. Microbiol.* **2016**, *6*, 109. [CrossRef]
254. Ulett, G.C.; Adderson, E.E. Regulation of Apoptosis by Gram-Positive Bacteria: Mechanistic Diversity and Consequences for Immunity. *Curr. Immunol. Rev.* **2006**, *2*, 119–141. [CrossRef]
255. Krakauer, T. Staphylococcal Superantigens: Pyrogenic Toxins Induce Toxic Shock. *Toxins* **2019**, *11*, 178. [CrossRef] [PubMed]
256. Grumann, D.; Scharf, S.S.; Holtfreter, S.; Kohler, C.; Steil, L.; Engelmann, S.; Hecker, M.; Völker, U.; Bröker, B.M.; Volker, U.; et al. Immune Cell Activation by Enterotoxin Gene Cluster (egc)-Encoded and Non-egc Superantigens from *Staphylococcus aureus*. *J. Immunol.* **2008**, *181*, 5054–5061. [CrossRef] [PubMed]
257. Islander, U.; Andersson, A.; Lindberg, E.; Adlerberth, I.; Wold, A.E.; Rudin, A. Superantigenic *Staphylococcus aureus* stimulates production of interleukin-17 from memory but not naive T cells. *Infect. Immun.* **2010**, *78*, 381–386. [CrossRef] [PubMed]
258. Jupin, C.; Anderson, S.; Damais, C.; Alouf, J.E.; Parant, M. Toxic shock syndrome toxin 1 as an inducer of human tumor necrosis factors and λ interferon. *J. Exp. Med.* **1988**, *167*, 752–761. [CrossRef]
259. Kum, W.W.S.; Cameron, S.B.; Hung, R.W.Y.; Kalyan, S.; Chow, A.W. Temporal Sequence and Kinetics of Proinflammatory and Anti-Inflammatory Cytokine Secretion Induced by Toxic Shock Syndrome Toxin 1 in Human Peripheral Blood Mononuclear Cells. *Infect. Immun.* **2001**, *69*, 7544–7549. [CrossRef]
260. Rust, C.J.; Koning, F. Gamma delta T cell reactivity towards bacterial superantigens. *Semin. Immunol.* **1993**, *5*, 41–46. [CrossRef]
261. Fisher, E.L.; Otto, M.; Cheung, G.Y.C. Basis of virulence in enterotoxin-mediated staphylococcal food poisoning. *Front. Microbiol.* **2018**, *9*. [CrossRef]
262. Brandes, M.; Willimann, K.; Moser, B. Immunology: Professional antigen-presentation function by human γδ cells. *Science* **2005**, *309*, 264–268. [CrossRef]
263. Maeurer, M.; Zitvogel, L.; Elder, E.; Storkus, W.J.; Lotze, M.T. Human intestinal V delta 1+ T cells obtained from patients with colon cancer respond exclusively to SEB but not to SEA. *Nat. Immun.* **1995**, *14*, 188–197. [PubMed]
264. Morita, C.T.; Li, H.; Lamphear, J.G.; Rich, R.R.; Fraser, J.D.; Mariuzza, R.A.; Lee, H.K. Superantigen recognition by gamma delta T cells: SEA recognition site for human Vgamma2 T cell receptors. *Immunity* **2001**, *14*, 331–344. [CrossRef]
265. Hayworth, J.L.; Mazzuca, D.M.; Vareki, S.M.; Welch, I.; McCormick, J.K.; Haeryfar, S.M. CD1d-independent activation of mouse and human i NKT cells by bacterial superantigens. *Immunol. Cell Biol.* **2012**, *90*, 699–709. [CrossRef] [PubMed]
266. Rieder, S.A.; Nagarkatti, P.; Nagarkatti, M. CD1d-independent activation of invariant natural killer T cells by Staphylococcal enterotoxin B through major histocompatibility complex class II/T cell receptor interaction results in acute lung injury. *Infect. Immun.* **2011**, *79*, 3141–3148. [CrossRef] [PubMed]
267. Szabo, P.A.; Rudak, P.T.; Choi, J.; Xu, S.X.; Schaub, R.; Singh, B.; McCormick, J.K.; Haeryfar, S.M.M. Invariant Natural Killer T Cells Are Pathogenic in the HLA-DR4-Transgenic Humanized Mouse Model of Toxic Shock Syndrome and Can Be Targeted to Reduce Morbidity. *J. Infect. Dis.* **2017**, *215*, 824–829. [CrossRef]

268. Stohl, W.; Elliott, J.E.; Linsley, P.S. Human T cell-dependent B cell differentiation induced by staphylococcal superantigens. *J. Immunol.* **1994**, *153*, 117–127.

269. Domiati-Saad, R.; Attrep, J.F.; Brezinschek, H.P.; Cherrie, A.H.; Karp, D.R.; Lipsky, P.E. Staphylococcal enterotoxin D functions as a human B cell superantigen by rescuing VH4-expressing B cells from apoptosis. *J. Immunol.* **1996**, *156*, 3608–3620.

270. Domiati-Saad, R.; Lipsky, P.E. Staphylococcal Enterotoxin A Induces Survival of V H 3 Expressing Human B Cells by Binding to the V H Region with Low Affinity. *J. Immunol.* **1998**, *161*, 1257–1266.

271. Gould, H.J.; Takhar, P.; Harries, H.E.; Chevretton, E.; Sutton, B.J. The allergic March from *Staphylococcus aureus* superantigens to immunoglobulin E. *Chem. Immunol. Allergy* **2007**, *93*, 106–136. [CrossRef]

272. Shin, S.Y.; Choi, G.S.M.; Lee, K.H.; Kim, S.W.; Cho, J.S.; Park, H.S. IgE Response to Staphylococcal Enterotoxins in Adenoid Tissues from Atopic Children. *Laryngoscope* **2009**, *119*, 171–175. [CrossRef]

273. Lotfi-Emran, S.; Ward, B.R.; Le, Q.T.; Pozez, A.L.; Manjili, M.H.; Woodfolk, J.A.; Schwartz, L.B. Human mast cells present antigen to autologous CD4+ T cells. *J. Allergy Clin. Immunol.* **2018**, *141*, 311–321.e10. [CrossRef]

274. Ono, H.K.; Nishizawa, M.; Yamamoto, Y.; Hu, D.L.; Nakane, A.; Shinagawa, K.; Omoe, K. Submucosal mast cells in the gastrointestinal tract are a target of staphylococcal enterotoxin type A. *FEMS Immunol. Med. Microbiol.* **2012**, *64*, 392–402. [CrossRef]

275. Scheuber, P.H.; Denzlinger, C.; Wilker, D.; Beck, G.; Keppler, D.; Hammer, D.K. Staphylococcal enterotoxin B as a nonimmunological mast cell stimulus in primates: The role of endogenous cysteinyl leukotrienes. *Int. Arch. Allergy Immunol.* **1987**, *82*, 289–291. [CrossRef]

276. Shaler, C.R.; Choi, J.; Rudak, P.T.; Memarnejadian, A.; Szabo, P.A.; Tun-Abraham, M.E.; Rossjohn, J.; Corbett, A.J.; McCluskey, J.; McCormick, J.K.; et al. MAIT cells launch a rapid, robust and distinct hyperinflammatory response to bacterial superantigens and quickly acquire an anergic phenotype that impedes their cognate antimicrobial function: Defining a novel mechanism of superantigen-induced immunopathology and immunosuppression. *PLoS Biol.* **2017**, *15*, e2001930. [CrossRef]

277. Godfrey, D.I.; Uldrich, A.P.; Mccluskey, J.; Rossjohn, J.; Moody, D.B. The burgeoning family of unconventional T cells. *Nat. Publ. Gr.* **2015**, *16*, 1114–1123. [CrossRef]

278. D'Souza, C.; Chen, Z.; Corbett, A.J. Revealing the protective and pathogenic potential of MAIT cells. *Mol. Immunol.* **2018**, *103*, 46–54. [CrossRef]

279. Godfrey, D.I.; Koay, H.F.; McCluskey, J.; Gherardin, N.A. The biology and functional importance of MAIT cells. *Nat. Immunol.* **2019**, *20*, 1110–1128. [CrossRef]

280. Langley, R.; Patel, D.; Jackson, N.; Clow, F.; Fraser, J.D. Staphylococcal superantigen super-domains in immune evasion. *Crit. Rev. Immunol.* **2010**, *30*, 149–165. [CrossRef]

281. Langley, R.J.; Ting, Y.T.; Clow, F.; Young, P.G.; Radcliff, F.J.; Choi, J.M.; Sequeira, R.P.; Holtfreter, S.; Baker, H.; Fraser, J.D. Staphylococcal enterotoxin-like X (SElX) is a unique superantigen with functional features of two major families of staphylococcal virulence factors. *PLoS Pathog.* **2017**, *13*, e1006549. [CrossRef]

282. Tuffs, S.W.; James, D.B.A.; Bestebroer, J.; Richards, A.C.; Goncheva, M.I.; O'Shea, M.; Wee, B.A.; Seo, K.S.; Schlievert, P.M.; Lengeling, A.; et al. The *Staphylococcus aureus* superantigen SElX is a bifunctional toxin that inhibits neutrophil function. *PLoS Pathog.* **2017**, *13*, e1006461. [CrossRef]

283. Spaan, A.N.; Surewaard, B.G.J.; Nijland, R.; van Strijp, J.A.G. Neutrophils versus *Staphylococcus aureus*: A Biological Tug of War. *Annu. Rev. Microbiol.* **2013**, *67*, 629–650. [CrossRef]

284. Chen, J.Y. Superantigens, superantigen-like proteins and superantigen derivatives for cancer treatment. *Eur. Rev. Med. Pharmacol. Sci.* **2021**, *25*, 1622–1630.

285. Liu, X.; Zeng, L.; Zhao, Z.; He, J.; Xie, Y.; Xiao, L.; Wang, S.; Zhang, J.; Zou, Z.; He, Y.; et al. PBMC activation via the ERK and STAT signaling pathways enhances the anti-tumor activity of Staphylococcal enterotoxin A. *Mol. Cell. Biochem.* **2017**, *434*, 75–87. [CrossRef]

286. Perabo, F.G.E.; Willert, P.L.; Wirger, A.; Schmidt, D.H.; Von Ruecker, A.; Mueller, S.C. Superantigen-activated Mononuclear Cells Induce Apoptosis in Transitional Cell Carcinoma. *Anticancer Res.* **2005**, *25*, 3565–3573.

287. Kodama, H.; Suzuki, M.; Katayose, Y.; Shinoda, M.; Sakurai, N.; Takemura, S.I.; Yoshida, H.; Saeki, H.; Ichiyama, M.; Tsumoto, K.; et al. Mutated SEA-D227A-conjugated antibodies greatly enhance antitumor activity against MUC1-expressing bile duct carcinoma. *Cancer Immunol. Immunother.* **2001**, *50*, 539–548. [CrossRef]

288. Zhang, J.; Cai, Y.M.; Xu, M.K.; Song, Z.H.; Li, C.Y.; Wang, H.R.; Dai, H.H.; Zhang, Z.P.; Liu, C.X. Anti-tumor activity and immunogenicity of a mutated staphylococcal enterotoxin C2. *Pharmazie* **2013**, *68*, 359–364. [CrossRef]

289. Forsberg, G.; Skartved, N.J.; Wallén-Öhman, M.; Nyhlén, H.C.; Behm, K.; Hedlund, G.; Nederman, T. Naptumomab estafenatox, an engineered antibody-superantigen fusion protein with low toxicity and reduced antigenicity. *J. Immunother.* **2010**, *33*, 492–499. [CrossRef]

290. Golob-Urbanc, A.; Rajčević, U.; Strmšek, Ž.; Jerala, R. Design of split superantigen fusion proteins for cancer immunotherapy. *J. Biol. Chem.* **2019**, *294*, 6294–6305. [CrossRef]

291. Knopick, P.; Terman, D.; Riha, N.; Alvine, T.; Larson, R.; Badiou, C.; Lina, G.; Ballantyne, J.; Bradley, D. Endogenous HLA-DQ8αβ programs superantigens (SEG/SEI) to silence toxicity and unleash a tumoricidal network with long-term melanoma survival. *J. Immunother. Cancer* **2020**, *8*, e001493. [CrossRef]

292. Antonoglou, M.B.; Sánchez Alberti, A.; Redolfi, D.M.; Bivona, A.E.; Fernández Lynch, M.J.; Noli Truant, S.; Sarratea, M.B.; López, L.V.I.; Malchiodi, E.L.; Fernández, M.M. Heterologous Chimeric Construct Comprising a Modified Bacterial Superantigen and a Cruzipain Domain Confers Protection Against Trypanosoma cruzi Infection. *Front. Immunol.* **2020**, *1*, 1279. [CrossRef]

Article

Clostridium botulinum C3 Toxin for Selective Delivery of Cargo into Dendritic Cells and Macrophages

Maximilian Fellermann [1], Mia Stemmer [1], Reiner Noschka [2], Fanny Wondany [3], Stephan Fischer [1], Jens Michaelis [3], Steffen Stenger [2] and Holger Barth [1,*]

[1] Institute of Experimental and Clinical Pharmacology, Toxicology and Pharmacology of Natural Products, University of Ulm Medical Center, 89081 Ulm, Germany
[2] Institute for Medical Microbiology and Hygiene, University of Ulm Medical Center, 89081 Ulm, Germany
[3] Institute of Biophysics, Ulm University, 89081 Ulm, Germany
* Correspondence: holger.barth@uni-ulm.de

Abstract: The protein toxin C3bot from *Clostridium botulinum* is a mono-ADP-ribosyltransferase that selectively intoxicates monocyte-derived cells such as macrophages, osteoclasts, and dendritic cells (DCs) by cytosolic modification of Rho-A, -B, and -C. Here, we investigated the application of C3bot as well as its non-toxic variant C3bot$_{E174Q}$ as transporters for selective delivery of cargo molecules into macrophages and DCs. C3bot and C3bot$_{E174Q}$ facilitated the uptake of eGFP into early endosomes of human-monocyte-derived macrophages, as revealed by stimulated emission depletion (STED) super-resolution microscopy. The fusion of the cargo model peptide eGFP neither affected the cell-type selectivity (enhanced uptake into human macrophages ex vivo compared to lymphocytes) nor the cytosolic release of C3bot. Moreover, by cell fractionation, we demonstrated that C3bot and C3bot$_{E174Q}$ strongly enhanced the cytosolic release of functional eGFP. Subsequently, a modular system was created on the basis of C3bot$_{E174Q}$ for covalent linkage of cargos via thiol–maleimide click chemistry. The functionality of this system was proven by loading small molecule fluorophores or an established reporter enzyme and investigating the cellular uptake and cytosolic release of cargo. Taken together, non-toxic C3bot$_{E174Q}$ is a promising candidate for the cell-type-selective delivery of small molecules, peptides, and proteins into the cytosol of macrophages and DCs.

Keywords: clostridial C3 toxin; *Clostridium botulinum*; C3bot; C3bot$_{E174Q}$; dendritic cells; macrophages; monocytes; stimulated emission depletion (STED); super-resolution microscopy

Key Contribution: C3bot and especially the non-toxic variant C3bot$_{E174Q}$ can be used as cell-type-selective molecular Trojan horses for transporting different cargo molecules into the cytosol of macrophages and dendritic cells.

Citation: Fellermann, M.; Stemmer, M.; Noschka, R.; Wondany, F.; Fischer, S.; Michaelis, J.; Stenger, S.; Barth, H. *Clostridium botulinum* C3 Toxin for Selective Delivery of Cargo into Dendritic Cells and Macrophages. *Toxins* **2022**, *14*, 711. https://doi.org/10.3390/toxins14100711

Received: 2 September 2022
Accepted: 14 October 2022
Published: 18 October 2022

Publisher's Note: MDPI stays neutral with regard to jurisdictional claims in published maps and institutional affiliations.

1. Introduction

C3bot is a protein toxin produced by *Clostridium botulinum* type C (*C. botulinum*) with a molecular weight of 23.5 kDa [1]. When C3bot reaches the cytosol of a target cell, it specifically mono-ADP-ribosylates Rho A, -B, and -C (in the following abbreviated as Rho) at position N41 [2–4]. For enzyme activity of C3bot, the amino acid E174 (without signal sequence) is essential, and the mutation E174Q leads to the enzymatically inactive and thus non-toxic variant C3bot$_{E174Q}$ [5,6]. As the first C3 toxin identified, C3bot is the prototype of the C3-like ADP-ribosyltransferase family [7], comprising at least nine different members produced by different organisms (*C. botulinum*, *C. limosum*, *Staphylococcus aureus*, *Bacillus cereus*, and *Paenibacillus larvae*) [1,7–15]. In contrast to classical AB-type protein toxins, most bacterial C3 toxins (except for PlxA and C3larvin [9,16,17]) do not comprise a binding and translocation B-component, resulting in poor uptake into most cell types [2,18]. However, it was demonstrated that the clostridial C3 toxins are efficiently internalized into monocytes, macrophages, osteoclasts, and neurons [18–22]. Recently, human monocyte-derived dendritic cells (DCs) were also identified as specific target cells for the clostridial C3 toxins [23]. The C3 catalyzed Rho-modification in these cells results in impaired cell functions such as adhesion [24,25]; endo-, phago-, or exocytosis [2,26–28]; cell migration [20,21,29,30]; or differentiation/maturation [19] (for a review, see [2,31]). Notably, inactivation of Rho also results in a characteristic change in cell morphology as determined by formation of long protrusions [32,33]. Despite the effect of Rho-ADP-ribosylation is well examined, the exact cell entry mechanism of the C3 toxins remains widely unknown. Comparable to bacterial AB toxins, the clostridial C3 toxins are also internalized into early endosomes of macrophages and immature or mature DCs [18,23], but the exact mechanism of endosomal escape is not understood.

Since C3bot is by nature cell-type-selective for monocyte-derived cells, its non-toxic variant C3bot$_{E174Q}$ represents a promising candidate as drug delivery tool. Cargo molecules can be attached to C3bot$_{E174Q}$ and specifically delivered into monocytic cells via the C3bot uptake mechanism without harming the cells. This strategy has already been examined in first attempts by delivering reporter enzymes into the cytosol of macrophages [34–36]. Two of the tested enzymes were subunits of other bacterial toxins, i.e., DTA (21 kDa) from diphtheria toxin [35] or C2I (50 kDa) from the binary *C. botulinum* C2 toxin [36]. Moreover, RNase A (14 kDa) [34] was delivered into the cytosol of macrophages by C3bot$_{E174Q}$. These cargo enzymes were attached via two strategies, i.e., generation of fusion proteins (C3bot$_{E174Q}$_C2I) [36] or by creating a modular system on the basis of the interaction of biotin (attached to cargo) and streptavidin (attached to C3bot$_{E174Q}$) as proven for delivery of DTA and RNase A [34,35]. Despite this strategy seeming quite promising for delivering foreign proteins into macrophages, non-covalent attachment of cargo via the streptavidin–biotin system comes with some disadvantages. The preparation of the streptavidin-based transporters was contaminated by unspecific side-products [34], and streptavidin reduces the cell type-selectivity of the C3bot$_{E174Q}$ transporters, potentially by influencing the interaction with cell surface proteins [34,35].

In the present study, we aimed to investigate the functions of C3bot and C3bot$_{E174Q}$ as cellular delivery systems in more detail and to extend their applications. Therefore, a fusion protein based on C3bot and the green fluorescent protein (eGFP) was generated. This fusion protein was cell-type selectively internalized into human macrophages as compared with lymphocytes of the same donor ex vivo. Fusion to C3bot$_{E174Q}$ strongly enhanced the internalization of the cargo model eGFP into early endosomes and cytosolic release of the cargo protein, as shown for macrophages and human DCs. Notably, fluorescence of the cytosolic eGFP was detected using a cell-fractionation assay verifying functionality of the released cargo. Moreover, on the basis of thiol–maleimide click chemistry, a novel modular system for fast, covalent, and specific attachment of cargo was created. Effective loading of the system with different cargo molecules was confirmed. It was demonstrated that the uptake of small molecules into human macrophages and DCs is strongly enhanced by their attachment to C3bot$_{E174Q}$. Finally, the novel C3bot$_{E174Q}$-based transport system enhanced the delivery of C2I into the cytosol of DCs and thereby proves the concept of the modular system with an established reporter enzyme. In conclusion, the C3bot$_{E174Q}$ transport system can increase the cell membrane penetration of cargo molecules. Notably, C3bot is by nature cell type selective for monocyte-derived immune cells and is therefore the ideal delivery system for targeting macrophages and DCs.

2. Results

2.1. C3bot and C3bot$_{e174q}$ Enhanced the Endosomal Uptake of eGFP

To demonstrate that C3bot serves as a delivery tool, the uptake of the model cargo eGFP into its target cells was investigated. The peptide eGFP was genetically fused to C3bot or C3bot$_{E174Q}$, and the uptake into human monocyte-derived macrophages analyzed by stimulated emission depletion (STED) super-resolution microscopy (Figure 1). Co-staining of the early endosomal antigen 1 (EEA1) was used as a marker for endosomes.

The fusion proteins [His]–eGFP_C3bot and [His]–eGFP_C3bot$_{E174Q}$ were both internalized into early endosomes, as indicated by surrounding of the eGFP-signals with EEA1 (Figure 1a).

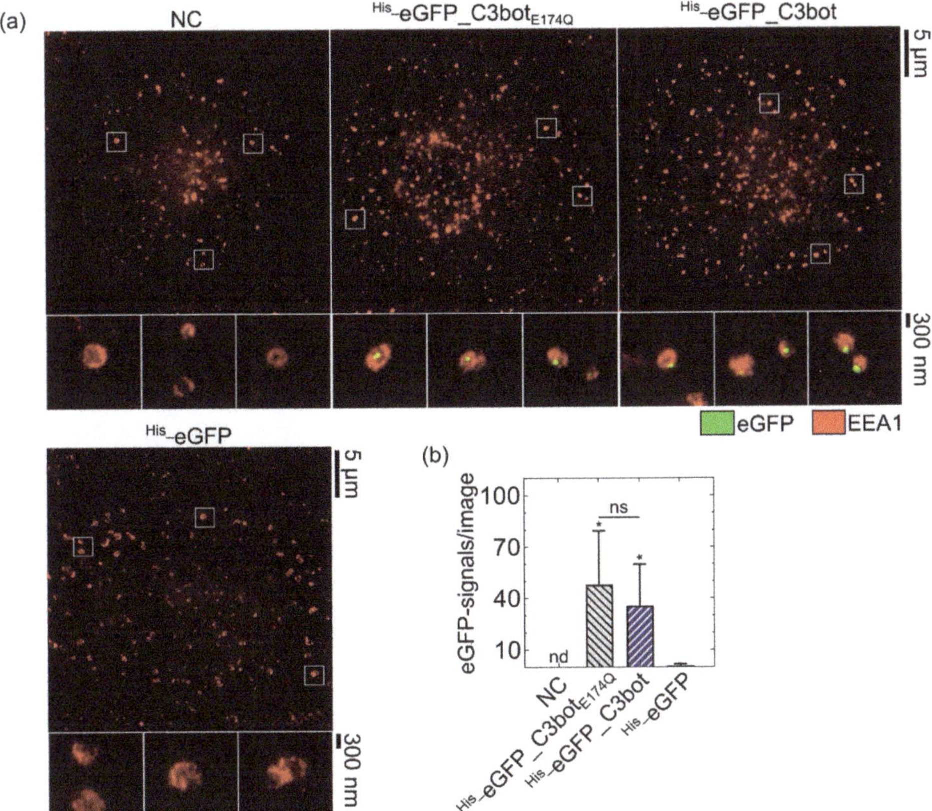

Figure 1. C3bot and C3bot$_{E174Q}$ facilitated the endosomal uptake of eGFP into human monocyte-derived macrophages ex vivo. Cells were treated at 37 °C for 30 min with 250 nM [His]–eGFP_C3bot, [His]–eGFP_C3bot$_{E174Q}$, or [His]–eGFP, or were left untreated (NC). After staining of EEA1 and eGFP, STED super-resolution microscopy was performed. (**a**) Representative super-resolution images are depicted with magnification marked by white squares. Scale bars on the right correspond to 5 µm or 300 nm and hold for all images. (**b**) The experiment was repeated with macrophages derived from five individual and independent donors (n = 5 donors). The detected eGFP signals were quantified for each treatment, and the means are depicted with standard deviations (± SD). For NC, no eGFP signals were detectable, as marked by "nd". Compared to the [His]–eGFP samples, significance testing was performed using Student's *t*-test (ns = not significant, * $p < 0.05$). By comparing the samples of [His]–eGFP_C3bot$_{E174Q}$ to [His]–eGFP_C3bot, no significant differences were found, as indicated by "ns" and the line above the respective columns. The figure is modified from [37] under the authors' rights.

These eGFP signals are condensed at specific locations within the endosomes either in the inner lumen or at the membrane of the vesicles (see magnifications). For [His]–eGFP_C3bot$_{E174Q}$, 81.5 ± 15.9% and for [His]–eGFP_C3bot 91.6 ± 4.5% of the detected eGFP signals were in close proximity (<250 nm) with EEA1 (see Table S2), indicating that most eGFP signals are associated with early endosomes. In contrast, [His]–eGFP alone was not efficiently endocytosed by the macrophages, and only a neglectable amount of green fluorescence was detected, as shown in the quantification of the STED images (Figure 1b).

All tested five blood monocyte donors showed an enhanced uptake of ^{His}-eGFP_C3bot$_{E174Q}$ and ^{His}-eGFP_C3bot compared to ^{His}-eGFP alone; however, the mean number of eGFP signals per image varies from donor to donor (see Table S1). These results thereby indicate that cellular uptake of eGFP is specific and significantly enhanced by coupling the cargo model to the transporter platform C3bot or C3bot$_{E174Q}$.

2.2. ^{His}-eGFP_C3bot$_{E174Q}$ Was Selectively Internalized into Human Macrophages Ex Vivo

After confirming that transport of cargo into macrophages can be enhanced by C3bot or C3bot$_{E174Q}$, we investigated whether ^{His}-eGFP_C3bot$_{E174Q}$ still possessed the cell-type selectivity of wild-type C3bot. Hence, cellular uptake into human-monocyte-derived macrophages was compared with uptake into lymphocytes of the same blood donor. Initially, the different cells were treated separately with ^{His}-eGFP_C3bot$_{E174Q}$, and internalization of eGFP was quantified in flow cytometry (Figure 2a). Since only internalized cargo should be analyzed, cell-surface-bound eGFP signals were quenched by using an established trypan-blue-based assay [38–40]. Uptake of ^{His}-eGFP_C3bot$_{E174Q}$ into macrophages was strongly enhanced in comparison with lymphocytes, which internalized only a minor portion of the fusion protein (Figure 2a). Notably, even with 800 nM of ^{His}-eGFP_C3bot$_{E174Q}$, only a minor portion was internalized by the lymphocytes (Figure 2a; for the experiment with a lower concentration, see Figure S1).

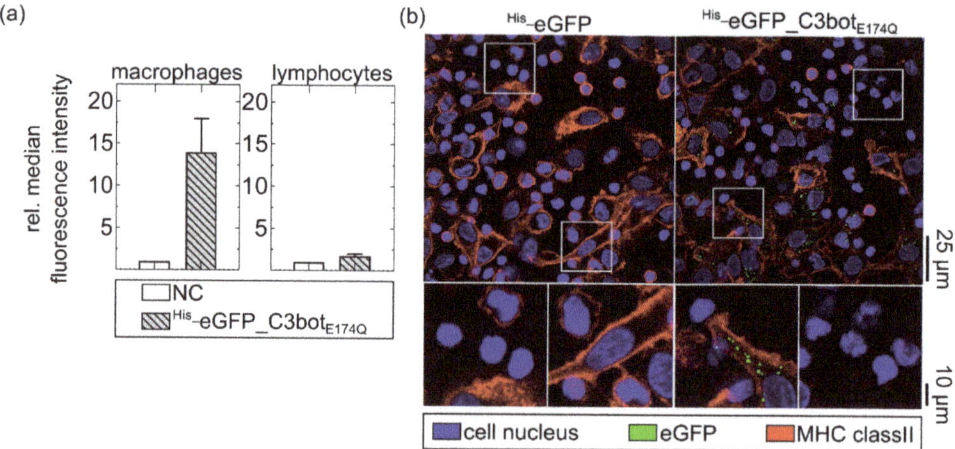

Figure 2. Cell-type-selective uptake of ^{His}-eGFP_C3bot$_{E174Q}$ into primary human macrophages compared to lymphocytes ex vivo. (**a**) Human-monocyte-derived macrophages and lymphocytes of the same donor were separately treated with 800 nM ^{His}-eGFP_C3bot$_{E174Q}$ for 20 min at 37 °C. The cells were washed with PBS, and extracellular or membrane-bound eGFP signals were quenched with trypan blue directly before flow cytometry measurement. The averaged relative median fluorescence intensity (normalized to untreated cells (NC)) is depicted in the column diagrams (mean ± SD, n = 5). (**b**) Monocyte-derived macrophages and lymphocytes of the same human blood donor were co-cultured (1:5 ratio) and treated with 400 nM ^{His}-eGFP_C3bot$_{E174Q}$ or ^{His}-eGFP at 37 °C for 1 h. The cells were fixed, and the cell nuclei (blue) were stained with DAPI. MHC class II (red) was stained as marker for macrophages. All in confocal microscopy detected fluorescence signals (eGFP, DAPI, and MHC class II) were merged, and representative images are depicted with magnification as indicated by the white squares. Scale bars on the right apply to the complete row of images (25 µm for full-size images, and 10 µm for magnifications). The figure is modified from [37] under the authors' rights.

On the basis of these findings, a co-culture of macrophages and lymphocytes of the same blood donor was treated with ^{His}-eGFP_C3bot$_{E174Q}$. For the control, the co-culture was treated with ^{His}-eGFP alone. The macrophage population was marked by staining of the major histocompatibility complex (MHC) class II on the cell surface (Figure 2b). In this co-culture, ^{His}-eGFP_C3bot$_{E174Q}$ was cell-type selectively internalized into human macrophages, as indicated by the green signals surrounded by red MHC class II staining. Moreover, about 70–90% of the total macrophage population internalized ^{His}-eGFP_C3bot$_{E174Q}$ (see Figure S2). Hence, direct coupling of cargo molecules to the transporter system did not affect the cell-type selectivity, and C3bot$_{E174Q}$ can be used to enhance the uptake of cargo molecules.

2.3. Functional eGFP Was Delivered into the Cytosol of DCs and Macrophages via C3bot and C3bot$_{E174Q}$

Next, we investigated whether the cargo model eGFP was released into the cytosol of target cells. Since eGFP is hardly detectable in the cytosol, even with state-of-the-art fluorescence microscopic techniques, at first, an indirect approach based on the cytotoxic effect of C3bot was used. We compared the morphological changes induced by C3bot with those induced by cargo-labeled ^{His}-eGFP_C3bot to evaluate whether coupling of cargo interfered with the uptake process of the toxin. Therefore, cells of a mouse macrophage cell line (J774A.1) were treated with increasing concentrations of C3bot or ^{His}-eGFP_C3bot (Figure 3). Treatment with ^{His}-eGFP_C3bot induced characteristic changes in cell macrophage morphology, i.e., formation of long cell protrusions, comparable to wild-type C3bot (Figure 3a). In contrast, ^{His}-eGFP_C3bot$_{E174Q}$ did not induce such characteristic changes in cell morphology, confirming that enzymatic activity of C3bot is required for this effect. Since the fusion proteins were produced in *Escherichia coli* (*E. coli*), the effect of lipopolysaccharides (LPS) was also tested. LPS endotoxins are common impurities in recombinant protein expression in Gram-negative bacteria that can activate immune cells, e.g., macrophages [41,42]. Despite treatment with LPS slightly influencing J774A.1 cell morphology and more intracellular inclusions being visible compared to the negative control (NC), the clear and striking formation of protrusions cannot be explained by a potential LPS contamination (Figure 3a). Hence, the results indicate that active C3bot is required for induction of the morphological change and, most importantly, eGFP labeling did not affect this effect. This becomes even more obvious by quantifying the cells with C3 morphology and comparing different concentrations in a time course (Figure 3b) or at defined time points (Figure 3c). Importantly, no significant differences were detected by comparing same concentrations of wild-type C3bot with ^{His}-eGFP_C3bot.

These results were furthermore confirmed by analyzing the Rho ADP-ribosylation status inside the cells. A sequential Rho ADP-ribosylation assay was used for detection of non-ADP-ribosylated Rho in intact cells (Rho$_{non-ADP-rib.}$) (Figure 4). Notably, in this assay, weak signals indicate strong ADP-ribosylation of Rho by the toxin inside the target cells. Equal protein loading was controlled by detection of heat shock protein 90 (HSP90). The detected Rho$_{non-ADP-rib.}$ decreased, i.e., Rho-ADP-ribosylation in the intact cells increased with the concentration of C3bot or ^{His}-eGFP_C3bot. In accordance with the morphological assays, eGFP labeling inhibited neither the Rho ADP-ribosyltransferase activity of C3bot in J774A.1 macrophages (Figure 4a) nor the activity in human DCs derived from a sarcoma cell line (U-DCS cells, see Figure 4b). Taken together, the results indicate that cytosolic release of C3bot into target cells (macrophages and DCs) is not inhibited by fusion with the cargo model eGFP.

Since this approach is quite indirect for the detection of cytosolic cargo release, alongside the fact that intracellular degradation of the fusion constructs cannot be excluded, we directly analyzed the presence of cytosolic eGFP with an established digitonin-based cell fractionation [34,36]. After treatment with the eGFP-labeled proteins (^{His}-eGFP, ^{His}-eGFP_C3bot, and ^{His}-eGFP_C3bot$_{E174Q}$), the cells were separated into cytosolic and membrane fractions.

For isolation of the cytosol, digitonin was used to form small pores into the cell membrane that are smaller than endosomes or other organelles. Hence, only the cytosolic proteins were able to leave the cells. Successful separation of the cytosolic fraction from the membrane fraction (containing endosomes) was confirmed by Western blot detection of EEA1, which was only present in the membrane fractions (Figure 5a). As a loading control for cytosolic proteins, HSP90 was detected.

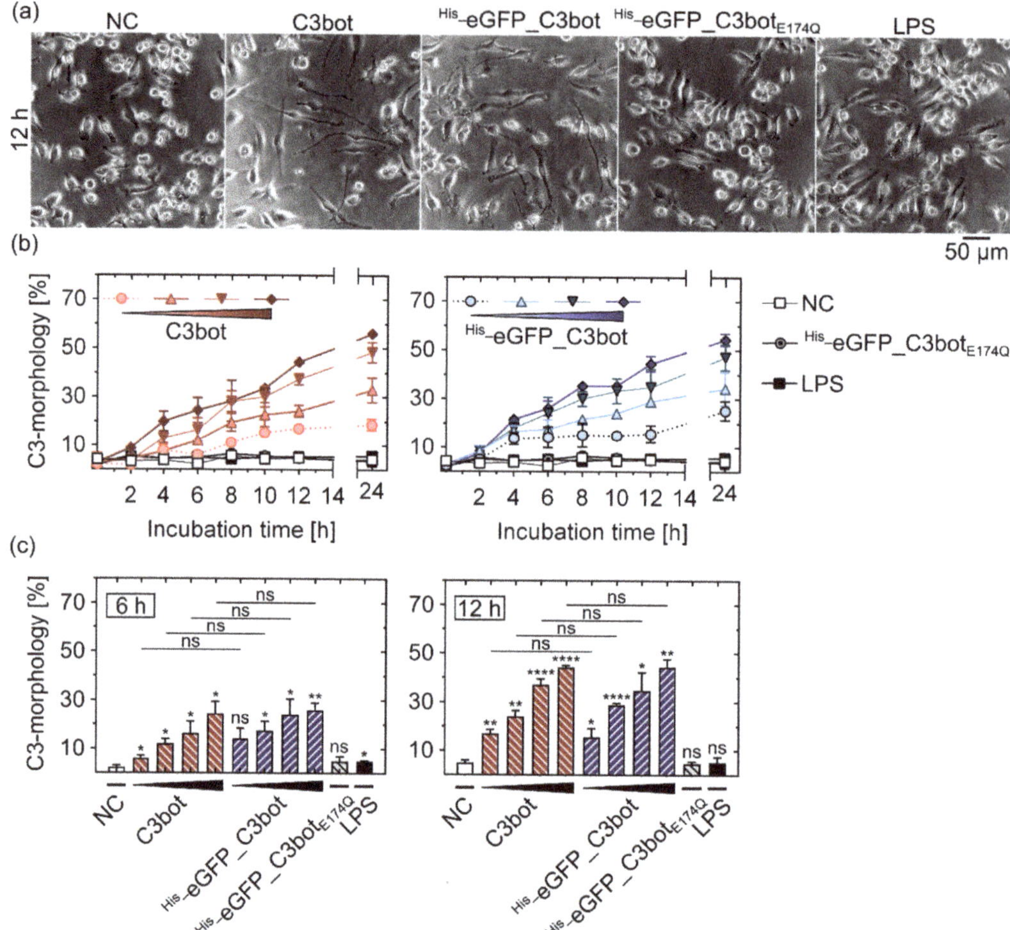

Figure 3. Fusion of eGFP to C3bot did not affect the cytosolic release of C3bot in J774A.1 macrophages. (**a**) J774A.1 cells were treated with C3bot, His–eGFP_C3bot, His–eGFP_C3bot$_{E174Q}$ (200 nM each), or 1 μg/mL LPS, or were left untreated (NC). The images show representative phase contrast images after an incubation time of 12 h with 50 μm scale bar. (**b,c**) J774A.1 macrophages were treated with increasing concentrations (25, 50, 100, 200 nM) of C3bot or His–eGFP_C3bot, with 200 nM His–eGFP_C3bot$_{E174Q}$, with 1 μg/mL LPS, or were left untreated (NC). Phase contrast images were taken after incubation times of 2, 4, 6, 8, 10, 12, and 24 h at 37 °C. The cell portion with characteristic stellate C3 morphology was quantified and divided by the total amount of cells per image. The ratios (mean ± SD, n = 3) are depicted in time course diagrams (**b**) and column diagrams after 6 or 12 h (**c**). (**c**) Compared to the NC, statistical significance was tested via Student's *t*-test, and columns are labeled according to following significance levels: ns $p > 0.05$, * $p < 0.05$, ** $p < 0.01$, **** $p < 0.0001$. Additionally, the same concentrations of C3bot and His–eGFP_C3bot were compared via Student's *t*-test, as indicated by the lines and significance levels above the representative columns. The figure is modified from [37] under the authors' rights.

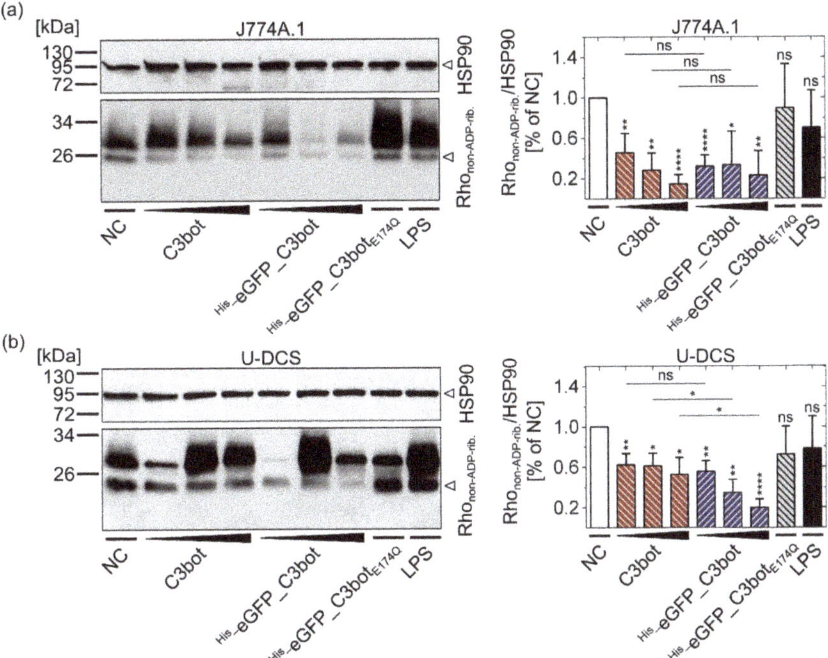

Figure 4. Fusion of eGFP to C3bot did not affect the cytosolic ADP-ribosylation of Rho in J774A.1 macrophages or U-DCS cells. J774A.1 cells (**a**) and U-DCS cells (**b**) were treated with increasing concentrations (50, 100, 200 nM) of C3bot or [His]-eGFP_C3bot, respectively, with 200 nM [His]-eGFP_C3bot$_{E174Q}$, with 1 µg/mL LPS, or they were left untreated (NC). After 7 h of incubation at 37 °C, the cells were washed and lysed, and then a sequential ADP-ribosylation assay was performed for detection of non-ADP-ribosylated Rho in intact cells (lower band in left lower panels). The relative integrated density values for Rho$_{non-ADP-rib.}$ were normalized to the HSP90 loading control (left upper panels) and depicted in a column diagram (mean ± SD, right panels). The detected bands above 26 kDa (upper band in left lower panels) were probably unspecific binding sites of the used streptavidin–peroxidase conjugate, which did not correlate with any treatment conditions. For J774A.1 macrophages in (**a**), five replicates (n = 5), and for U-DCS cells in (**b**), four replicates (n = 4) were averaged. Notably, weak signals in the sequential ADP-ribosylation assay indicated strong toxin activity in intact cells. Compared to the NC, statistical significance was tested via Student's *t*-test, and columns are labeled according to following significance levels: ns $p > 0.05$, * $p < 0.05$, ** $p < 0.01$, **** $p < 0.0001$. Additionally, the same concentrations of C3bot and [His]-eGFP_C3bot were compared via Student's *t*-test, as indicated by the lines and significance levels above the representative columns. The figure is modified from [37] under the authors' rights.

Already after 1 h of incubation with the constructs, signals for eGFP were detected in both fractions, indicating cytosolic release of eGFP (Figure 5a). For [His]-eGFP_C3bot and [His]-eGFP_C3bot$_{E174Q}$, the eGFP signals were detected at a height of about 55 kDa, corresponding to the size of full-length fusion proteins (calculated at 52 kDa). For [His]-eGFP, the signals were detected at about 34 kDa, which also fitted into the calculated molecular weight of about 28.4 kDa. These results indicate that eGFP-labeled proteins were not degraded during cellular uptake. Due to the nature of this assay, it is not possible to compare the protein amounts in one fraction with that in the other fraction, but rather the signals within one fraction should be compared. In both fractions, the detected eGFP signals were much stronger for the fusion constructs [His]-eGFP_C3bot and [His]-eGFP_C3bot$_{E174Q}$ compared to [His]-eGFP alone (Figure 5a), indicating that uptake into intracellular vesicles (membrane fraction) as well as cytosolic release of the cargo model eGFP were significantly enhanced by coupling to C3bot or C3bot$_{E174Q}$ (for Western blot quantification and statistical analysis, see Figure 5b).

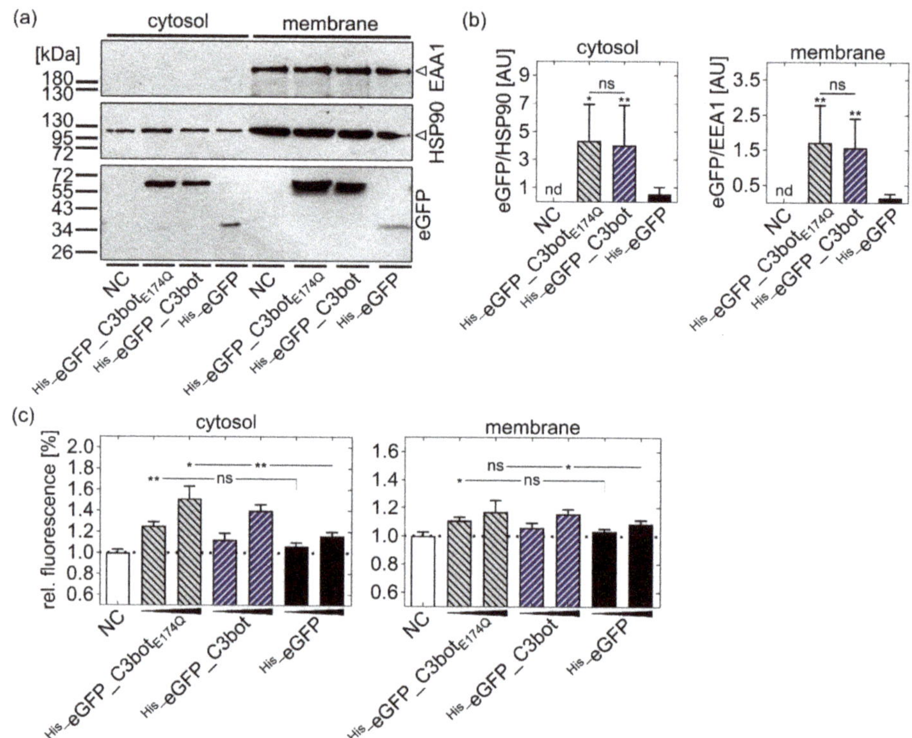

Figure 5. C3bot and C3bot$_{E174Q}$ increased the cytosolic release of the functional cargo model eGFP. (**a**,**b**) U-DCS cells were treated with ^{His}–eGFP_C3bot, ^{His}–eGFP_C3bot$_{E174Q}$, or ^{His}–eGFP (250 nM each), or were left untreated (NC) for 1 h at 37 °C. Subsequently, the cytosol was separated from the membrane fraction (containing endosomes) in a digitonin-based assay. (**a**) In a Western blot, the separation was confirmed by detection of EEA1 as a maker for endosomes that should not be present in the cytosolic fraction. HSP90 was detected as a cytosolic marker, present in both fractions, since the cytosol did not completely leave the cells (see Section 5.14). The presence of eGFP in the cytosol or membrane fraction was analyzed by using a primary anti-GFP antibody. (**b**) The Western blots were quantified, and the normalized integrated density values were plotted in a column diagram as mean ± SD (n = 7). For the cytosolic fraction, eGFP was normalized to the detected HSP90 signals, and for the membrane fraction, it was normalized to the EEA1 signals. The individual values for each repetition are available in the Supplementary Materials (Table S3 for the cytosolic fraction and Table S4 for the membrane fraction). (**c**) U-DCS cells were treated with ^{His}–eGFP_C3bot, ^{His}–eGFP_C3bot$_{E174Q}$, or ^{His}–eGFP (250 and 500 nM each), or left untreated (NC) for 5 h at 37 °C. Digitonin-based cell fractionation was performed as described above and controlled by Western blotting (data not shown). The functionality of eGFP in each fraction (cytosol in the left panel and membrane in the right panel) was analyzed by fluorescence detection at 488 nm and an emission wavelength of 510 nm in a microplate reader. (**b**,**c**) Compared to the respective ^{His}–eGFP treatment, statistical significance was tested via Student's *t*-test, and the following significance levels were defined: ns $p > 0.05$, * $p < 0.05$, ** $p < 0.01$. (**b**) Additionally, the same concentrations of C3bot and ^{His}–eGFP_C3bot were compared via Student's *t*-test, as indicated by the lines and significance levels above the representative columns.

For assay control, it was confirmed that the fusion proteins ^{His}-eGFP_C3bot and ^{His}-eGFP_C3bot$_{E174Q}$ were equally detected compared to ^{His}-eGFP by the GFP antibody (see Figure S3b,c). In this additional Western blot (Figure S3b,c), the detected signals for ^{His}-eGFP_C3bot were slightly weaker compared to ^{His}-eGFP and ^{His}-eGFP_C3bot$_{E174Q}$; however, this minor difference was neglected since it did not object to the final conclusion in the digitonin-based cell fractionation assay.

After confirming that cargo delivery can be enhanced by coupling to C3bot and C3bot$_{E174Q}$, we investigated whether the cytosolic eGFP was still functional. Therefore, fluorescence of the cytosolic and membrane fraction was analyzed with a microplate reader after treatment of cells with ^{His}-eGFP, ^{His}-eGFP_C3bot, or ^{His}-eGFP_C3bot$_{E174Q}$ and digitonin-based separation. Fluorescence signals thereby increased in a concentration-dependent manner, and cytosolic fluorescence was significantly enhanced by coupling eGFP to the C3bot transporters (Figure 5c, left panel). Taken together, the results indicated that the endosomal uptake and cytosolic release of the eGFP cargo model into macrophages and DCs can be strongly enhanced by using C3bot or C3bot$_{E174Q}$ as delivery tools and that the cargo remains functional inside these target cells.

2.4. Characterization of a Modular System for Fast Attachment of Cargo to C3bot$_{E174Q}$

After proofing the concept of using C3bot$_{E174Q}$ as a shuttle system for the cargo model eGFP, the next step was to generate a modular delivery system for fast attachment of various cargos, i.e., small molecules and proteins. Since the earlier published streptavidin-biotin system comes with the cost of reduced cell type selectivity [34,35], direct coupling of C3bot$_{E174Q}$ with cargo via thiol–maleimide click chemistry was investigated to possibly overcome this problem. By nature, C3bot$_{E174Q}$ does not contain any thiol group (no cysteine), and therefore this reaction group can simply be inserted side specifically by a point mutation. A single cysteine amino acid was inserted at the N-terminal end of C3bot$_{E174Q}$ (A1C), and this mutant will be further referred to as ^{Cys}-C3bot$_{E174Q}$. After protein purification, the thiol group can be loaded with different maleimide-labelled cargo molecules (see the cartoon in Figure S4a and example cargo in Figure S4b) by incubation for 2 h on ice. Successful protein loading was indicated by a shift to higher molecular weight, as observed by SDS-PAGE. As proof of concept, we loaded different maleimide-labeled cargos, i.e., a small molecule fluorophore (maleimide_Dylight 488 (mDL488), 0.8 kDa), larger fluorescein-isothiocyanate-labeled polyethylene glycol (mPEG_FITC, 5 kDa), or the reporter enzyme C2I (mC2I, 50 kDa). For all cargo molecules, a shift in molecular weight was detected as expected for successful coupling (Figure S4c).

After confirming the cargo coupling to the ^{Cys}-C3bot$_{E174Q}$-system, successful cellular delivery was tested in different assays. First, the uptake of the small molecule fluorophore mDL488 into macrophages (Figure 6a) and DCs (Figure 6b) was investigated. Internalization of mDL488 was only facilitated when the cargo was directly coupled to ^{Cys}-C3bot$_{E174Q}$. In contrast, neither for the treatment with cargo alone (mDL488) nor in combination with free uncoupled C3bot$_{E174Q}$ green internalization signals were detected. These results indicate that the generated modular system enhanced cellular uptake directly and not indirectly by simply enhancing endocytosis. Comparable results were obtained for the lager mPEG_FITC cargo model, which was also more efficiently internalized into target cells after coupling to ^{Cys}-C3bot$_{E174Q}$ (Figure S5).

Finally, we investigated whether the cytosolic release of cargo was also enhanced by the modular thiol–maleimide system. Therefore, the established reporter enzyme C2I was labeled with maleimide (mC2I) via the bifunctional linker m-maleimidobenzoyl-N-hydroxysuccinimide ester (MBS) and coupled to ^{Cys}-C3bot$_{E174Q}$. When C2I reached the cytosol, it ADP-ribosylated G-actin, leading to cell rounding without affecting the cell viability within 2 days of incubation [43,44]. Cell rounding also occurred when U-DCS cells were treated with C2I in combination with its activated B-component C2IIa, providing a robust and established control for successful cytosolic release of cargo (Figure 7a). In accordance with the literature [44,45], only minimal effects on cell viability/proliferation were detected for the delivered C2I (Figure S6).

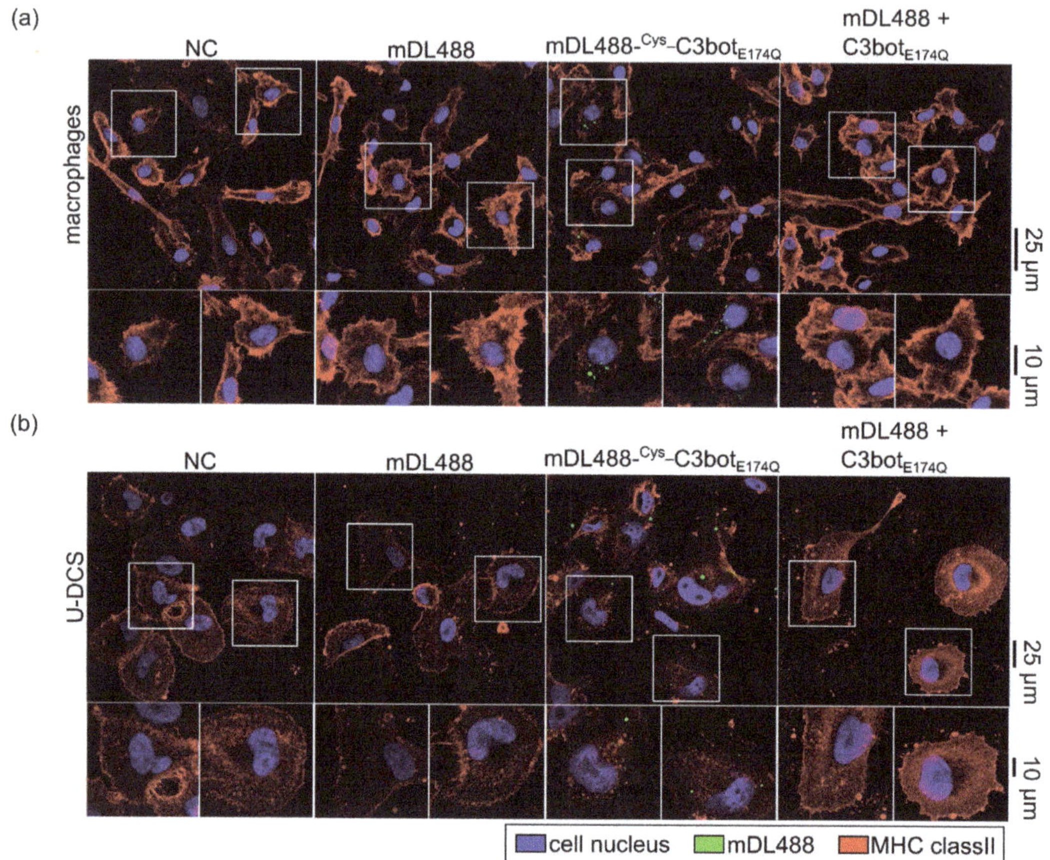

Figure 6. The small-molecule fluorophore was delivered into macrophages and DCs via ^{Cys}-C3bot$_{E174Q}$. Human-monocyte-derived macrophages (**a**) and U-DCS cells (**b**) were treated with free mDL488, mDL488-^{Cys}-C3bot$_{E174Q}$, or uncoupled mDL488 together with C3bot$_{E174Q}$ (250 nM each), or were left untreated (NC) for 30 min at 37 °C. The cells were fixed, and cell nuclei (DAPI) and MHC class II were stained as markers for cell shape and size. Confocal microscopy was used to detect fluorescence signals, and the signals for mDL488 (green), DAPI (blue), and MHC class II (red) were merged. Representative images are depicted with magnification as indicated by the white squares. Scale bars on the right are applied for the complete row of images (25 µm for full-size images, and 10 µm for magnifications).

Although mC2I alone applied in high concentrations (220 or 880 nM) for 27 or 50 h also induced some cell rounding of the dendritic sarcoma cells, this effect was significantly enhanced by coupling mC2I to ^{Cys}-C3bot$_{E174Q}$ (mC2I-^{Cys}-C3bot$_{E174Q}$), as shown in the quantification (Figure 7b). For proper control, 4.4-fold higher concentrations of mC2I were used in comparison to mC2I-^{Cys}-C3bot$_{E174Q}$ in order to compensate for free mC2I present in the coupling product (Figure S4b). Nevertheless, the ^{Cys}-C3bot$_{E174Q}$ transporter significantly increased the cytosolic release of mC2I, proving that the modular C3bot system can serve as selective and specific tool for delivery of cargo molecules into the cytosol of human monocytic cells.

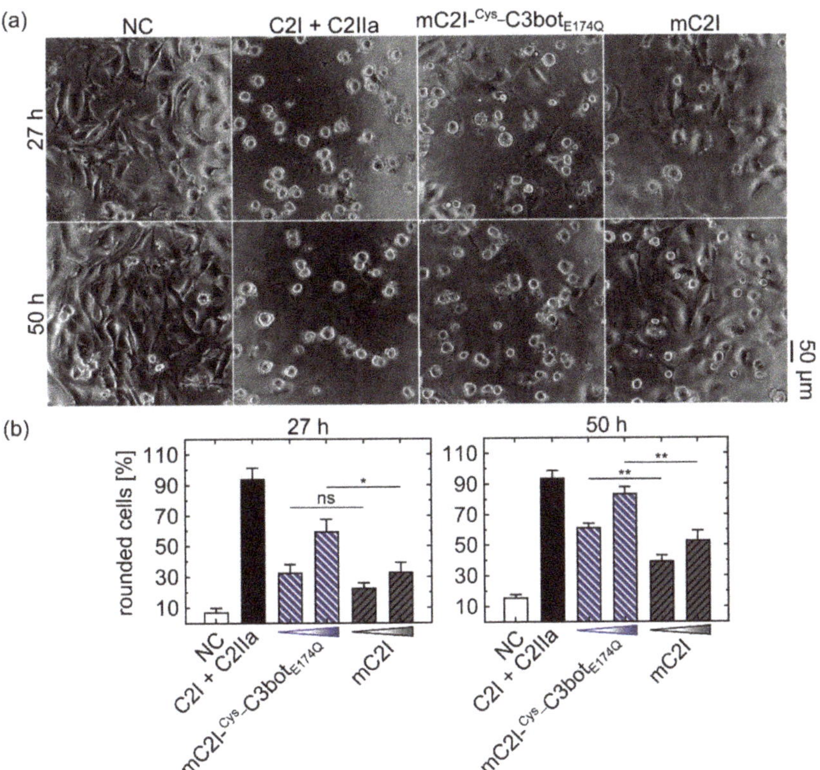

Figure 7. ^{Cys}-C3bot$_{E174Q}$ enhanced the cytosolic release of the reporter enzyme C2I. (**a**) U-DCS cells were treated with 200 nM mC2I-^{Cys}-C3bot$_{E174Q}$, 880 nM mC2I, or a combination of 1 nM C2I with 1.66 nM C2IIa, or were left untreated (NC). Phase contrast images were taken after 27 and 50 h of incubation, and representative micrographs are shown with a 50 µm scale bar. (**b**) U-DCS cells were treated with 50 or 200 nM mC2I-^{Cys}-C3bot$_{E174Q}$, 220 or 880 nM mC2I, or a combination of 1 nM C2I with 1.66 nM C2IIa, or were left untreated (NC). Phase contrast images were taken after 27 and 50 h of incubation. The number of rounded cells was counted and divided by the total cell number, and the resulting ratios are depicted in column diagrams as mean ± SD (n = 3) for the indicated time points. The values for mC2I-^{Cys}-C3bot$_{E174Q}$ were compared with the corresponding mC2I concentrations, and statistical significance was tested via Student's *t*-test (indicated with a line above columns) according to the following significance levels: ns $p > 0.05$, * $p < 0.05$, ** $p < 0.01$. The figure is modified from [37] under the authors' rights.

3. Discussion

The protein toxin C3bot selectively enters the cytosol of monocytic cells, including macrophages and DCs, by endocytosis, and inhibits Rho-dependent processes in such cells. Thereby, the toxin down-modulates essential functions of these important immune cells, which should be detrimental in the context of an infection with C3-toxin-producing bacteria. On the other hand, the cell-type selectivity of the C3 toxin can be exploited for the targeted pharmacological down-modulation of excessive pro-inflammatory activity of macrophages in the context of traumatic diseases [21]. Since hyperactive macrophages and DCs play crucial roles in several inflammatory diseases, they are important drug targets, and the C3 toxin should be a promising compound to selectively and specifically suppress excessive reactions of these innate immune cells. However, the cell-type selectivity of C3bot towards monocytic cells can be exploited for pharmacological purposes in a second way, i.e., for targeted delivery of therapeutic (macro-) molecules into these target cells and their controlled release into the cytosol. Various bacterial AB-type protein toxins have been used as drug delivery systems because of their unique mode of action, i.e., endocytic uptake and endosomal release of the therapeutic cargo molecules to reach cytosolic drug targets [46–48]. Since C3bot by nature enters monocyte-derived cells, a non-toxic variant of C3bot should represent an ideal molecule for targeted drug delivery into macrophages and DCs. Potentially, the cell penetration of novel therapeutics, e.g., peptides, proteins, nucleic acids, or cell-membrane-impermeable small molecules, can be facilitated, and/or selective targeting of macrophages and DCs as important immune cells can be improved by using C3bot$_{E174Q}$ as a drug delivery tool. The basic concepts for such a transport system were investigated in this study.

As proof of concept, the cargo model eGFP was genetically fused to both C3bot and C3bot$_{E174Q}$, and cellular uptake of the resulting fusion proteins into target cells was investigated in comparison to non-target human blood cells. The fusion protein ^{His}-eGFP_C3bot$_{E174Q}$ was selectively internalized into the human-monocyte-derived macrophages much stronger compared to lymphocytes of the same blood donor. Hence, the cell-type selectivity of the wild-type C3bot was maintained, despite the attachment of eGFP. The fluorescent protein eGFP is commonly used to follow uptake processes or to localize intracellular proteins. The generated fusion proteins (^{His}-eGFP_C3bot and ^{His}-eGFP_C3bot$_{E174Q}$) were specifically internalized into early endosomes of human-monocyte-derived macrophages, as unveiled by STED super-resolution microscopy. These results confirm earlier results obtained for the J774A.1 cell line [18] and provide more details about the localization within early endosomes. The eGFP signals are not evenly distributed over the complete vesicles, but rather are condensed at specific sites where potential translocation into the cytosol could take place. Earlier, our group observed a similar C3bot condensation in endosomes for human-monocyte-derived mature or immature DCs [23]. Only neglectable amounts of ^{His}-eGFP alone were internalized into human-monocyte-derived cells, indicating that the C3bot transporters strongly enhance the cellular uptake of this cargo model. After endocytosis, wild-type C3bot is released into the cytosol of macrophages or DCs [18,23]. However, due to the stable β-barrel structure, eGFP can potentially hinder the translocation process, resulting in lower toxin activity, as shown for the diphtheria toxin [49] or for the anthrax pore [50]. In the present study, we investigated whether this was also the case for C3bot translocation by comparing wild-type C3bot with ^{His}-eGFP_C3bot in intoxication assays. In contrast to the diphtheria toxin or the anthrax pore, not even the slightest reduction in toxin activity was detected for the fusion construct ^{His}-eGFP_C3bot. Hence, this could be the result of (i) a very effective delivery of the cargo together with C3bot into the cytosol, or (ii) very effective cleavage of the fusion construct in the early endosomes to only release C3bot into the cytosol. This second possibility, however, was ruled out by detection of cytosolic full-length ^{His}-eGFP_C3bot and ^{His}-eGFP_C3bot$_{E174Q}$ in a digitonin-based cell fractionation assay. Notably, no degradation products were detected in either the cytosol or the membrane fraction, excluding endosomal degradation. Importantly, cytosolic eGFP delivered by C3bot or C3bot$_{E174}$ was also proven to be functional by fluorescence detection. Since hypothesis (ii) could be rejected, the possibility of (i) being correct seems plausible, indicating that even relatively stable proteins such as eGFP are efficiently delivered into the cytosol of target cells via the toxins' uptake mechanism. It is possible that C3bot uses another translocation mechanism compared to the diphtheria or anthrax toxin. This could explain the preserved toxin activity of the eGFP fusion construct (^{His}-eGFP_C3bot). This seems reasonable since C3bot has no B-component that remains inside the endosomes [2,18], while this is the case for the classical AB-type protein toxins. Moreover, it was shown earlier that C3bot does not form pores but rather induces membrane flickering upon acidification [18]. By using C3bot or C3bot$_{E174Q}$ as a drug delivery tool, this special mechanism could potentially be utilized to deliver stable cargo molecules into the cytosol of C3bot-target cells. In the digitonin-based cell-fractionation assay, low amounts of cytosolic ^{His}-eGFP were detected when the cells were treated with cargo alone. This minor portion can be explained by unspecific uptake and release. Alternatively, the His-tag could potentially function as a very inefficient cell-penetrating peptide, as shown for cationic elastin-like polypeptide tags that enhance the cellular uptake of GFP in dependency on the number of positive charges [51]. However, in contrast to cells treated with ^{His}-eGFP_C3bot$_{E174Q}$ or ^{His}-eGFP_C3bot, this portion is neglectable, indicating that the C3bot-based transporters significantly enhanced the cytosolic release of cargo.

To evaluate drug delivery via C3bot in more detail, a modular system for fast attachment of cargo was generated on the basis of thiol–maleimide click chemistry. Three different cargo molecules were efficiently loaded onto the ^{Cys}-C3bot$_{E174Q}$-transporter, and uptake into DCs and macrophages was evaluated. Labeling with mDL488 was used to simulate small molecule cargo (0.8 kDa). A polyethylene glycol cargo model (mPEG_FITC) was used to simulate more bulky cargo molecules (5 kDa), and the reporter enzyme C2I was used as a model for larger protein cargo (50 kDa). For mDL488 and mPEG_FITC, uptake into macrophages and DCs was strongly enhanced by coupling to ^{Cys}-C3bot$_{E174Q}$. Increased cytosolic release into DCs was detected for mC2I. Thereby, the modular system was proven to be functional and can be used to deliver a wide range of different drugs into these target cells.

Besides the action on macrophages and DCs, earlier publications have shown that C3bot can independently of its enzymatic activity promote the outgrowth of neuronal cells after spinal cord injury [22]. Coupling C3bot$_{E174Q}$ with novel drugs could potentially enhance this effect. Moreover, C3bot$_{E174Q}$ could potentially also be used as a drug delivery tool for targeting neuronal cells upon local application. The thiol–maleimide system could serve as a platform for fast attachment of different drug candidates to C3bot$_{E174Q}$ and thereby accelerate such a drug development process.

By comparing this thiol–maleimide system with the earlier published biotin–streptavidin coupling system [19,34] for coupling of cargo molecules to C3bot$_{E174Q}$, this novel system provides some benefits but also comes with some costs. A disadvantage of the thiol–maleimide system is that the cargo molecules themselves should not contain thiol groups to ensure a specific and easy coupling process. Otherwise, maleimide-labeled cargo molecules could form multimers. This could be prevented by using protection groups, but this complicates the coupling process. In contrast, the main advantages of the thiol–maleimide system over the biotin–streptavidin coupling system are that streptavidin-induced multimerization can be excluded and that streptavidin cannot influence the natural cell-type selectivity of C3bot, which was reduced/lost, as shown in earlier publications for the biotin–streptavidin coupling system [34,35]. Internalization of the novel system only depends on ^{Cys}-C3bot$_{E174Q}$ and the attached cargo molecule.

Overall, the fully modular thiol–maleimide system will widely extend the applications of C3bot$_{E174Q}$ as a cell-type-selective drug delivery system for monocyte-derived cells, as well as allowing for fast and easy attachment of various therapeutic cargo molecules including macromolecules such as enzymes, peptides, or nucleic acids.

4. Conclusions

Taken together, we demonstrated that C3bot and its non-toxic variant C3bot$_{E174Q}$ can be used to deliver different cargo molecules into the cytosol of human macrophages and DCs. Direct fusion of eGFP as a cargo model to C3bot$_{E174Q}$ neither influenced the cell-type selectivity as evaluated for human blood cells, nor interfered with cytosolic release of the C3 toxin. C3bot and C3bot$_{E174Q}$ can therefore serve in the delivery of relatively stable functional proteins into the cytosol of their target cells. Additionally, thiol–maleimide technology was used for rapid and easy attachment of various maleimide-labeled cargo molecules to ^{Cys}-C3bot$_{E174Q}$ in order to generate a fully modular transport system for the targeted delivery and cytosolic release of molecules into macrophages and DCs.

5. Materials and Methods

5.1. Cell Culture

All cells were cultured at 37 °C at constant saturated humidity and 5% CO_2 up to passage number 25. For U-DCS cells, a medium mix of IMDM (Lonza, Basel, Switzerland) and RPMI 1640 (Gibco-Life Technologies, Carlsbad, CA, USA) in 4:1 ratio was used, and it was supplemented with 10% fetal calf serum (Gibco-Life Technologies, Carlsbad, CA, USA), 0.1% insulin–transferrin–sodium selenite supplement (Roche Diagnostics, Basel, Switzerland), 1% L-glutamine (Thermo Fisher Scientific, Waltham, MA, USA), and 100 U/mL (1%) penicillin–streptomycin (Gibco-Life Technologies, Carlsbad, CA, USA). Subconfluent U-DCS cells were passaged after trypsin detachment (Roche Diagnostics, Basel, SUI) every 2 to 3 days and split into ratios of 1:2 or 1:3, respectively. J774A.1 macrophages were cultivated in DMEM with 10% FCS, 1 nM sodium pyruvate, 1% penicillin–streptomycin, and 1% non-essential amino acids (all Gibco-Life Technologies, Carlsbad, CA, USA). Subconfluent J774A.1 cells were passaged after mechanical detachment with a cell scraper (Sarstedt, Nümbrecht, Germany) every 2 to 3 days and split into ratios of 1:4 or 1:10, respectively. During passaging, cells were seeded in microtiter plates and used for the experiments on the following or next but one day, respectively.

5.2. Differentiation of Human Monocytes into Macrophages and Co-Cultivation with Lymphocytes

Density gradient centrifugation (Ficoll-Paque™ Plus; GE Healthcare, Chicago, IL, USA) was used to isolate human peripheral blood mononuclear cells (PBMCs) from buffy coat preparations of anonymous healthy donors (Institute of Transfusion Medicine, Ulm University). PBMCs were allowed to adhere in a cell culture flask for 90 min in AIM V cell culture medium (Gibco-Life Technologies, Carlsbad, CA, USA). Non-adherent autologous cells (lymphocytes) were transferred to a fresh flask and kept in AIM V until their use in the incubator. Plastic-adherent monocytes were incubated with granulocyte-macrophage-colony-stimulating factor (10 ng/mL; Miltenyi Biotec, Bergisch Gladbach, Germany) in macrophage serum-free medium (Gibco-Life Technologies, Carlsbad, CA, USA) to generate primary human macrophages. After 6 days, macrophages were harvested using 1 mM EDTA/PBS (Sigma-Aldrich, St. Louis, MO, USA). If indicated, macrophages and autologous PBMCs were co-cultured in a 1:5 ratio in a poly-L-lysine-coated 8-chamber slide.

5.3. Protein Expression and Cell Lysis

The plasmids coding for ^{His}-eGFP_C3bot, ^{His}-eGFP_C3bot$_{E174Q}$, ^{His}-eGFP (for its origin, see [23]), C3bot, C3bot$_{E174Q}$, ^{Cys}-C3bot$_{E174Q}$ (for origin see [52]), C2I, or C2IIa were heat-shock transformed into competent *Escherichia coli* BL21. For the first preculture, 5 mL LB medium (1% tryptone, 0.5% yeast extract, 1% NaCl, 100 µg/mL ampicillin) was inoculated with a single colony for 5–8 h at 37 °C and at 180 rpm in a shaking incubator. A second 150 mL overnight culture in an Erlenmeyer flask was inoculated with the preculture. The main culture of 4 L LB medium was inoculated with the overnight preculture (30 mL per L) and incubated at 37 °C and 180 rpm until the OD_{600} reached 0.6–0.8. Protein expression was induced by adding 0.5 mM isopropyl β-D-1-thiogalactopyranoside (IPTG, Carl Roth, Karlsruhe, Germany). Subsequently, the incubation temperature was decreased to 16 °C (for eGFP-labeled proteins) or to 29 °C (GST-tagged proteins) for the main culture incubated at 180 rpm overnight. For harvesting, the cells were centrifuged at 5500 rcf and 4 °C for 10 min, and the pellet was resuspended in 40 mL buffer. GST-Lysis buffer (10 mM NaCl, 20 mM Tris, 1% Triton X-100, 1% phenylmethylsulfonyl fluoride (PMSF); pH 7.4) was used for GST-tagged proteins and NPI-20 buffer (50 mM NaH_2PO_4, 300 mM NaCl, 20 mM imidazole, 1% PMSF; pH 8.0) for His-tagged proteins. The cells were lysed by sonication (10 pulses each for 20 s with intermediate pauses of 30 s). Cell fragments were removed by centrifugation at 13,000 rcf and 4 °C for 30 min. The supernatant was filtered through 0.45 µm and 0.2 µm syringe filters.

5.4. Purification of GST-Tagged Proteins

C3bot, C3bot$_{E174Q}$, C2I, and ^{Cys}–C3bot$_{E174Q}$ were purified as GST-tagged proteins, as described previously [23]. The filtered cell lysates were incubated overnight at 4 °C with 1.2 mL Protino Glutathione Agarose 4B-beads (Macherey-Nagel., Düren, Germany), which were preequilibrated in PBS (137 mM NaCl, 2.7 mM KCl, 8 mM Na$_2$HPO$_4$, and 1.8 mM KH$_2$PO$_4$; pH 7.4). After centrifugation at 3000 rcf for 5 min, the beads were washed three times (twice with washing buffer (150 mM NaCl, 20 mM Tris–HCl; pH 7,4) and once with PBS). The proteins were eluted by removing the GST-tag with thrombin (80 NIH units, Amersham Biosciences, Little Chalfont, GBR) for 1 h at RT. The beads were removed by centrifugation for 30 s at 10,000 rcf and 4 °C. Thrombin was removed by incubating the supernatant with 120 μL Benzamidine–Sepharose 6B-beads (GE Healthcare, Chicago, IL, USA) for 10 min at RT. Centrifugation at 10,000 rcf and 4 °C for 30 s was used to remove the benzamidine beads, and the concentration of purified proteins was determined in SDS-PAGE by comparison to a BSA standard.

5.5. Purification of His-Tagged Proteins

The proteins ^{His}–eGFP_C3bot, ^{His}–eGFP_C3bot$_{E174Q}$, and ^{His}–eGFP were produced as described previously [23]. C2II was produced as a His-tagged protein, as described in [45]. In general, filtered cell lysate containing the His-tagged proteins was incubated with preequilibrated (in NPI-20, see Section 5.3) PureCube 100 INDIGO Ni-agarose (Cube Biotech, Monheim am Rhein, Germany) overnight at 4 °C. PureCube 1-step batch Midi Plus Columns (Cube Biotech, Monheim am Rhein, Germany) was used to collect the beads, and they were washed three times with 20 mL NPI-20. For elution of the His-tagged proteins, NPI-250 (50 mM NaH$_2$PO$_4$, 300 mM NaCl, 250 mM imidazole; pH 8.0) was used. Protein fractions were analyzed in SDS-PAGE, and the ones with the highest amount of target protein and the lowest content of impurities were collected. The buffer was exchanged with PBS by ultrafiltration (Vivaspin20 with 10 kDa molecular weight cutoff, Merk, Darmstadt, Germany). The protein solutions were stored at −80 °C, and the concentration of purified proteins was determined via SDS-PAGE by comparison to a BSA standard.

5.6. Coupling of Maleimide-Labeled Cargo to ^{Cys}–C3bot$_{E174Q}$

The modular ^{Cys}–C3bot$_{E174Q}$ system was loaded with maleimide-labeled cargo by incubation at 4 °C for 2 h. For mDL488 (Thermo Scientific, Waltham, MA, USA), ^{Cys}–C3bot$_{E174Q}$ and cargo were mixed at a 1:1 molar ratio. Due to the steric hindrance of PEG for mPEG_FITC (Nanocs, Boston, MA, USA), a 1:50 molar ratio (^{Cys}–C3bot$_{E174Q}$ to cargo) was needed. mC2I was prepared by labelling C2I at 4 °C for 2 h with a 10-fold molar excess of MBS (Thermo Scientific, Waltham, MA, USA) followed by removal of unbound MBS with Zeba Spin Desalting Columns (7 kDa molecular weight cutoff, Thermo Scientific, Waltham, MA, USA). For generation of mC2I-^{Cys}–C3bot$_{E174Q}$, ^{Cys}–C3bot$_{E174Q}$ and mC2I were mixed at a 1:1 molar ratio. Subsequent to coupling of the cargo, free maleimide-labeled small-molecule cargo (mDL488 and mPEG_FITC) was removed by two rounds of buffer exchange via the Zeba Spin Desalting Columns with 7 kDa molecular weight cutoff protocol. Finally, the concentration of loaded transporters was determined via SDS-PAGE by comparison to a BSA standard.

5.7. Flow Cytometry

Human-monocyte-derived macrophages were detached with 1 mM EDTA/PBS for 20 min at 37 °C, while lymphocytes were harvested by centrifugation (1300 rpm for 10 min). Per sample, either 2×10^5 macrophages or lymphocytes were incubated with ^{His}–eGFP or ^{His}–eGFP_C3bot$_{E174Q}$, or left untreated for 20 min at 37 °C. Afterwards, cells were washed with FACS buffer (1% FCS (Biochrom, Berlin, Germany) and 0.1% sodium azide (VWR, Radnor, PA, USA) in PBS (Gibco-Life Technologies, Carlsbad, CA, USA) and centrifuged for 10 min at 1300 rpm. The supernatant was discarded, and cells were incubated directly before measurement for 1 min with 50 μg/mL trypan blue to quench extracellular eGFP signals as described in [40]. Intracellular fluorescence was detected using a FACSCalibur™ flow cytometer (BD Biosciences, Heidelberg, Germany). Data analysis was performed using Flowing Software 2.5.1 (Turku Bioscience, Turku, Finland).

5.8. Phase Contrast Microscopy

Cells were seeded in 96-well microtiter (Corning Incorporated, Corning, NY, USA) plates and incubated with the indicated test substance at 37 °C and 5% CO_2 on the next day. One phase contrast image was taken per well (three wells for each treatment) with a LEICA DMi1 microscope connected to a MC 170 HD camera (both Leica Microsystems, Wetzlar, Germany). Representative images are depicted with scale bars.

5.9. Cell Viability and Proliferation Assay

Cells were seeded in a 96-well microtiter plate and treated as indicated. After the indicated incubation time, 10 µL CellTiter 96 AQueous One solution was added, containing 3-(4,5-dimethylthiazol-2-yl)-5-(3-carboxymethoxyphenyl)-2-(4-sulfophenyl)-2H-tetrazolium (MTS). Cells were incubated further for 1–2 h, and the absorbance at 492 nm was measured in a plate reader.

5.10. STED Super-Resolution Microscopy

STED microscopy was performed as described in [23,40]. A total of 10^5 monocyte-derived macrophages were seeded per well in an 8-well µ-slide with a glass bottom (ibidi GmbH, Gräfelfing, Germany). The cells were incubated with 250 nM of the indicated protein (^{His}-eGFP, ^{His}-eGFP_C3bot, or ^{His}-eGFP_C3bot$_{E174Q}$) for 30 min at 37 °C. Afterwards, the cells were washed twice with cold PBS and fixated with 3.2% paraformaldehyde in PBS (32% PFA aqueous solution, Electron Microscopy Sciences, Hatfield, PA, USA) for 20 min at RT. After washing the cells three times with PBS, the cells were permeabilized and blocked in 3% BSA and 0.3% TritonX-100 in PBS for 2 h. The samples were incubated overnight with 1 µg/mL of primary rabbit anti-EEA1 antibody (Thermo Scientific, Waltham, MA, USA) and 0.5 µg/mL Atto594-conjugated GFP-booster nanobody (Chromotek, Planegg-Martinsried, Germany) in 1:10 diluted blocking solution at 4 °C. The cells were washed three times and incubated with 1 µg/mL of the secondary Atto647N-conjugated goat anti-rabbit antibody (Sigma-Aldrich, St. Louis, MO, USA) dissolved in 1:10 diluted blocking solution. Unbound antibodies were removed by washing the samples three times with PBS. Before imaging, PBS was exchanged with 2,2'-thiodiethanol (97% solution in PBS, pH 7.5). A self-build dual-color 3D STED microscope [53] was used for image recording with an average power of 0.8 µW for each excitation beam and 1.3 mW for each depletion beam. The pixel size was 12.5 nm, and images were captured with 300 µs dwell time and approximately 150 counts as a typical peak photon number. The recorded pictures were analyzed with ImageJ (v1.52n, National Institute of Health, Bethesda, MD, USA). A Gaussian blur σ = 1 pixel and >20 count intensity threshold was applied for better visualization.

The eGFP signals and their co-localization with EEA1 signals were automatically quantified by a self-written search algorithm in Python 3.7. The algorithm loaded respective raw image file pairs, i.e., one image file containing Atto594-conjugated GFP-booster nanobody signal intensities and the belonging second image file representing signal intensities for EEA1. On both image files, a Gaussian blur σ = 1 pixel was applied to reduce background noise and to smoothen signals for later automated search. Next, a threshold of >35 counts was set for eGFP signal intensities and >50 counts for EEA1 signal intensities to further eliminate unwanted background. After thresholding, the algorithm horizontally searched for eGFP signals. Whenever a signal was found horizontally, the respective vertical coordinate was searched for. Each coordinate pair was saved for later use and counted. Thereby, for each donor, the mean signal number per image was calculated and averaged for five individual blood donors (n = 5). After every pixel was analyzed with regard to eGFP signals, the saved eGFP coordinate pairs were loaded to be compared with the image file containing EEA1 signal intensities. In more detail, within a radius of 250 nm (based on an estimated endosome diameter of 500 nm) from the previously found eGFP coordinate pair, the algorithm searched for EEA1 signals, indicating an EEA1-associated eGFP signal. Such co-localizing signals were successively counted. EEA1 signals above a radius of 250 nm were cut off as non-co-localizing eGFP signals. For each donor, the mean percentage of co-localizing eGFP signals was calculated (co-localizing eGFP signals divided by the total number of eGFP signals) and averaged for five individual blood donors (n = 5).

5.11. Immunofluorescence Staining for Confocal or Epifluorescence Microscopy

Cells were incubated as indicated at 37 °C in 5% CO_2. Afterwards, cells were washed twice and fixed (4% PFA, Sigma-Aldrich, St. Louis, MO, USA). For MHC class II staining, cells were incubated with 2% BSA in PBS and labeled with anti-HLA-DR antibody (1:200, L243, Leinco, St. Louis, MO, USA) for 30 min at RT. After three washing steps with PBS, MHC class II was detected by Cy5-conjugated goat anti-mouse antibody (1:250, Dianova, Hamburg, Germany). If indicated, cell nuclei were stained either with DAPI (1:200, Sigma-Aldrich, St. Louis, MO, USA) diluted in 1% BSA and 0.1% Triton X-100 in PBS, or with 5 µg/mL Hoechst33342 in PBS for 10 min at RT. Confocal images were acquired by using the inverted laser scanning confocal microscope LSM 710 (Zeiss, Oberkochen, Germany). Epifluorescence microscopic images were recorded using the iMIC digital microscope (FEI, Munich, Germany). Images were processed using ImageJ software (v1.51n, National Institute of Health, Bethesda, MD, USA).

5.12. SDS-PAGE and Western Blotting

For protein separation depending on molecular weight, SDS-PAGE was used with 12.5% acrylamide gels. Subsequently to electrophoresis, the proteins were transferred onto a nitrocellulose membrane by using semi-dry blotting, which was controlled by Ponceau S (AppliChem GmbH, Darmstadt, Germany) staining. Unspecific binding to the membrane was blocked by incubation in 5% skim milk powder diluted in PBS-T (137 mM NaCl, 2.7 mM KCl, 8 mM Na_2HPO_4, 1.8 mM KH_2PO_4, 0.1% Tween20; pH 7.4) for 1 h at RT. After washing with PBS-T, the membrane was incubated with the indicated antibody or streptavidin–peroxidase conjugate (1:5000; Sigma-Aldrich, St. Louis, MO, USA) diluted in PBS-T for 1 h at RT. For eGFP detection, 1:10,000 diluted anti-GFP antibody (ab290, Abcam, Cambridge, GBR) was used. HSP90 was detected with 1:500 diluted HSP90 α/β antibody (F-8, Santa Cruz Biotechnology, Dallas, TX, USA), while for EEA1-detection, 1:1000 diluted rabbit EEA1 polyclonal antibody (PA1-063A, Thermo Fisher Scientific, Waltham, MA, USA) was used. Unbound antibodies/proteins were removed with three washing steps with PBS-T for 5 min at RT on an orbital shaker. For the detection of eGFP and EEA1, 1:2500 diluted mouse anti-rabbit IgG-HRP (sc-2357, Santa Cruz Biotechnology, Dallas, TX, USA) was used. For detection of HSP90, 1:2500 diluted m-IgGκ BP-HRP (sc-516102, Santa Cruz Biotechnology, Dallas, TX, USA) was used. The peroxidase-labeled antibodies/proteins were detected with Pierce ECL Western blotting substrate (Thermo Fisher Scientific, Waltham, MA, USA) and X-ray films (AGFA Health Care, Mortsel, BEL).

5.13. Sequential ADP-Ribosylation Assay

Cells were seeded in 24-well microtiter plates and treated as indicated at 37 °C and 5% CO_2. Extracellular toxins were removed by washing the cells two times with PBS. The medium was removed, and the samples were frozen at −20 °C and thawed in ADP-ribosylation buffer (20 mM Tris-HCl, 1 mM EDTA, 1 mM DTT, 5 mM $MgCl_2$, cOmplete (1:50, freshly added); pH 7.5) to lyse the cells. The samples were collected, and 5 pmol fresh C3bot and 6-biotin-17-NAD^+ (10 µM) were added in excess. The sequential ADP-ribosylation reactions at 37 °C were started and stopped at the same time (after 30 min). Laemmli buffer (0.3 M Tris-HCl, 10% SDS, 37.5% glycerol, 0.4 mM bromophenol blue) was added, and the samples were heat denatured. Notably, in this sequential ADP-ribosylation reaction, only the non-ADP-ribosylated Rho can be biotin-labeled from 6-biotin-17-NAD^+. Hence, in SDS-PAGE and Western blotting (see Section 5.12), the non-ADP-ribosylated Rho in intact cells is detected with streptavidin–peroxidase conjugate. For densitometric analysis of the detected protein bands, ImageJ software (v1.51n, National Institute of Health, Bethesda, MD, USA) was used. Importantly, weak signals detected in this assay indicate strong toxin activity in intact cells.

5.14. Digitonin-Based Cell Fractionation Assay

A total of 10^6 U-DCS cells were seeded on a 24-well microtiter plate. After two days, the cells were incubated with the respective proteins (^{His}-eGFP_C3bot, ^{His}-eGFP_C3bot$_{E174Q}$, and ^{His}-eGFP) with indicated concentrations and time points at 37 °C. Subsequently, the cells were carefully washed twice with PBS and then incubated with digitonin (20 µg/mL in PBS) for 5 min at RT. Thereby, the cell membrane was permeabilized, and the cytosol left the cells through the previously formed pores during incubation for 25 min at 4 °C on ice. In this process, the cells were divided into two fractions: the supernatant (cytosol-only fraction) and the solid cellular portion (membrane fraction) containing cellular organelles, vesicles, cell membranes, and the remaining cytosolic proteins that did not flow out through the pores. The fluorescence of eGFP was analyzed in both fractions using a microplate reader at 488 nm excitation and 510 nm emission. SDS-PAGE and Western blot analysis (see Section 5.12) were performed to ensure clean separation of both fractions. The early endosomal marker EEA1 was only detectable in the membrane fraction. Cytosolic HSP90 can be detected in both fractions since it is not possible to force complete outflow of all cytosolic proteins. The presence of ^{His}-eGFP and ^{His}-eGFP-labeled proteins was analyzed in the respective cell fractions with the GFP antibody.

5.15. Data Analysis and Visualization

The depicted data points are provided as mean $\pm$ standard deviation ($\pm$ SD) with corresponding sample size (n). A two-tailed unpaired Student's *t*-test was used to compare the two treatment groups. The following significance levels were defined: not significant (ns) $p > 0.05$, * $p < 0.05$, ** $p < 0.01$, *** $p < 0.001$, **** $p < 0.0001$. Diagrams were generated by using GraphPad Prism (version 9.1.2 GraphPad Software, San Diego, CA, USA). Figures were assembled in Inkscape (version 0.92, Free Software Foundation, Boston, MA, USA).

Supplementary Materials: The following supporting information can be downloaded at https://www.mdpi.com/article/10.3390/toxins14100711/s1, Table S1. Quantification of the mean eGFP signals per image given for the individual human blood donors. Table S2. Co-localization analysis of eGFP and EEA1. Table S3. Quantification of the cytosol fraction from the digitonin-based cell-fractionation assay. Table S4. Quantification of the membrane fraction from the digitonin-based cell-fractionation assay. Figure S1. Cell-type selective uptake of ^{His}-eGFP_C3bot$_{E174Q}$ into primary human macrophages compared to lymphocytes ex vivo. Figure S2. Quantification of eGFP-positive macrophages from Figure 2b. Figure S3. Detection of the eGFP-labeled C3bot variants in SDS-PAGE and Western blotting. Figure S4. The modular thiol–maleimide system enabled attachment of cargo molecules to ^{Cys}-C3bot$_{E174Q}$. Figure S5. Internalization of mPEG_FITC into macrophages and DCs was strongly enhanced by coupling to ^{Cys}-C3bot$_{E174Q}$. Figure S6. The cell viability/proliferation of U-DCS cells was not or only minimally effected by the transported mC2I.

Author Contributions: Conceptualization, M.F., S.F. and H.B.; Funding acquisition, S.F., J.M., S.S. and H.B.; Investigation, M.F., M.S., R.N. and F.W.; Project administration, H.B.; Supervision, S.F., J.M., S.S. and H.B.; Visualization, M.F.; Writing—original draft, M.F., M.S. and R.N.; Writing—review & editing, M.F., M.S., R.N., F.W., S.F., J.M., S.S. and H.B. All authors have read and agreed to the published version of the manuscript.

Funding: This research was funded by the German Research Foundation (DFG) as part of the CRC 1279 (A01 and C02)—Project number 316249678—SFB 1279 and part of the CRC 1149 (A05)—Project number 251293561—SFB 1149 together with a CRC 1149 start-up grant (Stephan Fischer). Maximilian Fellermann, Mia Stemmer, and Reiner Noschka are members of the International Graduate School in Molecular Medicine Ulm (IGradU) and grateful thank the IGradU for its support.

Institutional Review Board Statement: Not applicable.

Informed Consent Statement: Not applicable.

Data Availability Statement: The data presented in this study is available on request from the corresponding author.

Conflicts of Interest: The authors declare no conflict of interest. The funders had no role in the study design; in collection, analyses, or interpretation of data; in writing of the manuscript, or in the decision to publish the results.

References

1. Popoff, M.R.; Hauser, D.; Boquet, P.; Eklund, M.W.; Gill, D.M. Characterization of the C3 Gene of *Clostridium botulinum* Types C and D and Its Expression in *Escherichia coli*. *Infect. Immun.* **1991**, *59*, 3673–3679. [CrossRef]
2. Aktories, K.; Wilde, C.; Vogelsgesang, M. Rho-Modifying C3-like ADP-Ribosyltransferases. In *Reviews of Physiology, Biochemistry and Pharmacology*; Springer: Berlin, Germany, 2004; Volume 152, pp. 1–22. [CrossRef]
3. Maehama, T.; Sekine, N.; Nishina, H.; Takahashi, K.; Katada, T. Characterization of Botulinum C3-Catalyzed ADP-Ribosylation of Rho Proteins and Identification of Mammalian C3-like ADP-Ribosyltransferase. *Mol. Cell. Biochem.* **1994**, *138*, 135–140. [CrossRef] [PubMed]
4. Sekine, A.; Fujiwara, M.; Narumiya, S. Asparagine Residue in the Rho Gene Product Is the Modification Site for Botulinum ADP-Ribosyltransferase. *J. Biol. Chem.* **1989**, *264*, 8602–8605. [CrossRef]
5. Aktories, K. Studies on the Active-Site Structure of C3-like Exoenzymes: Involvement of Glutamic Acid in Catalysis of ADP-Ribosylation. *Biochimie* **1995**, *77*, 326–332. [CrossRef]
6. Saito, Y.; Nemoto, Y.; Ishizaki, T.; Watanabe, N.; Morii, N.; Narumiya, S. Identification of Glu173 as the Critical Amino Acid Residue for the ADP-Ribosyltransferase Activity of *Clostridium botulinum* C3 Exoenzyme. *FEBS Lett.* **1995**, *371*, 105–109. [CrossRef]
7. Aktories, K.; Weller, U.; Chhatwal, G.S. *Clostridium botulinum* Type C Produces a Novel ADP-Ribosyltransferase Distinct from Botulinum C2 Toxin. *FEBS Lett.* **1987**, *212*, 109–113. [CrossRef]
8. Nemoto, Y.; Namba, T.; Kozaki, S.; Narumiya, S. *Clostridium botulinum* C3 ADP-Ribosyltransferase Gene. Cloning, Sequencing, and Expression of a Functional Protein in Escherichia Coli. *J. Biol. Chem.* **1991**, *266*, 19312–19319. [CrossRef]
9. Fünfhaus, A.; Poppinga, L.; Genersch, E. Identification and Characterization of Two Novel Toxins Expressed by the Lethal Honey Bee Pathogen Paenibacillus Larvae, the Causative Agent of American Foulbrood. *Environ. Microbiol.* **2013**, *15*, 2951–2965. [CrossRef]
10. Krska, D.; Ravulapalli, R.; Fieldhouse, R.J.; Lugo, M.R.; Merrill, A.R. C3larvin Toxin, an ADP-Ribosyltransferase from *Paenibacillus larvae*. *J. Biol. Chem.* **2015**, *290*, 1639–1653. [CrossRef]
11. Just, I.; Mohr, C.; Schallehn, G.; Menard, L.; Didsbury, J.R.; Vandekerckhove, J.; van Damme, J.; Aktories, K. Purification and Characterization of an ADP-Ribosyltransferase Produced by *Clostridium limosum*. *J. Biol. Chem.* **1992**, *267*, 10274–10280. [CrossRef]
12. Just, I.; Schallehn, G.; Aktories, K. ADP-Ribosylation of Small GTP-Binding Proteins by *Bacillus cereus*. *Biochem. Biophys. Res. Commun.* **1992**, *183*, 931–936. [CrossRef]
13. Inoue, S.; Sugai, M.; Murooka, Y.; Paik, S.Y.; Hong, Y.M.; Ohgai, H.; Suginaka, H. Molecular Cloning and Sequencing of the Epidermal Cell Differentiation Inhibitor Gene from *Staphylococcus aureus*. *Biochem. Biophys. Res. Commun.* **1991**, *174*, 459–464. [CrossRef]
14. Wilde, C.; Chhatwal, G.S.; Schmalzing, G.; Aktories, K.; Just, I. A Novel C3-like ADP-Ribosyltransferase from *Staphylococcus aureus* Modifying RhoE and Rnd3. *J. Biol. Chem.* **2001**, *276*, 9537–9542. [CrossRef] [PubMed]
15. Yamaguchi, T.; Hayashi, T.; Takami, H.; Ohnishi, M.; Murata, T.; Nakayama, K.; Asakawa, K.; Ohara, M.; Komatsuzawa, H.; Sugai, M. Complete Nucleotide Sequence of a *Staphylococcus aureus* Exfoliative Toxin B Plasmid and Identification of a Novel ADP-Ribosyltransferase, EDIN-C. *Infect. Immun.* **2001**, *69*, 7760–7771. [CrossRef]
16. Ebeling, J.; Fünfhaus, A.; Genersch, E. The Buzz about ADP-Ribosylation Toxins from *Paenibacillus larvae*, the Causative Agent of American Foulbrood in Honey Bees. *Toxins* **2021**, *13*, 151. [CrossRef] [PubMed]
17. Ebeling, J.; Knispel, H.; Fünfhaus, A.; Genersch, E. The Biological Role of the Enigmatic C3larvinAB Toxin of the Honey Bee Pathogenic Bacterium *Paenibacillus larvae*. *Environ. Microbiol.* **2019**, *21*, 3091–3106. [CrossRef] [PubMed]
18. Fahrer, J.; Kuban, J.; Heine, K.; Rupps, G.; Kaiser, E.; Felder, E.; Benz, R.; Barth, H. Selective and Specific Internalization of Clostridial C3 ADP-Ribosyltransferases into Macrophages and Monocytes. *Cell. Microbiol.* **2010**, *12*, 233–247. [CrossRef]
19. Tautzenberger, A.; Förtsch, C.; Zwerger, C.; Dmochewitz, L.; Kreja, L.; Ignatius, A.; Barth, H. C3 Rho-Inhibitor for Targeted Pharmacological Manipulation of Osteoclast-like Cells. *PLoS ONE* **2013**, *8*, e85695. [CrossRef]
20. Rotsch, J.; Rohrbeck, A.; May, M.; Kolbe, T.; Hagemann, S.; Schelle, I.; Just, I.; Genth, H.; Huelsenbeck, S.C. Inhibition of Macrophage Migration by C. Botulinum Exoenzyme C3. *Naunyn-Schmiedeberg's Arch. Pharmacol.* **2012**, *385*, 883–890. [CrossRef]
21. Martin, T.; Möglich, A.; Felix, I.; Förtsch, C.; Rittlinger, A.; Palmer, A.; Denk, S.; Schneider, J.; Notbohm, L.; Vogel, M.; et al. Rho-Inhibiting C2IN-C3 Fusion Toxin Inhibits Chemotactic Recruitment of Human Monocytes Ex Vivo and in Mice In Vivo. *Arch. Toxicol.* **2018**, *92*, 323–336. [CrossRef]
22. Just, I.; Rohrbeck, A.; Huelsenbeck, S.C.; Hoeltje, M. Therapeutic Effects of *Clostridium botulinum* C3 Exoenzyme. *Naunyn-Schmiedeberg's Arch. Pharmacol.* **2011**, *383*, 247–252. [CrossRef] [PubMed]

23. Fellermann, M.; Huchler, C.; Fechter, L.; Kolb, T.; Wondany, F.; Mayer, D.; Michaelis, J.; Stenger, S.; Mellert, K.; Möller, P.; et al. Clostridial C3 Toxins Enter and Intoxicate Human Dendritic Cells. *Toxins* **2020**, *12*, 563. [CrossRef]

24. Liu, L.; Schwartz, B.R.; Lin, N.; Winn, R.K.; Harlan, J.M. Requirement for RhoA Kinase Activation in Leukocyte De-Adhesion. *J. Immunol.* **2002**, *169*, 2330–2336. [CrossRef] [PubMed]

25. Laudanna, C.; Campbell, J.J.; Butcher, E.C. Role of Rho in Chemoattractant-Activated Leukocyte Adhesion through Integrins. *Science* **1996**, *271*, 981–983. [CrossRef] [PubMed]

26. Caron, E.; Hall, A. Identification of Two Distinct Mechanisms of Phagocytosis Controlled by Different Rho GTPases. *Science* **1998**, *282*, 1717–1721. [CrossRef] [PubMed]

27. Kippert, A.; Trajkovic, K.; Rajendran, L.; Ries, J.; Simons, M. Rho Regulates Membrane Transport in the Endocytic Pathway to Control Plasma Membrane Specialization in Oligodendroglial Cells. *J. Neurosci.* **2007**, *27*, 3560–3570. [CrossRef]

28. Norman, J.C.; Price, L.S.; Ridley, A.J.; Koffer, A. The Small GTP-Binding Proteins, Rac and Rho, Regulate Cytoskeletal Organization and Exocytosis in Mast Cells by Parallel Pathways. *Mol. Biol. Cell* **1996**, *7*, 1429–1442. [CrossRef]

29. Aepfelbacher, M.; Essler, M.; Huber, E.; Czech, A.; Weber, P.C. Rho Is a Negative Regulator of Human Monocyte Spreading. *J. Immunol.* **1996**, *157*, 5070–5075.

30. Worthylake, R.A.; Lemoine, S.; Watson, J.M.; Burridge, K. RhoA Is Required for Monocyte Tail Retraction during Transendothelial Migration. *J. Cell Biol.* **2001**, *154*, 147–160. [CrossRef]

31. Barth, H.; Fischer, S.; Möglich, A.; Förtsch, C. Clostridial C3 Toxins Target Monocytes/Macrophages and Modulate Their Functions. *Front. Immunol.* **2015**, *6*, 339. [CrossRef]

32. Chardin, P.; Boquet, P.; Madaule, P.; Popoff, M.R.; Rubin, E.J.; Gill, D.M. The Mammalian G Protein RhoC Is ADP-Ribosylated by *Clostridium botulinum* Exoenzyme C3 and Affects Actin Microfilaments in Vero Cells. *EMBO J.* **1989**, *8*, 1087–1092. [CrossRef] [PubMed]

33. Barth, H.; Hofmann, F.; Olenik, C.; Just, I.; Aktories, K. The N-Terminal Part of the Enzyme Component (C2I) of the Binary *Clostridium botulinum* C2 Toxin Interacts with the Binding Component C2II and Functions as a Carrier System for a Rho ADP-Ribosylating C3-like Fusion Toxin. *Infect. Immun.* **1998**, *66*, 1364–1369. [CrossRef]

34. Lillich, M.; Chen, X.; Weil, T.; Barth, H.; Fahrer, J. Streptavidin-Conjugated C3 Protein Mediates the Delivery of Mono-Biotinylated RNAse A into Macrophages. *Bioconjug. Chem.* **2012**, *23*, 1426–1436. [CrossRef]

35. Christow, H.; Lillich, M.; Sold, A.; Fahrer, J.; Barth, H. Recombinant Streptavidin-C3bot for Delivery of Proteins into Macrophages. *Toxicon* **2013**, *75*, 144–147. [CrossRef]

36. Dmochewitz, L.; Förtsch, C.; Zwerger, C.; Vaeth, M.; Felder, E.; Huber-Lang, M.; Barth, H. A Recombinant Fusion Toxin Based on Enzymatic Inactive C3bot1 Selectively Targets Macrophages. *PLoS ONE* **2013**, *8*, e54517. [CrossRef] [PubMed]

37. Fellermann, M. Characterization of Clostridial C3 Toxin and Diphtheria Toxin Mutant Proteins, Their Cellular Internalization and Application as Delivery Tools for Foreign Cargo Molecules. Ph.D. Thesis, Universität Ulm, Ulm, Germany, 2021.

38. Sahlin, S.; Hed, J.; Rundquist, I. Differentiation between Attached and Ingested Immune Complexes by a Fluorescence Quenching Cytofluorometric Assay. *J. Immunol. Methods* **1983**, *60*, 115–124. [CrossRef]

39. Fang, I.-J.; Trewyn, B.G. Application of Mesoporous Silica Nanoparticles in Intracellular Delivery of Molecules and Proteins. In *Methods in Enzymology*; Elsevier Inc.: Amsterdam, The Netherlands, 2012; Volume 508, pp. 41–59. ISBN 9780123918604.

40. Fellermann, M.; Wondany, F.; Carle, S.; Nemeth, J.; Sadhanasatish, T.; Frick, M.; Barth, H.; Michaelis, J. Super-Resolution Microscopy Unveils Transmembrane Domain-Mediated Internalization of Cross-Reacting Material 197 into Diphtheria Toxin-Resistant Mouse J774A.1 Cells and Primary Rat Fibroblasts In Vitro. *Arch. Toxicol.* **2020**, *94*, 1753–1761. [CrossRef] [PubMed]

41. Barth, H.; Stiles, B.G. Protein Toxins from Bacteria. In *Comprehensive Natural Products II*; Elsevier: Amsterdam, The Netherlands, 2010; pp. 149–173. ISBN 9780080453828.

42. Mamat, U.; Wilke, K.; Bramhill, D.; Schromm, A.B.; Lindner, B.; Kohl, T.A.; Corchero, J.L.; Villaverde, A.; Schaffer, L.; Head, S.R.; et al. Detoxifying *Escherichia coli* for Endotoxin-Free Production of Recombinant Proteins. *Microb. Cell Fact.* **2015**, *14*, 57. [CrossRef] [PubMed]

43. Aktories, K.; Bärmann, M.; Ohishi, I.; Tsuyama, S.; Jakobs, K.H.; Habermann, E. Botulinum C2 Toxin ADP-Ribosylates Actin. *Nature* **1986**, *322*, 390–392. [CrossRef]

44. Barth, H.; Klingler, M.; Aktories, K.; Kinzel, V. *Clostridium botulinum* C2 Toxin Delays Entry into Mitosis and Activation of P34cdc2 Kinase and Cdc25-C Phosphatase in HeLa Cells. *Infect. Immun.* **1999**, *67*, 5083–5090. [CrossRef]

45. Eisele, J.; Schreiner, S.; Borho, J.; Fischer, S.; Heber, S.; Endres, S.; Fellermann, M.; Wohlgemuth, L.; Huber-Lang, M.; Fois, G.; et al. The Pore-Forming Subunit C2IIa of the Binary *Clostridium botulinum* C2 Toxin Reduces the Chemotactic Translocation of Human Polymorphonuclear Leukocytes. *Front. Pharmacol.* **2022**, *13*, 810611. [CrossRef]

46. Barth, H.; Stiles, B.G. Binary Actin-ADP-Ribosylating Toxins and Their Use as Molecular Trojan Horses for Drug Delivery into Eukaryotic Cells. *Curr. Med. Chem.* **2008**, *15*, 459–469. [CrossRef] [PubMed]

47. Piot, N.; van der Goot, F.G.; Sergeeva, O.A. Harnessing the Membrane Translocation Properties of AB Toxins for Therapeutic Applications. *Toxins* **2021**, *13*, 36. [CrossRef] [PubMed]

48. Beilhartz, G.L.; Sugiman-Marangos, S.N.; Melnyk, R.A. Repurposing Bacterial Toxins for Intracellular Delivery of Therapeutic Proteins. *Biochem. Pharmacol.* **2017**, *142*, 13–20. [CrossRef] [PubMed]

49. Auger, A.; Park, M.; Nitschke, F.; Minassian, L.M.; Beilhartz, G.L.; Minassian, B.A.; Melnyk, R.A. Efficient Delivery of Structurally Diverse Protein Cargo into Mammalian Cells by a Bacterial Toxin. *Mol. Pharm.* **2015**, *12*, 2962–2971. [CrossRef] [PubMed]

50. Zornetta, I.; Brandi, L.; Janowiak, B.; Dal Molin, F.; Tonello, F.; Collier, R.J.; Montecucco, C. Imaging the Cell Entry of the Anthrax Oedema and Lethal Toxins with Fluorescent Protein Chimeras. *Cell. Microbiol.* **2010**, *12*, 1435–1445. [CrossRef]
51. Pesce, D.; Wu, Y.; Kolbe, A.; Weil, T.; Herrmann, A. Enhancing Cellular Uptake of GFP via Unfolded Supercharged Protein Tags. *Biomaterials* **2013**, *34*, 4360–4367. [CrossRef]
52. Kuan, S.L.; Fischer, S.; Hafner, S.; Wang, T.; Syrovets, T.; Liu, W.; Tokura, Y.; Ng, D.Y.W.; Riegger, A.; Förtsch, C.; et al. Boosting Antitumor Drug Efficacy with Chemically Engineered Multidomain Proteins. *Adv. Sci.* **2018**, *5*, 1701036. [CrossRef]
53. Osseforth, C.; Moffitt, J.R.; Schermelleh, L.; Michaelis, J. Simultaneous Dual-Color 3D STED Microscopy. *Opt. Express* **2014**, *22*, 7028–7039. [CrossRef]

Article

The Role of Plasma Membrane Pleiotropic Drug Resistance Transporters in the Killer Activity of *Debaryomyces hansenii* and *Wickerhamomyces anomalus* Toxins

Monika Czarnecka [1,*], Xymena Połomska [1], Cristina Restuccia [2] and Barbara Żarowska [1]

[1] Faculty of Biotechnology and Food Science, Wrocław University of Environmental and Life Sciences, ul. Chełmońskiego 37/41, 51-630 Wrocław, Poland; xymena.polomska@upwr.edu.pl (X.P.); barbara.zarowska@upwr.edu.pl (B.Ż.)

[2] Department of Agriculture, Food and Environment (Di3A), University of Catania, via Santa Sofia 100, 95123 Catania, Italy; crestu@unict.it

* Correspondence: monika.grzegorczyk89@gmail.com

Abstract: The killer strains of *Debaryomyces hansenii* and *Wickerhamomyces anomalus* species secrete antimicrobial proteins called killer toxins which are active against selected fungal phytopathogens. In our research, we attempted to investigate the role of plasma membrane pleiotropic drug resistance (PDR) transporters (Pdr5p and Snq2p) in the mechanism of defense against killer toxins. *Saccharomyces cerevisiae* mutant strains with strengthened or weakened pleiotropic drug resistance due to increased or reduced number of mentioned PDR efflux pumps were tested for killer toxin susceptibility. The present study demonstrates the influence of the Snq2p efflux pump in immunity to *W. anomalus* BS91 killer toxin. It was also shown that the activity of killer toxins of *D. hansenii* AII4b, KI2a, MI1a and CBS767 strains is regulated by other transporters than those influencing *W. anomalus* killer toxin activity. In turn, this might be related to the functioning of the Pdr5p transporter and a complex cross-talk between several regulatory multidrug resistance networks. To the best of our knowledge, this is the first study that reports the involvement of PDR transporters in the cell membrane of susceptible microorganisms in resistance to killer yeasts' toxins.

Keywords: *Debaryomyces hansenii*; *Wickerhamomyces anomalus*; *Saccharomyces cerevisiae*; PDR transporters; killer toxin

Key Contribution: This study reveals that PDR transporters in *S. cerevisiae* play a role in resistance to the activity of *D. hansenii* and *W. anomalus* killer toxin, wherein the *pdr1-3 S. cerevisiae* mutant was resistant to *W. anomalus* toxin and the Δ*pdr1*Δ*pdr3 S. cerevisiae* mutant was susceptible to it. Killer toxin activity of *W. anomalus* and *D. hansenii* was affected by Snq2p and Pdr5p transporters, respectively.

Citation: Czarnecka, M.; Połomska, X.; Restuccia, C.; Żarowska, B. The Role of Plasma Membrane Pleiotropic Drug Resistance Transporters in the Killer Activity of *Debaryomyces hansenii* and *Wickerhamomyces anomalus* Toxins. *Toxins* **2022**, *14*, 180. https://doi.org/10.3390/toxins14030180

Received: 5 February 2022
Accepted: 24 February 2022
Published: 28 February 2022

Publisher's Note: MDPI stays neutral with regard to jurisdictional claims in published maps and institutional affiliations.

1. Introduction

Killer strains of the species *Debaryomyces hansenii* and *Wickerhamomyces anomalus* exhibit antagonistic activity against several yeast species and fungal phytopathogens. Thus, they are considered promising biocontrol agents in plant protection [1–12]. Antagonistic activity of *D. hansenii* and *W. anomalus* killer strains can be manifested through various mechanisms, including competition for nutrients and space, mycoparasitism, biofilm formation, induction of plants' immune response to pathogens, and the secretion of antifungal substances, such as volatile organic compounds (VOCs), β-glucanases, and killer toxins [13–18]. The production of killer toxins, exhibiting antimicrobial activity, is a relatively common phenomenon in yeast. Strains with killer phenotypes produce killer toxins and are immune to their toxin. However, they may be susceptible to killer toxins produced by other killer yeasts [19,20]. The protein structure, specific receptors, and mode of action of killer toxins may differ significantly among yeast genera, species, and even strains. Most killer toxins exhibit a two-step killing mechanism. First, they bind to their specific receptors in the cell wall of susceptible microorganisms. Subsequently, toxins may exhibit killer mechanisms such as: blockage of DNA synthesis in the cell nucleus, cleavage of selected tRNAs, inhibition of β-1,3-glucan synthesis, disruption of electrochemical ion gradient across the plasma membrane, or enzymatic activity, leading to increased plasma membrane permeability with an eventual lethal effect [21–24].

The plasma membrane is essential for maintaining proper cell metabolism and osmotic pressure, for the exchange of genetic information and translocation of molecules across the plasma membrane. Compounds excreted from the cell are cellular metabolites, and xenobiotics, including antibiotics and antimicrobial substances. The main cellular elements responsible for the extrusion of numerous structurally unrelated molecules are the ATP-binding cassette (ABC) transporters, located in the cell membrane. These work as one-way pumps, thereby creating a multi-drug resistance (MDR) network. One of the main components of such a network in *S. cerevisiae* is the pleiotropic drug resistance (PDR) transporters system. PDR genes of *S. cerevisiae* share homology with mammalian MDR genes and some genes in other yeasts and fungi [25]. These homologies, with their ease of genome manipulation and simplicity of breeding, made *S. cerevisiae* a convenient and valuable tool for multidrug resistance research in medicine and agronomy [26].

There are three main regulatory networks involved in controlling the expression of multidrug resistance genes in *S. cerevisiae*, represented by different transcription factors: *PDR1* and *PDR3*, *YAP1* (yeast activator protein) and *YRR1* (yeast reveromycin resistance), from which *PDR1* and *PDR3* are the most characterized and studied [27]. *PDR5*, *SNQ2* (sensitivity to 4-Nitroquinoline N-oxide) and *YOR1* (yeast oligomycin resistance) genes encoding Pdr5p, Snq2p, and Yor1p efflux pumps in the plasma membrane, are positively regulated mainly by *PDR1* and *PDR3* transcription factors. However, they may be also activated by *YAP1* or *YRR1* (Figure 1). These pumps are responsible for resistance to a wide spectrum of substrates, including molecules commonly used in the chemical treatment of fungal and bacterial infections [25,28,29]. *PDR1* and *PDR3* are homologous genes, at which point mutations lead to the overexpression of downstream target genes [30]. Single point mutations in the *PDR1* locus (*pdr1-2*, *pdr1-3*, *pdr1-6*, *pdr1-7*, *pdr1-8*) increase the expression of *PDR5*, *SNQ2*, *YOR1*, *PDR10*, *PDR15*, and *PDR16* genes, whereas mutations in *PDR3* (*pdr3-2* to *pdr3-10*) activate the expression of *PDR5*, *SNQ2*, *PDR15*, and *PDR3* genes, encoding efflux pumps in the cell membrane. Therefore, these point mutations enhance multidrug resistance in microorganisms. In turn, disruption, or deletion of *PDR1* and/or *PDR3* genes cause hypersensitivity, which is more pronounced in *PDR1* mutants [31–35]. The regulation of the PDR genes expression in microorganisms and plants may be also triggered by the exposure of cells to PDR-specific substrates [36–39].

So far, it has been established that β-glucans in the cell wall of sensitive microorganisms are the primary receptors for killer toxins of *D. hansenii* and *W. anomalus* [6,14,16,40–43]. However, little is known about the further effect of the killer toxins on the plasma membrane of sensitive microorganisms and potential defence systems. In this research, we attempt to establish whether plasma membrane components, such as PDR transporters, play a role in resistance to *D. hansenii* and *W. anomalus* killer toxins. In comparative studies, we analyzed the sensitivity of *S. cerevisiae* wild type strains and their isogenic mutants with increased or reduced pleiotropic drug resistance to *D. hansenii* and *W. anomalus* killer toxins. To the best of our knowledge, this is the first report considering the correlation between killer toxins activity and the function of plasma membrane transporters which contributes to the overall research on killer toxins and killer yeast antagonistic activity in biocontrol of pathogenic fungi.

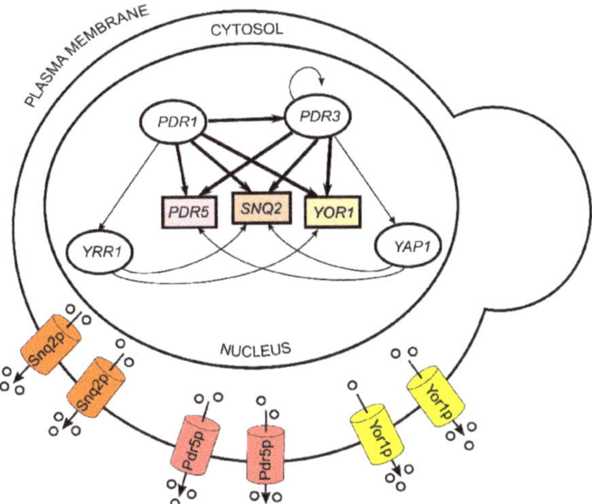

Figure 1. Positive regulation of *PDR5*, *SNQ2*, and *YOR1* genes encoding Pdr5p, Snq2p, and Yor1p efflux pumps, by prevailing transcription factors of PDR family: *PDR1* and *PDR3* and by *YAP1* and *YRR1*- transcription factors of interconnected with PDR drug resistance networks in *S. cerevisiae*.

2. Results

PDR pumps are responsible for the translocation of small molecule compounds. However, in the present study, only crude killer toxins preparations exhibited the inhibitory effect on tested S. cerevisiae strains, whereas the small-molecule fractions (below 10 kDa), obtained from killer yeast cultures did not. Also, the post-culture fluids obtained from the culture of non-killer control strain D. hansenii CLIB 545 did not inhibit the growth of S. cerevisiae, thus confirming that the differences in the susceptibility of tested S. cerevisiae strains are toxin-dependent since CLIB 545 strain was grown under the same conditions as killer strains but the post culture medium did not contain killer toxins (Figures 2–4).

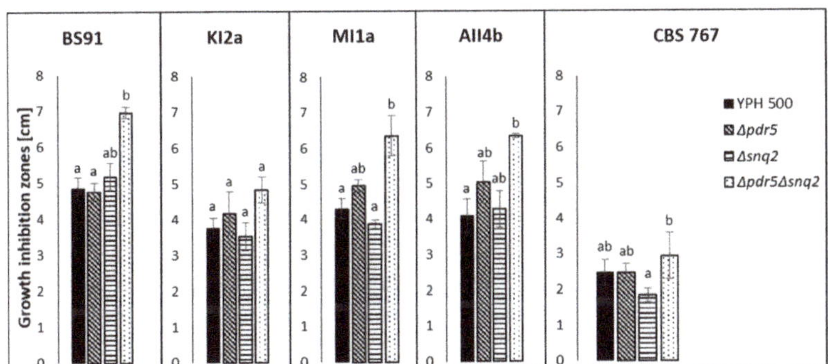

Figure 2. Growth inhibition zones of *S. cerevisiae* YPH500 wild type strain and its isogenic Δ*pdr5*; Δ*snq2* and Δ*pdr5*Δ*snq2* mutants in the presence of crude killer toxin preparations of *W. anomalus* BS91 and *D. hansenii* killer strains: KI2a, MI1a, AII4b, and CBS 767. Means and standard deviations of means are presented. The statistical significance of the differences between the means was tested separately for each killer toxin preparation (BS91, KI2a, MI1a, AII4b, and CBS 767) and marked with different letters. The study was performed with a Kruskal-Wallis one-way analysis of variance by ranks at *p* = 0.05. Means followed by different letters (a,b) are significantly different, while means marked by ab are not significantly different from groups a and b.

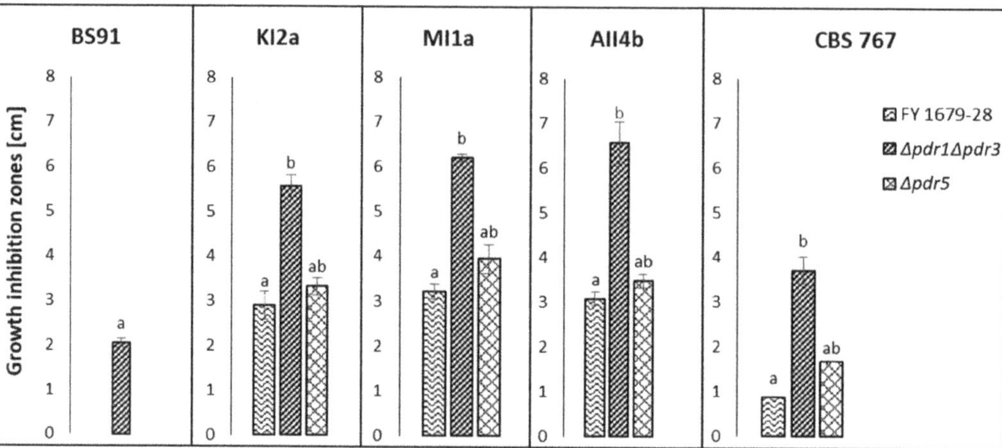

Figure 3. Growth inhibition zones of *S. cerevisiae* FY1679-28 wild type strain and its isogenic *Δpdr1Δpdr3* and *Δpdr5* mutants in the presence of crude killer toxin preparations of *W. anomalus* BS91 and *D. hansenii* killer strains: KI2a, MI1a, AII4b, and CBS 767. Means and standard deviations of means are presented. The statistical significance of the differences between the means was tested separately for each killer toxin preparation (BS91, KI2a, MI1a, AII4b, and CBS 767) and marked with different letters. The study was performed with a Kruskal-Wallis one-way analysis of variance by ranks at $p = 0.05$. Means followed by different letters (a,b) are significantly different, while means marked by ab are not significantly different from groups a and b.

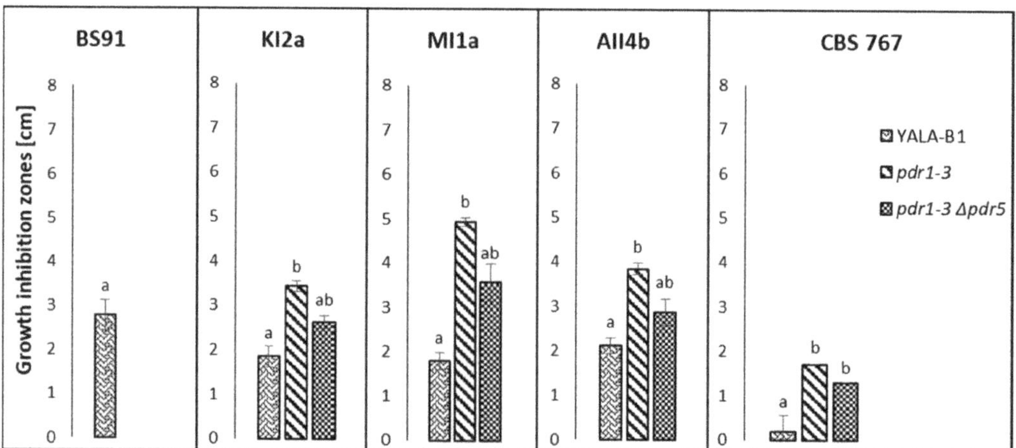

Figure 4. Growth inhibition zones of *S. cerevisiae* YALA-B1 wild type strain and its isogenic *pdr1-3* and *pdr1-3Δpdr5* mutants in the presence of crude killer toxin preparations of *W. anomalus* BS9 and *D. hansenii*: KI2a, MI1a, AII4b, and CBS 767. Means and standard deviations of means are presented. The statistical significance of the differences between the means was tested separately for each killer toxin preparation (BS91, KI2a, MI1a, AII4b, and CBS 767) and marked with different letters. The study was performed with a Kruskal-Wallis one-way analysis of variance by ranks at $p = 0.05$. Means followed by different letters (a,b) are significantly different, while means marked by ab are not significantly different from groups a and b.

The growth of *S. cerevisiae* wild type strain YPH 500 and its isogenic mutants with deleted transporter genes *PDR5*, *SNQ2*, and *PDR5* together with *SNQ2* was inhibited by crude killer toxins of all tested killer yeast strains (AII4b, KI2a, MI1a, CBS 767, and BS91 at varying levels (Figure 2)). In general, the double deletant (Δ*pdr5*Δ*snq2*) was more susceptible to killer toxins than the parental YPH 500 strain or the single knock-outs Δ*pdr5* and Δ*snq2* (Figure 2). Deletion of *PDR5* resulted in higher sensitivity to killer toxins of *D. hansenii* AII4b, KI2a, and MI1a strains, and not to the toxin of *D. hansenii* CBS 767 and *W. anomalus* BS91 (Figure 2). Interestingly, Δ*snq2* mutant was less sensitive to KI2a, MI1a, and CBS 767 in relation to isogenic wild type strain YPH 500, whereas in the presence of *W. anomalus* BS91 killer toxin, this strain was more sensitive than the wild type isogenic YPH 500 strain (Figure 2).

Double deletion of genes encoding two main transcription factors *PDR1* and *PDR3* in *S. cerevisiae* FY1679-28 mutant resulted in significantly higher sensitivity to crude killer toxins of all tested killer *D. hansenii* yeast strains: AII4b, KI2a, MI1a, CBS 767, as compared to the wild type strain (Figure 3). Interestingly, the growth of wild-type FY 1679-28 strain and its isogenic Δ*pdr5* mutant strain was not inhibited in the presence of *W. anomalus* BS9 killer toxin preparation, whereas double knock-out mutant (Δ*pdr1*Δ*pdr3*) appeared to be sensitive (Figure 3).

Unexpectedly, the sensitivity of *S. cerevisiae* hyper-resistant *pdr1-3* mutant to all toxins of *D. hansenii*, was significantly higher, as compared to that of the wild type strain YALA-B1 (Figure 4). However, deletion of the *PDR5* transporter gene in a mutant strain with retained *pdr1-3* gain-of-function mutation resulted in a phenotype that was less sensitive to all killer toxins of *D. hansenii*, still more sensitive than wild-type strain YALA-B1. On the contrary, the *pdr1-3* mutant strain was completely resistant to the killer toxin of *W. anomalus* BS91. Also, there were no growth inhibition zones observed for *pdr1-3*Δ*pdr5* isogenic mutant in the presence of BS91 toxin, whereas the growth of the control wild type YALA-B1 strain was inhibited by BS91 killer toxin (Figure 4).

3. Discussion

PDR5, *SNQ2*, and *YOR1* genes, under prevailing control of *PDR1* and *PDR3* transcription factors, encode Pdr5p, Snq2p, and Yor1p efflux pumps in the cell membrane (Figure 1). These pumps are one of the most characterized transporters in the ABC family and exhibit extremely broad, sometimes overlapping, substrate specificity. The joint and compensatory action of these efflux pumps is responsible for the extrusion of various undesired small molecules and peptides from cells [25,44]. In turn, killer toxins derived from *D. hansenii* and *W. anomalus* are proteins of the molecular mass exceeding 10 kDa, which exhibit antagonistic activity towards sensitive microorganisms upon binding to their cell wall [1,12,14,17,41]. So far, there have been no reports that PDR transporters could transport large proteins across the cell membrane. In our experiments, only a large molecule fraction containing killer toxin exhibited an inhibitory effect on tested *S. cerevisiae*, while the small molecular fraction containing peptides and other small molecules did not. Therefore, it can be inferred that PDR pumps may play a role as plasma membrane components that are involved in resistance to killer toxins. However, these pumps may not necessarily work as a killer toxin extrusion system, and may instead modulate their activity.

In the present study, the role of PDR pumps in the killer toxin effect was more pronounced for *W. anomalus* BS91 killer toxin, as compared to *D. hansenii* killer toxins (Figure 5).

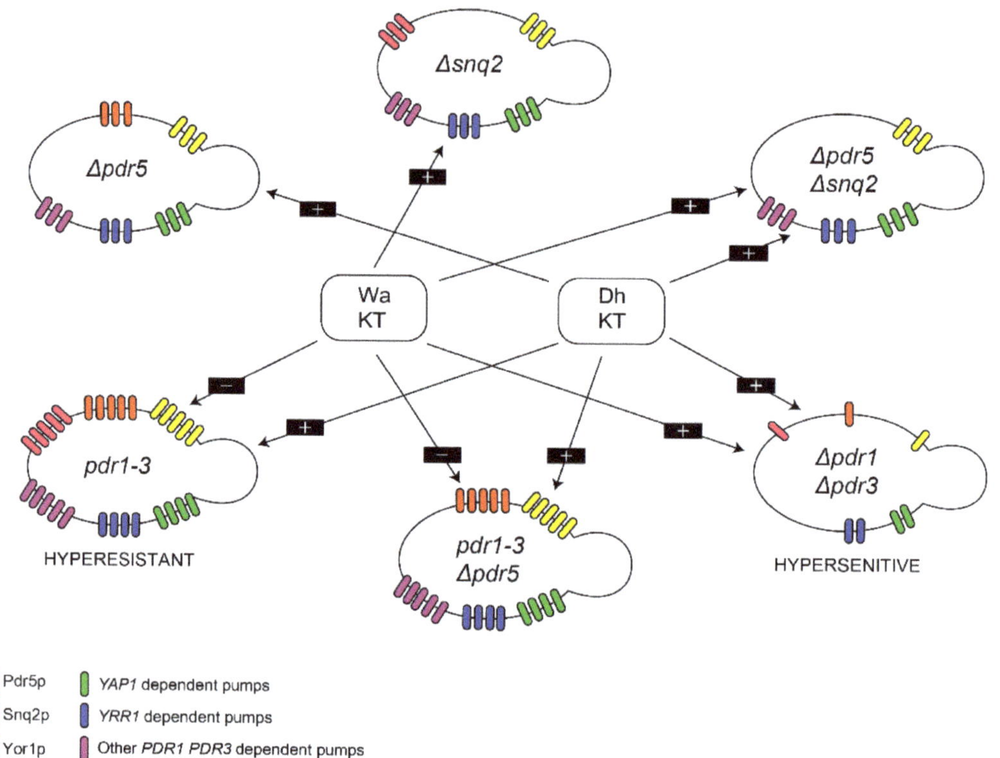

Figure 5. The schematic representation of the effect of type and quantity of transmembrane efflux pumps in *S. cerevisiae* PDR mutants on the sensitivity to *W. anomalus* (Wa) and *D. hansenii* (Dh) killer toxins (KT). The signs "+" and "−" in the black rectangle mean respectively: enhanced and decreased sensitivity to killer toxins.

The activity of *W. anomalus* BS91 killer toxin towards *S. cerevisiae* mutants with a single deletion of *SNQ2* gene as well as deletion of both *SNQ2* and *PDR5* genes was enhanced as compared to the wild type strain YPH 500 and its single *PDR5* deletant suggesting the possible role of Snq2p in resistance to BS91 killer toxin. In turn, in *S. cerevisiae pdr1-3* hyper-resistant mutant, exhibiting enhanced expression of *PDR5*, *SNQ2* and *YOR1*, the activity of BS91 killer toxin was extinguished. The deletion of the *PDR5* gene in the *pdr1-3* genetic background did not result in regaining of susceptibility of cells to BS91 killer toxin suggesting that the Pdr5p transporter may not be involved in the inhibition of BS91 killer toxin activity, since the lack of this protein did not change the resistance of *pdr1-3* mutant to BS91 killer toxin. This was also shown in FY 1679-28 wild type strain and its *PDR5* single deletant. None of them was sensitive to BS91 toxin. A double deletant in *PDR1* and *PDR3* was significantly more sensitive to BS91 killer toxin than the isogenic wild type *S. cerevisiae* strain, again indicating the involvement of Snq2p and possibly Yor1p transporters in resistance to BS91 toxin (Figure 5).

In turn, killer toxins of *D. hansenii* exhibited varying activity towards *S. cerevisiae* PDR mutants and the tendency in their performance differed from that of *W. anomalus* BS91 killer toxin. As was expected, double deletant in both major transcription factors *PDR1* and *PDR3* was significantly more susceptible to all killer toxins of *D. hansenii*. Also, the single deletion of *PDR5* and the double deletion of *PDR5* and *SNQ2* affected the activity of most *D. hansenii* killer toxins as compared to the wild type YPH 500 strain. Whereas, single deletion of *SNQ2* did not, suggesting the role of Pdr5p in resistance to *D. hansenii* killer toxins in susceptible cells. However, the gain-of-function *pdr1-3* mutant, bearing an armour of PDR efflux pumps including Pdr5p, Snq2p, and Yor1p in the plasma membrane, was unexpectedly more sensitive to all *D. hansenii* killer toxins than the wild type isogenic strain. Single deletion of the *PDR5* gene in the *pdr1-3* background resulted in a phenotype that was more sensitive to *D. hansenii* killer toxins than the wild type strain. However, it was still significantly less sensitive than the hyper-resistant *pdr1-3* mutant (Figure 5). It could be therefore inferred that the *D. hansenii* killer toxin may be either recognized and affected by Pdr5p transporter in the cell membrane with a regular density of PDR transporters, and not with an excess of efflux pumps as it is in the *pdr1-3* mutant, or that killer toxin activity may be influenced by other than tested efflux pumps in the cell membrane (possibly under the control of *YAP1* or *YRR1* transcription factors). Also, according to Kolaczkowska et al. [44], there exists a compensatory activation of one multidrug transporter upon disruption of genes encoding different multidrug transporters, which is accompanied by increased efflux of substrates specific for the activated ones. According to their studies, upon the disruption of the *PDR5* gene, the resistance to Yor1p- and Snq2p-specific substrates increases. Since *D. hansenii* killer toxin activity did not seem to depend on Snq2p in YPH 500 background, the role of Yor1p should be considered. Yor1p is, among others, the most important pump in the extrusion of Aureobasidin A– a cyclic depsipeptide produced by killer yeast *Aurobasidium pullulans*, toxic to several yeast species and filamentous fungi, such as *Botrytis cinerea*, *Monilinia* sp. and *Penicillium* sp. [45,46], towards which *D. hansenii* also exhibits antagonistic activity [16,47,48].

4. Conclusions

The present research, with the use of *S. cerevisiae* mutants in terms of PDR transporters responsible for the extrusion of substances of low molecular mass from cells, demonstrated that the killer effect may depend on the presence and density of PDR pumps in the plasma membrane of sensitive to killer toxin microorganisms, including Pdr5p and Snq2p (Figure 5). It also pointed out a new difference in killer phenomenon between killer strains of *D. hansenii* and *W. anomalus* species, where *W. anomalus* killer toxin was inhibited by the PDR transporters under the control of *PDR1* and *PDR3* transcription factors, whereas *D. hansenii* killer toxins activity was dependent on different transporters in plasma membrane than BS91 killer toxin, which could not be unambiguously defined. In previous studies, it was already proven that killer toxins derived from *D. hansenii* and *W. anomalus* exhibited varying antagonistic activity against several fungal plant pathogens through different mechanisms of action [15,16,18]. A comparative study of killer yeast strains of *D. hansenii* and *W. anomalus* species revealed their common feature, which is the recognition of β-glucans in the cell wall of the attacked microorganisms. It also pointed out a differentiating feature, which is the formation and release of antimicrobial volatile organic compounds, that was observed only in *W. anomalus*, and not in *D. hansenii* killer strains in combating pathogenic filamentous fungi [9,16]. This biocontrol mechanism of *W. anomalus* might be in turn attributed to the production of ethyl acetate [13,49] and 2-phenylethanol [50], from which the latter one may be ejected from *S. cerevisiae* cells by Pdr12p efflux pump, whose *PDR12* gene stays under the control of *WAR1* transcription factor [51,52]. In relation to biocontrol of phytopathogenic fungi, that contain MDR pumps in plasma membrane responsible for acquired resistance to chemical treatment of crops with azoles, strobilurins and succinate dehydrogenase inhibitors [53–55], this may elucidate the varying susceptibility of fungal pathogenic strains to killer toxins of *W. anomalus* and *D. hansenii*. As was shown in this study the inhibition of killer toxin produced by these yeast species relies on different PDR transporters. These findings shed new light on the perception of the mechanisms of the killer effect, understood as the antagonistic effect of killer yeasts against sensitive, often pathogenic yeast and microscopic fungi. Our novel findings constitute the foundations for research on the involvement of pleiotropic drug resistance efflux pumps of different families in the antagonistic effect of killer yeasts on sensitive microorganisms.

5. Materials and Methods

5.1. Yeast Strains

Killer strains of *D. hansenii* KI2a, MI1a and AII4b were isolated from blue-veined Rokpol cheese [56] and belong to the Culture Collection of the Department of Biotechnology and Food Microbiology at the Wroclaw University of Environmental and Life Sciences (Wroclaw, Poland). Another killer strain of *D. hansenii*, CBS 767 was obtained from CBS-KNAW Culture Collection. *W. anomalus* BS91 killer yeast was isolated from naturally fermented olive brine [57] and belongs to the Di3A Culture Collection (University of Catania, Italy). Killer phenotypes of these strains are presented in Table 1. *D. hansenii* CLIB 545, obtained from Centre International de Ressources Microbiennes (CIRM-Levures, http://www.inra.fr/cirmlevures, 16 August 2021), was used as a control non-killer, non-resistant to killer toxins strain (Table 1). *S. cerevisiae* yeast strains were chosen based on the characteristics of drug resistance associated with PDR transporters in their plasma membrane and their genotypes are presented in Table 2. YPH500 is an isogenic wild type strain to YKKB-13 strain with a single *PDR5* knock-out, YYM5 strain with a single *SNQ2* knock-out, and YYM3—with a double *PDR5* and *SNQ2* knock-outs. FY-1679-28C is an isogenic wild type strain to hyper-sensitive FY Δpdr1Δpdr3 double knock-out strain and a single *PDR5* knock-out strain FY-WT/Δpdr5-2. YALA-B1 is an isogenic strain to YALA-G4 hyper-resistant mutant (*pdr1-3*) and YZGA 278 hyper-resistant mutant (*pdr1-3*) with a single *PDR5* knock-out.

Table 1. Killer phenotypes of *D. hansenii*, *W. anomalus*, and *S. cerevisiae* isogenic yeast strains used in this study.

Yeast Strain	Killer Phenotype
D. hansenii KI2a	K$^+$R$^+$
D. hansenii MI1a	K$^+$R$^+$
D. hansenii AII4b	K$^+$R$^+$
D. hansenii CBS 767	K$^+$R$^-$
D. hansenii CLIB 545	K$^-$R$^-$
W. anomalus BS91	K$^+$R$^+$
S. cerevisiae YPH 500	K$^-$R$^{-D,-W}$
S. cerevisiae FY 1679-28	K$^-$R$^{-D,+W}$
S. cerevisiae YALA-B1	K$^-$R$^{-D,-W}$

"K"—Killer; "R"—Resistant; "D"—*D. hansenii* killer toxins; "W"—*W. anomalus* killer toxin. K$^+$R$^+$—produces a killer toxin and is resistant to killer toxins; K$^+$R$^-$—produces a killer toxin and is sensitive to killer toxins produced by other killer strains; K$^-$R$^-$—does not produce a killer toxin and is sensitive to killer toxins.

Table 2. Genotype characterization of *S. cerevisiae* strains used in this study.

Strain *	Genotype	Reference
YPH 500	*MATα ura3-52 his3-Δ200 leu2-Δ1 trp1-Δ63 lys2-801^{amber} ade2-101^{ochre}*	[34]
YKKB-13	*MATα ura3-52 his3-Δ200 leu2-Δ1 trp1-Δ63 lys2-801 ade2-101 pdr5Δ::TRP1*	[58]
YYM 5	*MATα ura3-52 his3-Δ200 leu2-Δ1 trp1-Δ63 lys2-801 ade2-101 snq2Δ::hisG*	[34]
YYM 3	*MATα ura3-52 his3-Δ200 leu2-Δ1 trp1-Δ63 lys2-801 ade2-101 pdr5Δ::TRP1 snq2Δ::hisG*	[35]

Table 2. *Cont.*

Strain *	Genotype	Reference
FY-1679-28C	*MATa ura3-53, leu2-Δ1, trp1-Δ63, his3Δ200, GAL2⁺*	[30]
FY Δpdr1Δpdr3	*MATa ura3-53, leu2-Δ1, trp1-Δ63, his3Δ200, GAL2⁺ pdr1-Δ2::TRP1 pdr3-1Δ::HIS3*	[30]
FY-WT/Δpdr5-2	*MATa ura3-53, leu2-Δ1, trp1-Δ63, his3Δ200, GAL2⁺ pdr5Δ::URA3*	[59]
YALA-B1	*MATa ura3-52 leu2-3,112 his3-11,115 trp1-1*	[35]
YALA-G4	*MATa ura3-52 leu2-3,112 his3-11,115 trp1-1 pdr1-3*	[35]
YZGA 278	*MATa ura3-52, leu2-3,112 his3-11,115 trp1-1 pdr1-3 pdr5Δ::hisG*	[59]

* The strains in bold are parental strains to the mutants listed below them in the table.

5.2. Killer Toxin Production

Yeasts were cultured on yeast peptone dextrose agar plates (YPDA; yeast extract, 10 g; peptone, 10 g; dextrose, 20 g; agar, 20 g per litre of distilled water) at 25 °C for 48 h. Then, the single colony biomass of each yeast strain was inoculated to 50 mL of YPD broth of pH adjusted to 4.5 with McIlvaine buffer and incubated either at 14 °C for 48 h (*D. hansenii* strains) or 25 °C (*W. anomalus*) for 24 h on a rotary shaker at 160 rpm to provide favourable conditions for killer toxin production. Yeast cells from liquid cultures were pelleted at 8000 g for 10 min. The supernatant was filtrated using a sterile syringe filter with a 0.22 μm pore size (Merck Millipore, Burlington, MA, USA, Milex-GP 33 mm, PES membrane). The obtained post-culture media served as crude toxin preparations. In addition, the small-molecule fractions (<10 kDa) were separated from the post-culture media using a filter with an appropriate cut-off point (Merck Millipore, Amicon® Ultra-15, 10 kDa). Both crude killer toxin preparations and small-molecule fractions were used in the diffusion assay.

5.3. Killer Toxin Diffusion Assay

The killer activity of *D. hansenii* and *W. anomalus* toxins was tested in agar diffusion bioassay according to the method described by Żarowska [12]. Twenty-five mL of YPDA medium, supplemented with 0.03% (*w/v*) methylene blue and 4% (*w/v*) of NaCl, was adjusted to pH 4.5 with McIlvaine buffer and used in the assay for the evaluation of killer toxin activity. Fifty μL of each tested sample (i) crude killer toxin preparation, (ii) small-molecule fraction (<10 kDa), was added into wells, sterilely cut in YPDA plates containing *S. cerevisiae* cells at a final concentration of 5×10^5 per mL of the assay medium (Table 2). The zones of *S. cerevisiae* growth inhibition [cm] around wells were measured after 48 h of incubation at 20 °C. The experiment was repeated three times with three replicates for each tested sample.

5.4. Statistical Analysis

In all repeated experiments, the arithmetic means were calculated and analyzed using Kruskal-Wallis one-way analysis of variance by ranks at $p = 0.05$. Post-culture media derived from the cultures of *D. hansenii* and *W. anomalus* were treated as individual preparations, the activity of which was tested against *S. cerevisiae* strains. The significant differences in growth inhibition zones between each *S. cerevisiae* wild type strain (YPH 500; FY-1679-28C and YALA-B1) and its isogenic PDR-mutants (YKKB-13, YYM 5, YMM3; FYΔpdr1Δpdr3, FY-WT/Δpdr5-2 and YALA-G4, YALA 278) were analyzed separately for each killer toxin preparation to demonstrate the role of PDR pumps on the toxic effect of all tested killer toxin preparations. The strength of the killer effect was not compared among killer strains. The results of mean killer activity of each preparation in relation to the tested *S. cerevisiae* strains, significantly different from each other, were grouped into homogenous groups a and b, while the results without statistically significant differences were assigned to the ab group, which was not significantly different from groups a or b. Data from experiments were analyzed using Statistica package software (Version 12; Statsoft Inc., Tulsa, OK, USA).

Author Contributions: Conceptualization, M.C.; funding acquisition, X.P. and B.Ż.; investigation, M.C.; Resources, X.P., C.R. and B.Ż.; writing—original draft, M.C.; writing—review & editing, M.C., X.P., B.Ż. and C.R. All authors have read and agreed to the published version of the manuscript.

Funding: This work was funded by research grant POIG.01.03.01-02-080/12 "The use of yeast *Y. lipolytica* and *D. hansenii*, their enzymes and toxins for application in industry and agriculture" co-financed by the European Union from the European Regional Development Found. The APC/BPC is financed/co-financed by Wroclaw University of Environmental and Life Sciences.

Institutional Review Board Statement: Not applicable.

Informed Consent Statement: Not applicable.

Data Availability Statement: The dataset used and/or analyzed during the current study are publicly available.

Acknowledgments: The authors would like to thank Serge Casaregola for kindly sharing *D. hansenii* CLIB 545 strain from the deposit of Centre International de Ressources Microbiennes (CIRM-Levures, http://www.inra.fr/cirmlevures).

Conflicts of Interest: The authors declare no conflict of interest.

References

1. Al-Qaysi, S.A.S.; Al-Haideri, H.; Thabit, Z.A.; Al-Kubaisy, W.H.A.A.-R.; Ibrahim, J.A.A.-R. Production, Characterization, and Antimicrobial Activity of Mycocin Produced by *Debaryomyces hansenii* DSMZ70238. *Int. J. Microbiol.* **2017**, *2017*, 2605382. [CrossRef]
2. Banjara, N.; Nickerson, K.W.; Suhr, M.J.; Hallen-Adams, H.E. Killer toxin from several food-derived *Debaryomyces hansenii* strains effective against pathogenic *Candida* yeasts. *Int. J. Food Microbiol.* **2016**, *222*, 23–29. [CrossRef]
3. Breuer, U.; Harms, H. *Debaryomyces hansenii*—An extremophilic yeast with biotechnological potential. *Yeast* **2006**, *23*, 415–437. [CrossRef]
4. Droby, S.; Chalutz, E.; Wilson, C.L.; Wisniewski, M. Characterization of the biocontrol activity of *Debaryomyces hansenii* in the control of *Penicillium digitatum* on grapefruit. *Can. J. Microbiol.* **1989**, *35*, 794–800. [CrossRef]
5. Hernández-Montiel, L.G.; Ochoa, J.L.; Troyo-Diéguez, E.; Larralde-Corona, C.P. Biocontrol of postharvest blue mold (*Penicillium italicum* Wehmer) on Mexican lime by marine and citrus *Debaryomyces hansenii* isolates. *Postharvest Biol. Technol.* **2010**, *56*, 181–187. [CrossRef]
6. Hernandez-Montiel, L.G.; Gutierrez-Perez, E.D.; Murillo-Amador, B.; Vero, S.; Chiquito-Contreras, R.G.; Rincon-Enriquez, G. Mechanisms employed by *Debaryomyces hansenii* in biological control of anthracnose disease on papaya fruit. *Postharvest Biol. Technol.* **2018**, *139*, 31–37. [CrossRef]
7. Kharchoufi, S.; Parafati, L.; Licciardello, F.; Muratore, G.; Hamdi, M.; Cirvilleri, G.; Restuccia, C. Edible coatings incorporating pomegranate peel extract and biocontrol yeast to reduce Penicillium digitatum postharvest decay of oranges. *Food Microbiol.* **2018**, *74*, 107–112. [CrossRef]
8. Lima, J.R.; Gondim, D.M.F.; Oliveira, J.T.A.; Oliveira, F.S.A.; Gonçalves, L.R.B.; Viana, F.M.P. Use of killer yeast in the management of postharvest papaya anthracnose. *Postharvest Biol. Technol.* **2013**, *83*, 58–64. [CrossRef]
9. Parafati, L.; Vitale, A.; Restuccia, C.; Cirvilleri, G. Biocontrol ability and action mechanism of food-isolated yeast strains against *Botrytis cinerea* causing post-harvest bunch rot of table grape. *Food Microbiol.* **2015**, *47*, 85–92. [CrossRef]
10. Platania, C.; Restuccia, C.; Muccilli, S.; Cirvilleri, G. Efficacy of killer yeasts in the biological control of *Penicillium digitatum* on Tarocco orange fruits (*Citrus sinensis*). *Food Microbiol.* **2012**, *30*, 219–225. [CrossRef]
11. Prista, C.; Michan, C.; Miranda, I.; Ramos, J. The halotolerant *Debaryomyces hansenii*, the Cinderella of non-conventional yeasts. *Yeast* **2016**, *33*, 523–533. [CrossRef] [PubMed]
12. Żarowska, B.; Wojtatowicz, M.; Połomska, X.; Juszczyk, P.; Chrzanowska, J. Factors affecting killer activity of some yeast species occurring in Rokpol cheese. *Folia Microbiol.* **2004**, *49*, 713–717. [CrossRef] [PubMed]
13. Contarino, R.; Brighina, S.; Fallico, B.; Cirvilleri, G.; Parafati, L.; Restuccia, C. Volatile organic compounds (VOCs) produced by biocontrol yeasts. *Food Microbiol.* **2019**, *82*, 70–74. [CrossRef]
14. Çorbacı, C.; Uçar, F.B. Purification, characterization and in vivo biocontrol efficiency of killer toxins from *Debaryomyces hansenii* strains. *Int. J. Biol. Macromol.* **2018**, *119*, 1077–1082. [CrossRef]
15. Czarnecka, M.; Żarowska, B.; Połomska, X.; Restuccia, C.; Cirvilleri, G. Role of biocontrol yeasts *Debaryomyces hansenii* and *Wickerhamomyces anomalus* in plants' defence mechanisms against *Monilinia fructicola* in apple fruits. *Food Microbiol.* **2019**, *83*, 1–8. [CrossRef]
16. Grzegorczyk, M.; Żarowska, B.; Restuccia, C.; Cirvilleri, G. Postharvest biocontrol ability of killer yeasts against *Monilinia fructigena* and *Monilinia fructicola* on stone fruit. *Food Microbiol.* **2017**, *61*, 93–101. [CrossRef] [PubMed]
17. Guo, F.-J.; Ma, Y.; Xu, H.-M.; Wang, X.-H.; Chi, Z.-M. A novel killer toxin produced by the marine-derived yeast *Wickerhamomyces anomalus* YF07b. *Antonie Van Leeuwenhoek* **2013**, *103*, 737–746. [CrossRef] [PubMed]

18. Parafati, L.; Vitale, A.; Polizzi, G.; Restuccia, C.; Cirvilleri, G. Understanding the mechanism of biological control of postharvest phytopathogenic moulds promoted by food isolated yeasts. *Acta Hortic.* **2016**, *1144*, 93–100. [CrossRef]
19. Rogers, D.; Bevan, E.A. Group Classification of Killer Yeasts Based on Cross-reactions between Strains of Different Species and Origin. *J. Gen. Microbiol.* **1978**, *155*, 199–202. [CrossRef]
20. Woods, D.R.; Ross, I.W.; Hendry, D.A. A New Killer Factor Produced by a Killer/Sensitive Yeast Strain. *J. Gen. Microbiol.* **1974**, *81*, 285–289. [CrossRef]
21. El-Banna, A.A.; Malak, A.; El-Sahn; Shehata, M.G. Yeasts Producing Killer Toxins: An Overview. *Alex. J. Food. Sci. Technol.* **2011**, *2*, 41–53.
22. Hodgson, V.J.; Button, D.; Walker, G.M. Anti-*Candida* activity of a novel killer toxin from the yeast *Williopsis mrakii*. *Microbiology* **1995**, *141*, 2003–2012. [CrossRef] [PubMed]
23. Sugisaki, Y.; Gunge, N.; Sakaguchi, K.; Yamasaki, M.; Tamura, G. Characterization of a novel killer toxin encoded by a double-stranded linear DNA plasmid of *Kluyveromyces lactis*. *Eur. J. Biochem.* **1984**, *141*, 241–245. [CrossRef]
24. Woods, D.R.; Bevan, E.A. Studies on the Nature of the Killer Factor Produced by *Saccharomyces cerevisiae*. *Microbiology* **1968**, *51*, 115–126. [CrossRef]
25. Kolaczkowska, A.; Goffeau, A. Regulation of pleiotropic drug resistance in yeast. *Drug Resist. Updates* **1999**, *2*, 403–414. [CrossRef] [PubMed]
26. Golin, J.; Ambudkar, S.V. The multidrug transporter Pdr5 on the 25th anniversary of its discovery: An important model for the study of asymmetric ABC transporters. *Biochem. J.* **2015**, *467*, 353–363. [CrossRef] [PubMed]
27. Prasad, R.; Gaur, N.A.; Gaur, M.; Komath, S.S. Efflux pumps in drug resistance of *Candida. Infect. Disord. Drug Targets* **2006**, *6*, 69–83. [CrossRef] [PubMed]
28. Balzi, E.; Chen, W.; Ulaszewskit, S.; Capieaux, E.; Goffeaust, A. The Multidrug Resistance Gene PDRl from *Saccharomyces cerevisiae*. *J. Biol. Chem.* **1987**, *262*, 16871–16879. [CrossRef]
29. Rogers, B.; Decottignies, A.; Kolaczkowski, M.; Carvajal, E.; Balzi, E.; Goffeau, A. The pleiotropic drug ABC transporters from *Saccharomyces cerevisiae*. *J. Mol. Microbiol. Biotechnol.* **2001**, *3*, 207–214.
30. Delaveau, T.; Delahodde, A.; Carvajal, E.; Subik, J.; Jacq, C. PDR3, a new yeast regulatory gene, is homologous to PDR1 and controls the multidrug resistance phenomenon. *Mol. Gen. Genet. MGG* **1994**, *244*, 501–511. [CrossRef]
31. Carvajal, E.; van den Hazel, H.B.; Cybularz-Kolaczkowska, A.; Balzi, E.; Goffeau, A. Molecular and phenotypic characterization of yeast PDR1 mutants that show hyperactive transcription of various ABC multidrug transporter genes. *Mol. Gen. Genet.* **1997**, *256*, 406–415. [CrossRef] [PubMed]
32. Decottignies, A.; Kolaczkowski, M.; Balzi, E.; Goffeau, A. Solubilization and characterization of the overexpressed PDR5 multidrug resistance nucleotide triphosphatase of yeast. *J. Biol. Chem.* **1994**, *269*, 12797–12803. [CrossRef]
33. Decottignies, A.; Lambert, L.; Catty, P.; Degand, H.; Epping, E.A.; Moye-Rowley, W.S.; Balzi, E.; Goffeau, A. Identification and characterization of SNQ2, a new multidrug ATP binding cassette transporter of the yeast plasma membrane. *J. Biol. Chem.* **1995**, *270*, 18150–18157. [CrossRef] [PubMed]
34. Mahé, Y.; Lemoine, Y.; Kuchler, K. The ATP Binding Cassette Transporters Pdr5 and Snq2 of *Saccharomyces cerevisiae* Can Mediate Transport of Steroids In Vivo. *J. Biol. Chem.* **1996**, *41*, 25167–25172. [CrossRef] [PubMed]
35. Mahé, Y.; Parle-McDermott, A.; Nourani, A.; Delahodde, A.; Lamprecht, A.; Kuchler, K. The ATP-binding cassette multidrug transporter Snq2 of *Saccharomyces cerevisiae*: A novel target for the transcription factors Pdr1 and Pdr3. *Mol. Microbiol.* **1996**, *20*, 109–117. [CrossRef]
36. de Waard, M.A.; Andrade, A.C.; Hayashi, K.; Schoonbeek, H.; Stergiopoulos, I.; Zwiers, L.-H. Impact of fungal drug transporters on fungicide sensitivity, multidrug resistance and virulence. *Pest Manag. Sci.* **2006**, *62*, 195–207. [CrossRef]
37. Hahn, M. The rising threat of fungicide resistance in plant pathogenic fungi: *Botrytis* as a case study. *J. Chem. Biol.* **2014**, *7*, 133–141. [CrossRef]
38. Hershkovitz, V.; Sela, N.; Taha-Salaime, L.; Liu, J.; Rafael, G.; Kessler, C.; Aly, R.; Levy, M.; Wisniewski, M.; Droby, S. De-novo assembly and characterization of the transcriptome of *Metschnikowia fructicola* reveals differences in gene expression following interaction with *Penicillium digitatum* and grapefruit peel. *BMC Genom.* **2013**, *14*, 168. [CrossRef]
39. Nuruzzaman, M.; Zhang, R.; Cao, H.-Z.; Luo, Z.-Y. Plant Pleiotropic Drug Resistance Transporters: Transport Mechanism, Gene Expression, and Function. *J. Integr. Plant Biol.* **2014**, *56*, 729–740. [CrossRef]
40. Izgu, F.; Altinbay, D. Isolation and Characterization of the K5-Type Yeast Killer Protein and Its Homology with an Exo-β-1,3-glucanase. *Biosci. Biotechnol. Biochem.* **2004**, *68*, 685–693. [CrossRef]
41. Muccilli, S.; Wemhoff, S.; Restuccia, C.; Meinhardt, F. Exoglucanase-encoding genes from three *Wickerhamomyces anomalus* killer strains isolated from olive brine. *Yeast* **2013**, *30*, 33–43. [CrossRef]
42. Santos, A.; Marquina, D.; Barroso, J.; Peinado, J.M. (1→6)-β-D-glucan as the cell wall binding site for *Debaryomyces hansenii* killer toxin. *Lett. Appl. Microbiol.* **2002**, *34*, 95–99. [CrossRef]
43. Żarowska, B. *Biosynthesis and Characteristics of Debaryomyces hansenii Killer Toxin, Monograph ed.*; University of Environmental and Life Sciences Publishing House: Tokyo, Japan, 2012.
44. Kolaczkowska, A.; Kolaczkowski, M.; Goffeau, A.; Moye-Rowley, W.S. Compensatory activation of the multidrug transporters Pdr5p, Snq2p, and Yor1p by Pdr1p in *Saccharomyces cerevisiae*. *FEBS Lett.* **2008**, *582*, 977–983. [CrossRef]

45. Ogawa, A.; Hashida-Okado, T.; Endo, M.; Yoshioka, H.; Tsuruo, T.; Takesako, K.; Kato, I. Role of ABC Transporters in Aureobasidin A Resistance. *Antimicrob. Agents Chemother.* **1998**, *42*, 755–761. [CrossRef]
46. Takesako, K.; Ikai, K.; Haruna, F.; Endo, M.; Shimanaka, K.; Sono, E.; Nakamura, T.; Kato, I.; Yamaguchi, H. Aureobasidins, new antifungal antibiotics. Taxonomy, fermentation, isolation, and properties. *J. Antibiot.* **1991**, *44*, 919–924. [CrossRef]
47. Chalutz, E. Postharvest biocontrol of green and blue mold and sour rot of citrus fruit by *Debaryomyces hansenii*. *Plant Dis.* **1990**, *74*, 135–137. [CrossRef]
48. Santos, A.; Sánchez, A.; Marquina, D. Yeasts as biological agents to control *Botrytis cinerea*. *Microbiol. Res.* **2004**, *159*, 331–338. [CrossRef]
49. Hua, S.S.T.; Beck, J.J.; Sarreal, S.B.L.; Gee, W. The major volatile compound 2-phenylethanol from the biocontrol yeast, *Pichia anomala*, inhibits growth and expression of aflatoxin biosynthetic genes of *Aspergillus flavus*. *Mycotoxin Res.* **2014**, *30*, 71–78. [CrossRef]
50. Druvefors, U.A.; Passoth, V.; Schnurer, J. Nutrient Effects on Biocontrol of *Penicillium roqueforti* by *Pichia anomala* J121 during Airtight Storage of Wheat. *Appl. Environ. Microbiol.* **2005**, *71*, 1865–1869. [CrossRef]
51. Hazelwood, L.A.; Tai, S.L.; Boer, V.M.; de Winde, J.H.; Pronk, J.T.; Daran, J.M. A new physiological role for Pdr12p in *Saccharomyces cerevisiae*: Export of aromatic and branched-chain organic acids produced in amino acid catabolism. *FEMS Yeast Res.* **2006**, *6*, 937–945. [CrossRef]
52. Kren, A.; Mamnun, Y.M.; Bauer, B.E.; Schüller, C.; Wolfger, H.; Hatzixanthis, K.; Mollapour, M.; Gregori, C.; Piper, P.; Kuchler, K. War1p, a novel transcription factor controlling weak acid stress response in yeast. *Mol. Cell. Biol.* **2003**, *23*, 1775–1785. [CrossRef]
53. Morschhäuser, J. Regulation of multidrug resistance in pathogenic fungi. *Fungal Genet. Biol.* **2010**, *47*, 94–106. [CrossRef] [PubMed]
54. Price, C.L.; Parker, J.E.; Warrilow, A.G.; Kelly, D.E.; Kelly, S.L. Azole fungicides—Understanding resistance mechanisms in agricultural fungal pathogens. *Pest Manag. Sci.* **2015**, *71*, 1054–1058. [CrossRef] [PubMed]
55. Sanglard, D. Emerging Threats in Antifungal-Resistant Fungal Pathogens. *Front. Med.* **2016**, *3*, 11. [CrossRef] [PubMed]
56. Wojtatowicz, M.; Chrzanowska, J.; Juszczyk, P.; Skiba, A.; Gdula, A. Identification and biochemical characteristics of yeast microflora of Rokpol cheese. *Int. J. Food Microbiol.* **2001**, *69*, 135–140. [CrossRef]
57. Muccilli, S.; Caggia, C.; Randazzo, C.L.; Restuccia, C. Yeast dynamics during the fermentation of brined green olives treated in the field with kaolin and Bordeaux mixture to control the olive fruit fly. *Int. J. Food Microbiol.* **2011**, *148*, 15–22. [CrossRef]
58. Bissinger, P.H.; Kuchler, K. Molecular cloning and expression of the *Saccharomyces cerevisiae* STS1 gene product. A yeast ABC transporter conferring mycotoxin resistance. *J. Biol. Chem.* **1994**, *11*, 4180–4186. [CrossRef]
59. Kugler, G.K.; Jandric, Z.; Beyer, R.; Klopf, E.; Glaser, W.; Lemmens, M.; Shams, M.; Mayer, K.; Adam, G.; Schuller, C. Ribosome quality control is a central protection mechanism for yeast exposed to deoxynivalenol and trichothecin. *BMC Genom.* **2016**, *17*, 417. [CrossRef]

Article

Gambierol Blocks a K⁺ Current Fraction without Affecting Catecholamine Release in Rat Fetal Adrenomedullary Cultured Chromaffin Cells

Evelyne Benoit [1,2], Sébastien Schlumberger [2], Jordi Molgó [1,2,*], Makoto Sasaki [3], Haruhiko Fuwa [4] and Roland Bournaud [2,*]

[1]   Service d'Ingénierie Moléculaire pour la Santé (SIMoS), Département Médicaments et Technologies pour la Santé (DMTS), Institut des Sciences du Vivant Frédéric Joliot, Université Paris-Saclay, CEA, INRAE, ERL CNRS 9004, F-91191 Gif-sur-Yvette, France; evelyne.benoit@cea.fr

[2]   CNRS, Laboratoire de Neurobiologie Cellulaire et Moléculaire-UPR 9040, F-91198 Gif-sur-Yvette, France; sebastienschlumberger@hotmail.com

[3]   Graduate School of Life Sciences, Tohoku University, Sendai 980-8577, Japan; masasaki@tohoku.ac.jp

[4]   Department of Applied Chemistry, Faculty of Science and Engineering, Chuo University, Tokyo 112-8551, Japan; hfuwa.50m@g.chuo-u.ac.jp

*   Correspondence: jordi.molgo@cea.fr (J.M.); bournaud.roland@gmail.com (R.B.)

Abstract: Gambierol inhibits voltage-gated K⁺ (K_V) channels in various excitable and non-excitable cells. The purpose of this work was to study the effects of gambierol on single rat fetal (F19–F20) adrenomedullary cultured chromaffin cells. These excitable cells have different types of K_V channels and release catecholamines. Perforated whole-cell voltage-clamp recordings revealed that gambierol (100 nM) blocked only a fraction of the total outward K⁺ current and slowed the kinetics of K⁺ current activation. The use of selective channel blockers disclosed that gambierol did not affect calcium-activated K⁺ (K_Ca) and ATP-sensitive K⁺ (K_ATP) channels. The gambierol concentration necessary to inhibit 50% of the K⁺ current-component sensitive to the polyether (IC_{50}) was 5.8 nM. Simultaneous whole-cell current-clamp and single-cell amperometry recordings revealed that gambierol did not modify the membrane potential following 11s depolarizing current-steps, in both quiescent and active cells displaying repetitive firing of action potentials, and it did not increase the number of exocytotic catecholamine release events, with respect to controls. The subsequent addition of apamin and iberiotoxin, which selectively block the K_Ca channels, both depolarized the membrane and enhanced by 2.7 and 3.5-fold the exocytotic event frequency in quiescent and active cells, respectively. These results highlight the important modulatory role played by K_Ca channels in the control of exocytosis from fetal (F19–F20) adrenomedullary chromaffin cells.

Keywords: fetal adrenomedullary chromaffin cell; gambierol; potassium currents; calcium-activated K⁺ channels; ATP-sensitive K⁺ channels; catecholamine release

Key Contribution: The study enhances the knowledge we have on the several types of K⁺ channels contributing to the total outward current of rat fetal (F19–F20) chromaffin cells lacking splanchnic innervation, and shows that gambierol, a dinoflagellate polyether toxin, affects only a K⁺ current fraction, distinct to the K_Ca and K_ATP currents, and has no action on Ca²⁺-dependent catecholamine secretion. The results further highlight the key modulatory role played by K_Ca currents in the control of exocytosis at this fetal stage.

Citation: Benoit, E.; Schlumberger, S.; Molgó, J.; Sasaki, M.; Fuwa, H.; Bournaud, R. Gambierol Blocks a K⁺ Current Fraction without Affecting Catecholamine Release in Rat Fetal Adrenomedullary Cultured Chromaffin Cells. *Toxins* **2022**, *14*, 254. https://doi.org/10.3390/toxins14040254

Received: 4 February 2022
Accepted: 30 March 2022
Published: 2 April 2022

Publisher's Note: MDPI stays neutral with regard to jurisdictional claims in published maps and institutional affiliations.

1. Introduction

Marine dinoflagellates are the source of a well-documented and distinctive group of bioactive polycyclic ether natural products, showing numerous associated ether rings in a ladder shape. These ladder-shaped polyethers are mainly found in marine microorganisms and are considered as secondary dinoflagellate metabolites constituting a rich source of complex compounds (for reviews see [1–3]). The demanding steps for their isolation and purification, and the small quantities obtained, have severely limited their structural and bioactivity characterization. Fortunately, some of these complex polyether molecules have been amenable to total organic chemical synthesis (for reviews see [4–6]). Among these marine polyethers, gambierol and analogs have been synthesized by means of diverse syntheses strategies [7–10], allowing detailed studies of their actions in a number of biology models.

Gambierol, distinguished by a transfused octacyclic polyether core (Figure 1), was the first toxin isolated and characterized from cultured *Gambierdiscus toxicus* dinoflagellates gathered from the Rangiroa peninsula in French Polynesia [11,12]. The genus' *Gambierdiscus* and *Fukuyoa* produce numerous ladder polycyclic ether compounds, including the well-identified ciguatoxins. This family of toxins is responsible for ciguatera poisoning, a seafood-borne disease resulting from the consumption of fish from tropical or temperate waters, or marine invertebrates that have bioaccumulated ciguatoxins (reviewed in [13–16]). It has been suggested that gambierol participates in ciguatera fish poisoning, but direct proof for this assumption has not yet been given, and to the best of our knowledge, the polyether toxin has not yet been identified in ciguateric fish, maybe because it is present at very low concentrations.

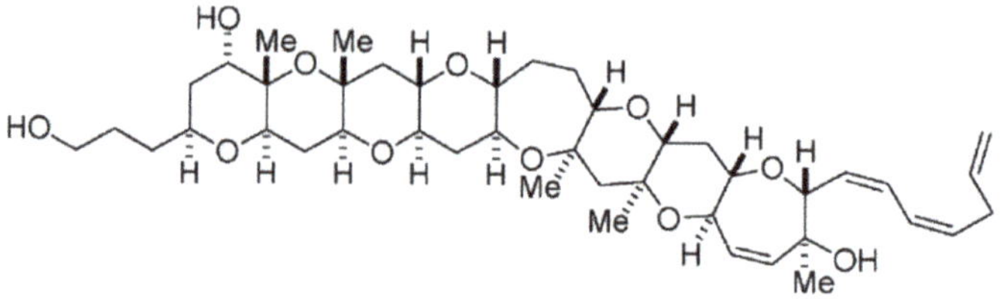

Figure 1. Chemical structure of gambierol.

Nanomolar concentrations of gambierol and analogs inhibit voltage-gated K^+ (K_V) channels in various excitable cells [17–19], including cells expressing $K_V1.1$–$K_V1.5$ channels [20–22] or $K_V3.1$ channels [23], and in frog and mouse motor nerve terminals [24,25].

Adrenomedullary chromaffin cells are known to generate action potentials [26], and to display voltage-gated Na^+ (Na_V) channels that are sensitive to tetrodotoxin (TTX) [27], and they are also involved in regulating the firing rate of action potentials [28]. The abundant diversity of K_V channels in chromaffin cells highlights their fundamental role in the control of the electrical properties of these cells, including the speed of action potential repolarization, the duration of the after-hyperpolarization, the firing rate, and the resting membrane potential (reviewed in [29]). In addition, individual chromaffin cells, depending on the animal species and stage of development, express distinct subtypes of voltage-gated Ca^{2+} (Ca_V) channels, including low-voltage-activated T-type (Ca_V3) channels [30], high-voltage-activated channels comprising L-type ($Ca_V1.2$ and $Ca_V1.3$), P/Q-type ($Ca_V2.1$), N-type ($Ca_V2.2$) [31] and R-type ($Ca_V2.3$) channels (for reviews see [29,32,33]).

The adrenomedullary chromaffin cells secrete catecholamines in response to various stressors, including acute hypoxia [34]. In adult mammals, the catecholamine secretion is triggered by the sympathetic nervous system that supplies, via the splanchnic nerve, the cholinergic innervation to the cells. In the perinatal period, the splanchnic innervation in the adrenal gland is either immature or absent and remains non-functional until the first postnatal week. At birth, hypoxia triggers adrenal catecholamine secretion by a non-neurogenic mechanism that is vital for adapting to the extra-uterine life. The replacement of the non-neurogenic adrenomedullary responses by the neurogenic mechanism is accurately connected to the beginning of the splanchnic secretory-induced nerve function and occurs at the postnatal period [34–36]. Interestingly, in rat fetal cells, hypoxia-induced catecholamine release was reported to be shaped by modulating the functioning of calcium-activated potassium (K_{Ca}) channels, and ATP-sensitive potassium (K_{ATP}) channels [37,38].

The aims of the present study, on cultured rat fetal adrenal medulla chromaffin (AMC) cells, were firstly, to determine the action of gambierol on outward K^+ currents, using perforated whole-cell voltage-clamp recordings, and secondly, to investigate whether gambierol by itself affects the release of catecholamines using simultaneous current-clamp and single-cell amperometry recordings.

2. Results

2.1. Effects of Gambierol on Outward K^+ Current in Rat Fetal Adrenomedullary Chromaffin Cells

In cultured AMC cells, the action of gambierol on K^+ currents was studied using the perforated whole-cell voltage-clamp configuration. Cells were continuously bathed in a standard physiological solution containing 1 μM tetrodotoxin (TTX) to block the Na_V channels, and depolarizing steps (90-ms duration) from a holding potential of -70 mV were delivered to specified membrane potentials. To study the involvement of different K^+ current components in the total outward K^+ current, we used the peptide neurotoxins apamin and iberiotoxin which selectively block the small-conductance Ca^{2+}-activated K^+ (SK_{Ca}) channels ($K_{Ca}2.1$–2.3, SK1-3 isoforms) [39] and the large-conductance Ca^{2+}-activated K^+ (BK_{Ca}) channels ($K_{Ca}1.1$, Slo1) [40], respectively. In addition, glibenclamide was used to target the ATP-sensitive K^+ (K_{ATP}) channel isoforms in neonatal AMC cells [37,41].

As shown on the superimposed recordings of Figure 2(Aa), the addition of apamin (400 nM), iberiotoxin (100 nM) and glibenclamide (200 μM) to the extracellular medium consistently reduced the total outward K^+ current. Under these conditions, the remaining K^+ current was further reduced by adding gambierol (100 nM) and was completely inhibited by further addition of the voltage-gated K^+ channel inhibitors tetraethylammonium chloride (TEA, 10 mM) and 3,4-diaminopyridine (3,4-DAP, 500 μM) to the external solution (Figure 2(Aa)). As depicted in Figure 2(Ba), gambierol (100 nM) only partly inhibited the total K^+ current when added before the K_{Ca} and K_{ATP} blockers and TEA and 3,4-DAP.

The normalized current-voltage relationships of steady-state K^+ current amplitudes in the presence of the various agents studied are shown in Figure 2(Ab,Bb). The columns relating the percentage of K^+ current block induced by the pharmacological agents used, are represented in Figure 2(Ac,Bc). Interestingly, the percentage of K^+ current inhibition by apamin, iberiotoxin and glibenclamide did not differ significantly in the presence ($\Delta 2 = 58.32 \pm 2.82\%$; $n = 4$), and in the absence ($\Delta 1 = 57.14 \pm 4.25\%$; $n = 4$) of 100 nM gambierol ($p = 0.8481$) (Figure 2(Ac,Bc)).

Taken as a whole, these results indicate that (i) AMC cells are endowed with several types of K^+ channels contributing to the total outward current, and (ii) gambierol only partly inhibited the total K^+ current when added after or before the K_{Ca} and K_{ATP} blockers. Therefore, these results strongly suggest that the polyether toxin affects neither the K_{Ca} nor K_{ATP} channels.

Because before the addition of K_{Ca} and K_{ATP} blockers the fraction of the total current blocked by gambierol is too low (approximatively 10%) to allow a proper detection of change in the voltage- and time-dependence of the K^+ current activation, additional studies, using the perforated whole-cell voltage-clamp configuration, were performed in AMC cells bathed with an external solution containing 1 μM TTX to block the Na_V channels, 200 μM glibenclamide to block the K_{ATP} channels, and 1 mM Cd^{2+} to inhibit the K_{Ca} channel activation via Ca^{2+} influx through the Ca_V channels. Under these conditions, the voltage-dependence of K^+ current activation was determined in the absence and presence of gambierol. As shown in Figure 3, the gambierol (100 nM) induced a slight (approximately 5 mV), but significant ($p = 0.028$), negative shift of the voltage-dependence of the K^+ current activation. Hence, the voltages corresponding to 50% maximal current ($V_{50\%}$), before and after gambierol action, were 14.25 ± 0.83 mV ($n = 4$) and 8.93 ± 0.76 mV ($n = 4$), respectively. In addition, the curve slope factor (k) was also slightly, but significantly ($p = 0.004$), higher in the presence than in the absence of the polyether, i.e., 10.17 ± 0.49 mV^{-1} ($n = 4$) and 8.07 ± 0.57 mV^{-1} ($n = 4$), respectively.

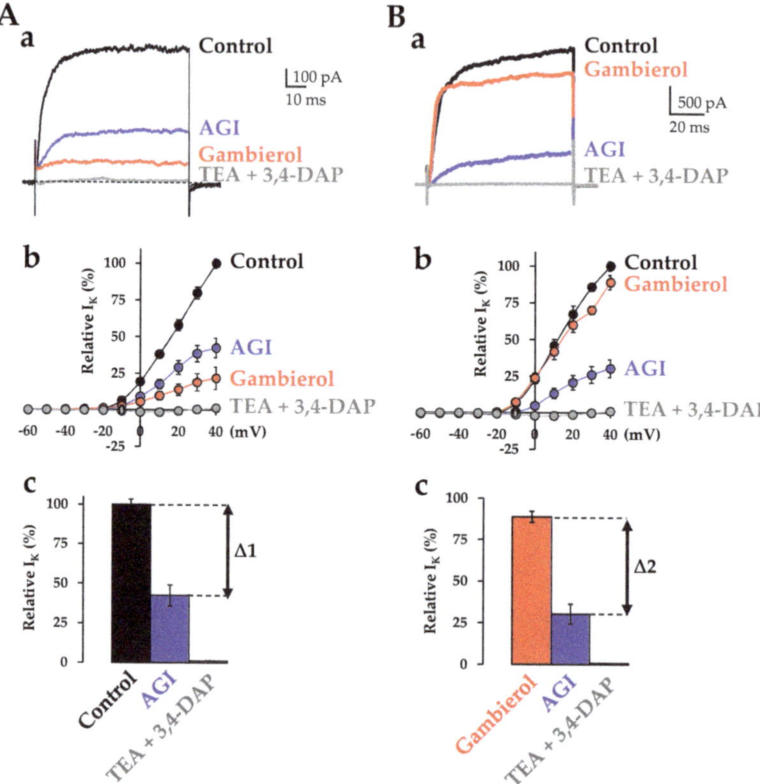

Figure 2. The contribution of various current components (K_{Ca}, K_{ATP}) and the action of gambierol (100 nM) on total K^+ current of rat fetal AMC cells. The K^+ current was recorded using the perforated whole-cell voltage-clamp technique, during 90 ms depolarizing steps from a holding potential of -70 mV, under the conditions indicated in the figure (**A,B**). Note in (**A,B**) the different sequences of gambierol addition to the external medium. (**a**) Superimposed traces of outward K^+ currents recorded under control conditions and after the perfusion of the different pharmacological agents indicated. AGI denotes the simultaneous perfusion of 400 nM apamin, 200 µM glibenclamide and 100 nM iberiotoxin. (**b**) Current–voltage relationships. The current was measured at the end of the depolarizing steps and expressed as a percentage of control values for the indicated agents. (**c**) Relative outward K^+ current contribution (expressed as a percentage of the total current) for the pharmacological treatments indicated. The K^+ current was measured at the end of the depolarizing steps to +40 mV and expressed as a percentage of control values. Note that the percentage of blocked K_{Ca} and K_{ATP} current components were not significantly different whatever the order of gambierol-induced channel inhibition was. In (**b,c**), data represent the mean $\pm$ SEM of 4 different experiments.

The K^+ current evoked by depolarizing pulses to +40 mV (90 ms duration, from a holding potential of -70 mV) reached a steady-state level (about 40% of control values) within about 4 min after addition of 20 nM gambierol (Figure 4A). Further increase in the gambierol concentration did not promote a supplementary reduction in the outward K^+-current. When individual K^+ current traces, recorded every 10 s pulsing, were analyzed before and during the gambierol (20 nM) perfusion, it was clear that gambierol not only reduced the amplitude of the K^+-current, but also slowed the kinetics of K^+ current activation by 75.4 $\pm$ 10.1% ($n = 4$), with respect to the control ($p = 0.031$) (Figure 4B,C). Hence, before and after gambierol action, the activation time constants of K^+ current were 3.82 $\pm$ 0.39 ms ($n = 4$) and 6.80 $\pm$ 1.02 ms ($n = 4$), respectively.

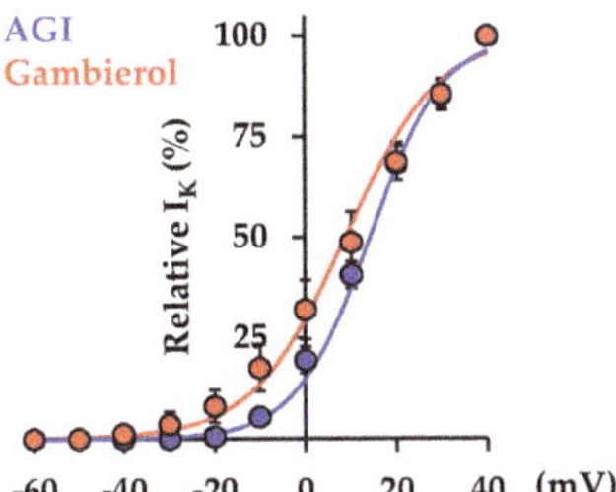

Figure 3. Action of gambierol on the voltage-dependence of K$^+$ current activation in rat fetal AMC cells. The current was measured at the end of 90 ms depolarizing steps from a holding potential of -70 mV, expressed as a percentage of its maximal value at +40 mV, and plotted as a function of membrane potential during depolarizing pulses, in the absence (AGI, in blue) and presence (in red) of 100 nM of gambierol. Data represent the mean $\pm$ SEM of 4 different experiments. The theoretical curves correspond to data point fits according to the Boltzmann equation, as described in Materials and Methods, with $V_{50\%}$ and k values of 14.25 mV and 8.07 mV^{-1} (R^2 = 0.9989), respectively, for AGI, and 8.93 mV and 10.17 mV^{-1} (R^2 = 0.9997), respectively, in the presence of gambierol. AGI denotes the simultaneous perfusion of 400 nM apamin, 200 μM glibenclamide and 100 nM iberiotoxin.

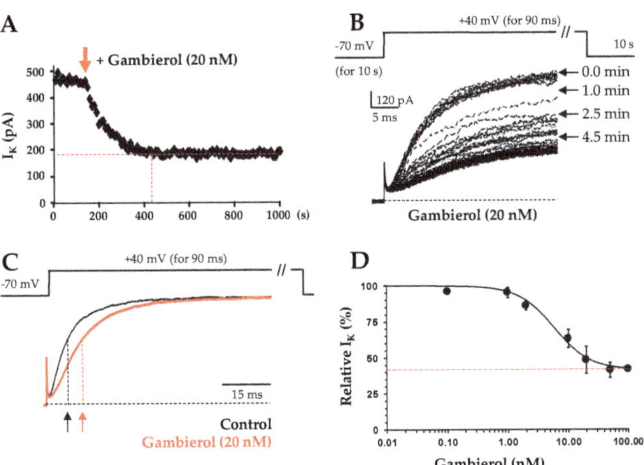

Figure 4. Time-course, kinetics and concentration-dependent action of gambierol on outward K$^+$ current in rat fetal AMC cells. (**A**) Time course of 20 nM gambierol action on the K$^+$ current measured at the end of 90 ms depolarizing pulses to +40 mV from a holding potential of -70 mV, applied every 10 s (see schema in (**B**)). The red arrow indicates the time of polyether addition to the bath. (**B**) Superimposed K$^+$ current recorded every 10-s pulsing, before and after the perfusion of gambierol (20 nM), during 90 ms depolarizing steps to +40 mV from a holding potential of -70 mV (schema). (**C**) Averaged normalized K$^+$ current recorded during 90 ms depolarizing steps to +40 mV from a holding potential of -70 mV (schema), under control conditions (black tracing), and after 20 nM gambierol (red tracing), note the slowing of the K$^+$ current activation. Data obtained from the same cell. The arrows indicate the activation time constants. (**D**) Concentration-response curve for the effect of gambierol on the steady-state K$^+$ current, measured after 90 ms depolarization steps to +40 mV from -70 mV. Each value, determined in the presence of 0.1–100 nM gambierol and normalized to its control value, represents the mean $\pm$ SEM of data obtained from 3–4 experiments. The theoretical curve was calculated as described in the text, with I_{SS}, IC_{50} and n_H values of 42%, 5.8 nM and 0.89 (r^2 = 0.982), respectively. The external medium in A–D contained 1 μM TTX, 200 μM glibenclamide and 1 mM Cd^{2+} to block, respectively, the Na$_V$, K$_{ATP}$ and Ca$_V$ channels and prevent, indirectly, the activation of K$_{Ca}$ channels.

The concentration-response relationship of the gambierol effect on the K^+ current was established by plotting the steady-state current amplitude, measured in the presence of gambierol (I_G) and expressed as a percentage of the value obtained before toxin application (I_C), as a function of the gambierol concentration ([gambierol]). The theoretical concentration-response curve was calculated from a typical sigmoid non-linear regression through data points (correlation coefficient = r^2), according to the Hill's equation (using GraphPad Prism v.5 software): $I_G/I_C = \{(1 - I_{ss})/[1 + ([gambierol]/IC_{50})^{nH}]\} + I_{ss}$, where I_{ss} is the current fraction remaining at high toxin concentrations (Figure 4D, dotted red line), IC_{50} is the toxin concentration necessary to inhibit 50% of the maximal current blocked by gambierol, and n_H is the Hill number. Under these conditions, the I_{ss}, IC_{50} and n_H values were $41.64 \pm 0.21\%$, 5.81 ± 1.56 nM and 0.89 ± 0.32 ($r^2 = 0.982$, $n = 4$), respectively (Figure 4D).

2.2. Effect of Gambierol on Cathecholamine Release

The use of single-cell amperometry is a valuable quantitative electrochemical method [37,38] to investigate the catecholamine secretion from AMC cells. The simultaneous whole-cell current-clamp and single-cell amperometry combination permitted us to investigate the action of gambierol on catecholamine secretion from fetal AMC cells. Exocytosis events were distinctly detected by positioning a carbon electrode (polarized to +650 mV to allow the oxidation of released catecholamine) as close as possible to the AMC cell, as shown in Figure 5A. Using the current-clamp configuration, the rat fetal AMC cells studied had a mean membrane resting potential of -51.8 ± 3.1 mV ($n = 32$) with a mean coefficient of variation of 0.32 (standard deviation/mean). Eleven of these cells were quiescent and no spontaneous action potential firing was observed while in twenty-one of the cells, spontaneous spike activity was present. In quiescent cells, the gambierol (50 nM) did not change significantly the number of amperometric spikes, related to catecholamine secretion, during 11-s depolarizing current-steps, as illustrated by a representative recording in Figure 5B. It is worth noting that in amperometric recordings, there was a delay between the pulse delivery and the first amperometric spike signal (Figure 5B, lower tracings). Furthermore, there was no correspondence between the recorded action potential (phasic with the current pulse, middle tracings, Figure 5B) and the amperometric signals, in good agreement with previously published data in which single action potentials were ineffective in triggering phasic secretion [42].

It has been previously reported that apamin and iberiotoxin induce catecholamine release in cultured rat fetal AMC cells from F19–F20 fetuses [37]. Therefore, it was of interest to test whether the subsequent addition of apamin (400 nM) and iberiotoxin (100 nM) to the extracellular medium containing the gambierol enhanced the catecholamine secretion. As exemplified in Figure 5B (middle and lower right tracing), during the action of the SK_{Ca} and BK_{Ca} channel blockers (apamin and iberiotoxin, respectively), a significant increase in the frequency of recorded amperometric spike events was detected (Figure 6A), concomitant with a sustained significant membrane depolarization (13.2 ± 0.5 mV, $n = 3$; $p < 0.05$). Further experiments were performed on fetal AMC cells that spontaneously fired action potentials. In those cells, long duration depolarizing current pulses triggered phasic action potentials exhibiting an overshoot of about 30 mV (Figure 5C, middle traces). Under control conditions, such phasic action potentials were followed by repetitive action potentials devoid of overshoot, whose frequency declined (accommodation) during the sustained depolarizing current, as shown in typical recordings (Figure 5C, middle left trace). Under control conditions, the mean action potential frequency during the 11-s current pulse was 2.50 ± 0.62 Hz with a mean coefficient of variation of 0.43 ($n = 3$). The addition of 50 nM gambierol to the medium increased significantly the frequency of action potentials, during the 11-s current pulse, to 4.5 ± 0.22 Hz ($p = 0.002$; $n = 3$) (Figure 5C, middle center trace), when compared to the control, but did not change significantly the amperometric spike events (Figure 5C, lower blue center trace; Figure 6B).

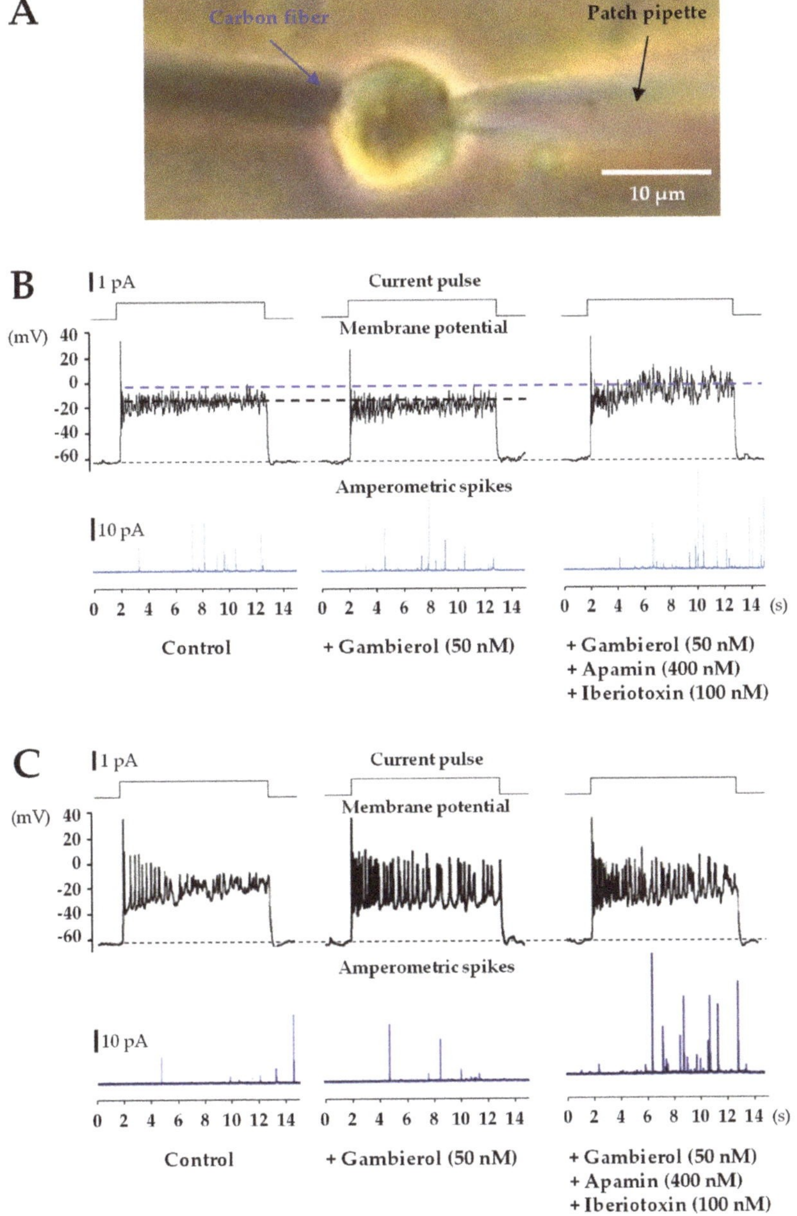

Figure 5. Combined whole-cell current-clamp and amperometric recordings in single rat fetal AMC cells under control conditions, and during the action of gambierol and K_{Ca} blockers. (**A**) Microphotograph of a typical recording configuration for single-cell amperometry, using a polarized carbon fiber electrode to detect exocytotic events, related to catecholamine release, and a whole-cell current-clamp pipette to control the level of membrane depolarization imposed to the cell membrane. (**B**,**C**) Whole-cell current-clamp recordings showing in the upper tracing the current

pulse depolarization used to trigger the changes in membrane potential (middle tracings), and the amperometric recording (lower tracings in blue), under control conditions (left), during gambierol action (center), and during the action of gambierol and the K_{Ca} channel blockers indicated (right). (**B**) Example taken from the same quiescent AMC cell (that had no spontaneous action potentials), while the addition of K_{Ca} channel blockers (in the continuous presence of gambierol) enhanced membrane depolarization during the current pulse, by about 14 mV (dotted blue line), and the frequency of exocytotic spikes related to catecholamine secretion events. (**C**) Example taken from an active AMC cell (that had spontaneous action potentials). Note that gambierol, as well as the addition of K_{Ca} channel blockers (in the continuous presence of gambierol), enhanced the frequency of repetitive action potentials during the current pulse depolarization, with respect to the control. Note also, that the frequency of exocytotic spikes, related to catecholamine secretion events, was only increased after the addition of K_{Ca} channel blockers.

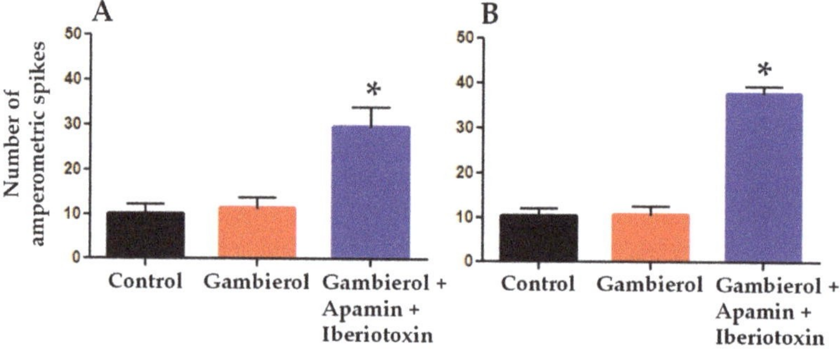

Figure 6. Number of amperometric spikes related to catecholamine release under control conditions, and during gambierol (50 nM), apamin (400 nM) and iberiotoxin (100 nM) applied cumulatively to the external medium by perfusion. Data obtained from fetal AMC cells showing initially either no action potential firing (**A**) or spontaneous spike activity (**B**), during stimulation with depolarizing current pulses of 11-s duration. Each column represents the mean ± SEM of 3 different experiments. Note that gambierol did not modify the number of release events, while a significant 2.7-fold (**A**) and 3.5-fold (**B**) increase occurred after the addition of K_{Ca} blockers. *: $p \leq 0.005$ versus gambierol.

In cells exhibiting repetitive action potentials, blockade of the SK_{Ca} and BK_{Ca} channels by apamin and iberiotoxin, respectively, further enhanced the action potential frequency by about 20% (5.38 ± 0.22 Hz, $p = 0.023$; $n = 3$), and significantly enhanced the amperometric spike events related to catecholamine release (Figure 5C, lower blue right trace; Figure 6B), in a similar manner as in quiescent cells.

On the whole, these results indicate that the specific inhibition of K_V channels induced by gambierol does not affect catecholamine release from quiescent or active rat fetal AMC cells. The increased amperometric spike number following the addition of K_{Ca} channel blockers may be the consequence of the increased depolarization induced by the SK_{Ca} and BK_{Ca} channel blockers. Furthermore, the recordings in Figure 5B (lower tracings) clearly show that the release of catecholamines by fetal AMC cells was not dependent on the action potential triggered by the current pulse of long duration but was controlled by the level of membrane depolarization. The frequency of the amperometric events increased when the depolarization of the membrane was larger, probably because more Ca_V channels were recruited. In active AMC cells (exhibiting spontaneous action potentials), it is likely that the K^+ current blocked by gambierol is involved in the control of the action potential firing (Figure 5C, middle center trace).

3. Discussion

K^+ channels constitute an important family of ion channels in excitable neuroendocrine cells and are involved in a number of physiological functions. To the best of our knowledge, this is the first time that the octacyclic polyether toxin gambierol, at nanomolar concentrations, is reported as blocking the K_V channels in rat fetal AMC cells. These cells express several types of K^+ channels contributing to the total outward current, as revealed by using the selective K_{Ca} channel blockers apamin [39] and iberiotoxin [40], and the K_{ATP} channel blocker glibenclamide [41]. Gambierol only partly inhibited the total K^+ current when added after, or before these channel blockers, suggesting that the polyether affects neither the K_{Ca} nor K_{ATP} channels (Figure 2).

In addition, gambierol slowed the kinetics of K^+ current activation in fetal AMC cells, implying a delayed opening of the K_V channels upon membrane depolarization (Figure 4C). In agreement with previous reports of such an effect [17,43,44], this strongly suggests that the polyether has a greater affinity for the channel resting state [45]. This particularity distinguishes the gambierol action from that of other lipophilic polyether toxins, such as Pacific ciguatoxin-1 (P-CTX-1) which also blocks K_V channels in rat myotubes and sensory neurons [46,47]. Interestingly, gambierol action on the K_V channels also differs from that of P-CTX-4B (the dinoflagellate-derived precursor of P-CTX-1), which is produced by the same dinoflagellate (*Gambierdiscus toxicus*), and blocks K_V channels in myelinated axons, without altering K^+ current activation [48].

Gambierol and synthetic analogues were previously reported to inhibit K_V channels in various cells and tissues, including neurosensory mouse taste cells [17], *Xenopus* skeletal myocytes [18], murine cerebellar neurons [19], human $K_V1.3$ channels from T-lymphocytes [22], mammalian $K_V1.1$–$K_V1.5$ channels expressed in *Xenopus* oocytes [20], and $K_V3.1$ channels expressed in both mouse fibroblasts [45] and Chinese hamster ovary (CHO) cells [21]. Gambierol was suggested to affect K_V channels by a new mechanism, interacting through a lipid-uncovered binding region of the channel [45]. Electrophysiological work, together with the use of expressed chimeric channels (between $K_V3.1$ and $K_V2.1$ channels) and homology modelling, revealed that gambierol high-affinity binding occurred in the resting state (when the channel is closed), by disturbing the gate opening and movements of the voltage-sensing domain [23,45,49]. The channel transitions between the resting and the open state require, initially, the dissociation of gambierol that is possible because the polyether has a considerably lower affinity for the open state. In the present work, an approximately 5 mV negative shift of the voltage-dependence of the K^+ current activation, associated with an increase in the curve slope factor, was detected in the presence of gambierol (Figure 3). These slight modifications in the voltage-dependence of K^+ current activation are in agreement with the assumption that the polyether is a gating modifier having a putative binding site on K_V channels equivalent to that of ciguatoxins, i.e., a cleft between the S5 and S6 segments [23,45,49]. It is worth noting that, in quiescent and active fetal AMC cells, gambierol, in the range of concentrations studied (0.1–100 nM), had no detectable activity on Na_V channels, as revealed by action potential recordings, which is in good agreement with previous reports on native [17,18,43] and expressed [20] channels.

The use of simultaneous whole-cell current-clamp and single-cell amperometry allowed for controlling the membrane potential and detecting exocytosis events. Catecholamine secretion in rat fetal AMC cells, lacking splanchnic innervation, is Ca^{2+}-dependent [37]. Gambierol did not modify significantly the number of amperometric spike-events triggered by current pulses of long-duration (11 s), causing measurable membrane depolarization that was not significantly different from that of controls (Figure 5B, middle and lower tracings, left and center). The fact that gambierol did not increase catecholamine release in quiescent fetal AMC cells can be explained by the following points: (i) the membrane potential was little affected by the polyether toxin; (ii) during the depolarizing current pulse, the membrane potential was unable to reach the threshold for activating the opening of voltage-gated calcium channels; (iii) other K_V channel subtypes remaining unaffected by gambierol, in particular the large-conductance (BK) and small conductance (SK) Ca^{2+}-activated K_V channel subtypes, curtailed membrane depolarization and voltage-gated Ca^{2+} entry, and therefore catecholamine secretion. Interestingly, block of the K_{Ca} channels, in the continuous presence of gambierol, enhanced membrane depolarization by about 13 mV (Figure 5B, during the 11 s current step), and at the same time, increased significantly the number of exocytotic events related to catecholamine secretion. Such enhanced depolarization is likely to bring the membrane potential above the activation threshold of high-voltage activated Ca_V channels, triggering both Ca^{2+} influx and subsequent catecholamine secretion.

In active cells, displaying spontaneous action potentials, gambierol was found to enhance the frequency of action potentials during the 11 s current pulse (Figure 5C) suggesting that the K^+ current blocked by the polyether may play a role in the control of the action potential firing in fetal AMC cells; however, despite this increase in action potential frequency by the gambierol, no enhancement of the amperometric events was detected, probably because of functional K_{Ca} channels' activity. The block of the K_{Ca} channels markedly increased the amperometric events, related to catecholamine secretion.

It was surprising to discover that gambierol did not increase catecholamine secretion from rat fetal AMC cells. The action of gambierol at the cellular level depends on the subtype of K_V channels that are expressed in a particular cell, their relative proportion, and finally their sensitivity to the polyether toxins. In neurosecretory fetal chromaffin cells, the proportion of K_{ATP} and K_{Ca} channels varied depending on the fetal development stage (F15 or F19–F20) [37,38], and present results. The results obtained in the fetal AMC cells, were quite distinct from previous ones in which gambierol was reported to block a fast K^+ current in motor nerve terminals, which lengthened the presynaptic action potential duration and thereby increased the amount of Ca^{2+} entry into the terminals and consequently the amount of acetylcholine quanta released upon nerve stimulation [24,25].

Overall, the pharmacological dissection of the several types of K^+ channels contributing to the total outward current of rat fetal chromaffin cells enhances the knowledge we have on gambierol action, showing that this phycotoxin affects only a fraction of the total K^+-current component distinct to the K_{Ca} and K_{ATP} currents. Some specific questions related to the type of K^+ current blocked in fetal AMC cells by gambierol remain unanswered and could motivate forthcoming studies. In addition, our results may help in understanding fetal viability, since gambierol, like other polyether toxins (e.g., brevetoxin-3 and ciguatoxin-1), likely crosses over the mammalian maternofetal barrier. Furthermore, our results show that the K^+-current block by gambierol in fetal AMC cells lacking splanchnic innervation has no effect on catecholamine secretion, emphasizing the key modulatory role of K_{Ca} currents in controlling exocytosis at this fetal stage (F19–F20).

4. Conclusions

In conclusion, (i) several types of K^+ channels contribute to the total outward current of rat fetal AMC cells; (ii) gambierol only partly inhibits the total K^+ current when added after or before K_{Ca} and K_{ATP} blockers, and affects neither the K_{Ca} nor K_{ATP} channels; (iii) after blocking the Na_V and K_{ATP} channels, and preventing activation of the K_{Ca} channels, gambierol blocks K^+ currents with a mean IC_{50} of 5.8 nM; (iv) in contrast to ciguatoxins, gambierol slows the kinetics of K^+-current activation; (v) gambierol does not modify the number of secretory events, related to catecholamine secretion and caused by long-lasting depolarizing pulses; (vi) gambierol increases the frequency of action potentials during a long-lasting current stimulation in cells exhibiting spontaneous action potentials; (vii) surprisingly the Ca^{2+}-dependent electrically-elicited catecholamine secretion is not affected by gambierol, but the subsequent block of K_{Ca} channels enhances membrane depolarization, the frequency of action potentials and increases the exocytotic event frequency, highlighting the modulatory role played by K_{Ca} channels in the control of exocytosis from rat fetal AMC cells.

The detailed mechanism of action of gambierol on the various types of transmembrane ion channels, the complexity of ion conductances, and firing activities in excitable rat fetal AMC cells still remain to be further explored.

5. Materials and Methods

5.1. Chemicals and Toxins

Gambierol was synthesized as described by Fuwa et al. in 2002 [7] and had a purity >97%. The synthetic gambierol was spectroscopically (NMR ^{13}C and ^{1}H, MS, IR) identical to natural gambierol. Due to the lipophilic nature of gambierol, stock solutions were prepared in dimethyl sulfoxide (DMSO) and diluted with the external physiological solution. The total DMSO concentration in the test solution did not exceed 0.1%. Tetrodotoxin, apamin, iberiotoxin, and glibenclamide were purchased from Alomone Labs (Jerusalem, Israel), and Latoxan (Portes-lès-Valence, France). All other chemicals, including cell culture media, reagents, tetraethylammonium chloride and 3,4-diaminopyridine, were purchased from Sigma-Aldrich (Saint-Quentin-Fallavier, France).

5.2. Animals

Adult pregnant Wistar rats were obtained from Elevage Janvier (Le Genest-Saint-Isle, France) and were acclimatized at the animal house for at least 48 h before experiments. Live animals were treated according to the European Community guidelines for laboratory animal handling, and to the guidelines established by the French Council on animal care "Guide for the Care and Use of Laboratory Animals" (EEC86/609 Council Directive; Decree 2001-131). All efforts were made to minimalize animal suffering and to reduce the number of animals used.

Animal care and surgical procedures were performed according to the Directive 2010/63/EU of the European Parliament, which had been approved by the French Ministry of Agriculture. The project was submitted to the French Ethics Committee CEEA (Comité d'Ethique en Expérimentation Animale) and obtained the authorization APAFIS#4111-2016021613253432 v5.

All experiments were performed in accordance with relevant named guidelines and regulations. Pregnant rats were housed in individual cages at constant temperature and a standard light cycle (12 h light/12 h darkness) and had food and water *ad libitum*. On the morning after an overnight breeding, the fetuses were considered to be at day 0.5 of gestation. At day 19 or 20 (F19 or F20), timed-pregnant rats were euthanized by carbon dioxide gas inhalation, and fetuses were rapidly collected and decapitated with scissors according to the guidance of the European Committee DGXI concerning animal experimentation.

5.3. Rat Fetal Adrenomedullary Chromaffin Cell Cultures

Fetal adrenomedullary chromaffin cells were obtained from adrenal glands, removed from fetuses collected at F19–F20. The method for chromaffin cell culture is detailed elsewhere [30]. Briefly, the adrenal glands removed from eight to ten rat fetuses were placed in an ice-cold phosphate buffer solution (PBS) to dissect with forceps the capsule and cortex of the adrenal glands under a binocular dissecting microscope.

The isolated medulla was cut into small pieces and treated at 37 °C with 5 mL Ca^{2+}-free digestion solution containing collagenase (0.2%, type IA), 0.1% hyaluronidase (type I–S) and 0.02% deoxyribonuclease (type I), to obtain dissociated chromaffin cells. After 30 min of tissue digestion, the enzymatic activity was stopped by adding 400 µL fetal bovine serum.

The digested tissue was rinsed with PBS, three times, and gently grinded with a Pasteur pipette. The cells were re-suspended in 5 mL Dulbecco's modified Eagle's medium (DMEM) supplemented with 7.5% fetal bovine serum, 50 IU/mL penicillin and 50 µg/mL streptomycin. Cells were plated onto 35 mm poly-L-lysine-coated dishes (2 mL volume) and kept at 37 °C in a controlled atmosphere (95% CO$_2$) for up to 2–4 days before the experiments.

5.4. Whole-Cell Voltage- and Current-Clamp Recordings

Membrane currents (under voltage-clamp conditions) and action potentials (under current-clamp conditions in the zero current mode) were recorded from fetal chromaffin cells using the perforated whole-cell technique [50], as previously described [30]. Briefly, recording microelectrodes were pulled from micro-hematocrit capillary tubes with a vertical microelectrode puller (PB-7-Narishige, Tokyo, Japan). Microelectrodes were coated with sticky wax (S.S. White, Gloucester, UK). Pipette resistance had typically 2–5 MΩ when filled with the internal solution. The seal resistance was typically 2–10 GΩ, and about 75–80% of the series resistance (ranging from 12 to 50 MΩ) was compensated electronically, under voltage-clamp conditions. Membrane currents and potentials were monitored with an RK-400 amplifier (Biologic, Claix, France), were filtered at 1–3 kHz (Frequency device, Haverhill, MA, USA), digitized with a DigiData-1200 interface (Axon Instruments, Union city, CA, USA), and stored on the hard disk of a PC computer.

Data acquisition and analyses were performed using the pCLAMP-v.8.0 software (Axon Instruments). The kinetic of K^+ current activation was determined by measuring the time constant (τ) of the exponential increase, both under control conditions (τ_C) and in the presence of 20 nM of gambierol (τ_G). Then, the percentage of change was calculated as $[(\tau_G/\tau_C) - 1] \times 100$. The voltage-dependence of the K^+ current activation was established by plotting the current (I_K), expressed as a percentage of its maximal value at +40 mV (I_{Kmax}), as a function of membrane potential (V) during 90 ms depolarizing pulses, in the absence and presence of 100 nM of gambierol, as elaborated previously by Hsu et al. in 2017 [44]. The theoretical curve corresponded to data point fits, according to the Boltzmann equation (GraphPad Prism v.5 software): $I_K/I_{Kmax} = 1 - [1/(1 + \exp((V - V_{50\%})/k))]$, where $V_{50\%}$ is the voltage corresponding to 50% maximal current, and k is the curve slope factor.

The standard external solution contained (in mM): 135 NaCl, 5 KCl, 2 $CaCl_2$, 2 $MgCl_2$, 10 glucose, and 10 HEPES (adjusted to pH 7.4 and an osmolarity of 300 mOsm). When necessary, 1 μM tetrodotoxin (TTX) was added to the external solution. The micropipette solution contained (in mM): 105 K^+ gluconate, 30 KCl, 0.1 $CaCl_2$ and 10 HEPES (adjusted to pH 7.2 and an osmolarity of 280 mOsm) and was added with amphotericin-B (24 μg/mL). Neither ATP nor EGTA were added in the internal solution to avoid the catecholamine release being affected, and thus the interpretation of results being complicated, since catecholamine release is calcium-dependent in fetal chromaffin cells. TTX, apamin, iberiotoxin and glibenclamide were added to the external solution. The solutions containing these drugs were freshly made from stock solutions, just before each experiment, and were applied by a custom-made gravity-fed micro-flow perfusion system, positioned as close as possible to the recorded cell. It is worth noting that glibenclamide was reported to have no significant effect on outward K^+ current under normoxia conditions in neonatal AMCs from P0 rat pups [51]. All experiments were performed at a constant room temperature (21 °C).

5.5. Amperometric Recordings from Single Cells

Electrochemical recordings of exocytotic events from single rat fetal AMC cells were performed, as described previously [37]. Cells were visualized with an inverted Olympus microscope. Catecholamine secretion was detected employing 5-μm diameter carbon fiber microelectrodes (purchased from ALA Scientific Instruments, Westbury, NY, USA), and prepared as previously described [49,52].

For the amperometry, a DC potential was applied to the carbon-fiber microelectrode, which appears at the interface between the carbon-fiber and the external solution bathing the cell. If the potential is much greater than the redox potential for a given transmitter, then catecholamines molecules diffuse to the carbon surface and are rapidly oxidized, yielding a current that can be measured. In our experiments, the carbon-fiber microelectrode was positioned adjacent to the individual cell, with the help of a micromanipulator, and a holding voltage of +650 mV was applied between the carbon fiber tip and the Ag/AgCl reference-electrode present in the bath to permit the oxidation of released catecholamines.

Electrochemical currents were filtered at 10 kHz (through a low pass filter) and amplified with a VA-10 current amplifier-system (NPI Electronic GmbH, Tamm, Germany). For peak detection, we used a threshold that was around four times the noise level (around

0.6 pA in our experiments). Amperometric spikes were identified, and carefully inspected offline on a personal computer, using Mini Analysis 5.1 (Synaptosoft, Leonia, NJ, USA) software. All spikes identified by the program were visually examined, and coinciding amperometric events were manually excluded from data sets. Current digitization and storage, data acquisition, as well as the standard external solution, were as described above. All experiments were carried out at a controlled room temperature (21 °C).

5.6. Statistics and Data Processing

Data are presented as the mean $\pm$ SEM. Comparison between data was completed using a Student's t test. Differences were considered to be statistically significant at $p < 0.05$. The number of experiments (n) refers to data obtained from the cells of different rat donors.

Author Contributions: Conceptualization, J.M., R.B. and E.B.; methodology, R.B., S.S., J.M. and E.B.; software, R.B., S.S. and E.B.; validation, E.B., S.S., J.M., M.S., H.F. and R.B.; formal analysis, S.S., R.B., J.M. and E.B.; investigation, S.S., R.B., E.B. and J.M.; resources, M.S. and H.F.; data curation, E.B., R.B. and J.M.; writing—original draft preparation, S.S., E.B., J.M. and R.B.; writing—review and editing, J.M., E.B. and R.B.; visualization, S.S., R.B., E.B., M.S., H.F. and J.M.; supervision, R.B., J.M. and E.B.; project administration, J.M., R.B. and. E.B.; funding acquisition, J.M., E.B. and R.B. All authors have read and agreed to the published version of the manuscript.

Funding: This study was supported in part by the grant ALERTOXNET (EAPA_317/2016) funded by the Interreg Atlantic program, and in part by the CNRS (through own resources from European contracts to J.M. and E.B.). S.S. was supported by a fellowship from the Délégation Générale pour l'Armement.

Institutional Review Board Statement: The animal study protocol was approved by the French Ethics Committee CEEA (Comité d'Ethique en Expérimentation Animale) (protocol code APAFIS#26651-2020072011192542v1 authorized on 20 July 2020).

Informed Consent Statement: Not applicable.

Data Availability Statement: Data is contained within the article.

Acknowledgments: We thank Gilles Ouanounou for helpful comments during perforated whole-cell voltage-clamp experiments.

Conflicts of Interest: The authors declare no conflict of interest. The funders had no role in the design of the study; in the collection, analyses, or interpretation of data; in the writing of the manuscript, or in the decision to publish the results.

References

1. Yasumoto, T. The chemistry and biological function of natural marine toxins. *Chem. Rec.* **2001**, *1*, 228–242. [CrossRef] [PubMed]
2. Shmukler, Y.B.; Nikishin, D.A. Ladder-Shaped Ion Channel Ligands: Current State of Knowledge. *Mar. Drugs* **2017**, *15*, 232. [CrossRef] [PubMed]
3. Sasaki, M.; Fuwa, H. Convergent strategies for the total synthesis of polycyclic ether marine metabolites. *Nat. Prod. Rep.* **2008**, *25*, 401–426. [CrossRef] [PubMed]
4. Nicolaou, K.C.; Frederick, M.O.; Aversa, R.J. The continuing saga of the marine polyether biotoxins. *Angew. Chem. Int. Ed. Engl.* **2008**, *47*, 7182–7225. [CrossRef] [PubMed]
5. Mori, Y. Development of New Synthetic Methods Using Oxiranyl Anions and Application in the Syntheses of Polycyclic Ether Marine Natural Products. *Chem. Pharm. Bull.* **2019**, *67*, 1–17. [CrossRef]
6. Wan, X.; Yao, G.; Liu, Y.; Chen, J.; Jiang, H. Research Progress in the Biosynthetic Mechanisms of Marine Polyether Toxins. *Mar. Drugs* **2019**, *17*, 594. [CrossRef]
7. Fuwa, H.; Sasaki, M.; Satake, M.; Tachibana, K. Total synthesis of gambierol. *Org. Lett.* **2002**, *4*, 2981–2984. [CrossRef]
8. Johnson, H.W.; Majumder, U.; Rainier, J.D. Total synthesis of gambierol: Subunit coupling and completion. *Chemistry* **2006**, *12*, 1747–1753. [CrossRef]
9. Fuwa, H.; Sasaki, M. Recent advances in the synthesis of marine polycyclic ether natural products. *Curr. Opin. Drug Discov. Devel.* **2007**, *10*, 784–806.
10. Alonso, E.; Fuwa, H.; Vale, C.; Suga, Y.; Goto, T.; Konno, Y.; Sasaki, M.; LaFerla, F.M.; Vieytes, M.R.; Giménez-Llort, L.; et al. Design and synthesis of skeletal analogues of gambierol: Attenuation of amyloid-β and tau pathology with voltage-gated potassium channel and N-methyl-D-aspartate receptor implications. *J. Am. Chem. Soc.* **2012**, *134*, 7467–7479. [CrossRef]

11. Satake, M.; Murata, M.; Yasumoto, T. Gambierol: A new toxic polyether compound isolated from the marine dinoflagellate *Gambierdiscus toxicus*. *J. Am. Chem. Soc.* **1993**, *115*, 361–362. [CrossRef]

12. Morohashi, A.; Satake, M.; Yasumoto, T. The absolute configuration of gambierol, a toxic marine polyether from the dinoflagellate, *Gambierdiscus toxicus*. *Tetrahedron Lett.* **1999**, *40*, 97–100. [CrossRef]

13. Chinain, M.; Gatti, C.M.I.; Darius, H.T.; Quod, J.P.; Tester, P.A. Ciguatera poisonings: A global review of occurrences and trends. *Harmful Algae* **2021**, *102*, 101873. [CrossRef] [PubMed]

14. Vetter, I.; Zimmerman, K.; Lewis, R.J. Ciguatera toxins: Pharmacology, toxicology, and detection. In *Seafood and Freshwater Toxins: Pharmacology, Physiology, and Detection*, 3rd ed.; CRC Press: Boca Raton, FL, USA, 2014; Chapter 32; pp. 925–950.

15. Friedman, M.A.; Fernandez, M.; Backer, L.C.; Dickey, R.W.; Bernstein, J.; Schrank, K.; Kibler, S.; Stephan, W.; Gribble, M.O.; Bienfang, P.; et al. An Updated Review of Ciguatera Fish Poisoning: Clinical, Epidemiological, Environmental, and Public Health Management. *Mar. Drugs* **2017**, *15*, 72. [CrossRef]

16. Chinain, M.; Gatti, C.M.; Roué, M.; Darius, H.T. Ciguatera-causing dinoflagellates in the genera *Gambierdiscus* and *Fukuyoa*: Distribution, ecophysiology and toxicology. In *Dinoflagellates: Morphology, Life History and Ecological Significance*; Subba Rao, D.V., Ed.; Nova Science: New York, NY, USA, 2020; pp. 405–457. ISBN 978-1-53617-888-3.

17. Ghiaroni, V.; Sasaki, M.; Fuwa, H.; Rossini, G.P.; Scalera, G.; Yasumoto, T.; Pietra, P.; Bigiani, A. Inhibition of voltage-gated potassium currents by gambierol in mouse taste cells. *Toxicol. Sci.* **2005**, *85*, 657–665. [CrossRef]

18. Schlumberger, S.; Ouanounou, G.; Girard, E.; Sasaki, M.; Fuwa, H.; Louzao, M.C.; Botana, L.M.; Benoit, E.; Molgó, J. The marine polyether gambierol enhances muscle contraction and blocks a transient K^+ current in skeletal muscle cells. *Toxicon* **2010**, *56*, 785–791. [CrossRef]

19. Pérez, S.; Vale, C.; Alonso, E.; Fuwa, H.; Sasaki, M.; Konno, Y.; Goto, T.; Suga, Y.; Vieytes, M.R.; Botana, L.M. Effect of gambierol and its tetracyclic and heptacyclic analogues in cultured cerebellar neurons: A structure-activity relationships study. *Chem. Res. Toxicol* **2012**, *25*, 1929–1937. [CrossRef]

20. Cuypers, E.; Abdel-Mottaleb, Y.; Kopljar, I.; Rainier, J.D.; Raes, A.L.; Snyders, D.J.; Tytgat, J. Gambierol, a toxin produced by the dinoflagellate Gambierdiscus toxicus, is a potent blocker of voltage-gated potassium channels. *Toxicon* **2008**, *51*, 974–983. [CrossRef]

21. Konoki, K.; Suga, Y.; Fuwa, H.; Yotsu-Yamashita, M.; Sasaki, M. Evaluation of gambierol and its analogs for their inhibition of human Kv1.2 and cytotoxicity. *Bioorg. Med. Chem. Lett.* **2015**, *25*, 514–518. [CrossRef]

22. Rubiolo, J.A.; Vale, C.; Martín, V.; Fuwa, H.; Sasaki, M.; Botana, L.M. Potassium currents inhibition by gambierol analogs prevents human T lymphocyte activation. *Arch. Toxicol.* **2015**, *89*, 1119–1134. [CrossRef]

23. Kopljar, I.; Labro, A.J.; de Block, T.; Rainier, J.D.; Tytgat, J.; Snyders, D.J. The ladder-shaped polyether toxin gambierol anchors the gating machinery of Kv3.1 channels in the resting state. *J. Gen. Physiol.* **2013**, *141*, 359–369. [CrossRef] [PubMed]

24. Molgó, J.; Schlumberger, S.; Sasaki, M.; Fuwa, H.; Louzao, M.C.; Botana, L.M.; Servent, D.; Benoit, E. Gambierol Potently Increases Evoked Quantal Transmitter Release and Reverses Pre- and Post-Synaptic Blockade at Vertebrate Neuromuscular Junctions. *Neuroscience* **2020**, *439*, 106–116. [CrossRef] [PubMed]

25. Molgó, J.; Schlumberger, S.; Sasaki, M.; Fuwa, H.; Louzao, M.C.; Botana, L.M.; Servent, D.; Benoit, E. Gambierol enhances evoked quantal transmitter release and blocks a potassium current in motor nerve terminals of the mouse neuromuscular junction. In *Harmful Algae 2018–From Ecosystems to Socio Ecosystems. Proceedings of the 18th International Conference on Harmful Algae*; Hess, P., Ed.; International Society for the Study of Harmful Algae: Nantes, France, 2020; pp. 145–148.

26. Kidokoro, Y.; Ritchie, A.K. Chromaffin cell action potentials and their possible role in adrenaline secretion from rat adrenal medulla. *J. Physiol.* **1980**, *307*, 199–216. [CrossRef] [PubMed]

27. Fenwick, E.M.; Marty, A.; Neher, E. Sodium and calcium channels in bovine chromaffin cells. *J. Physiol.* **1982**, *331*, 599–635. [CrossRef]

28. Martinez-Espinosa, P.L.; Neely, A.; Ding, J.; Lingle, C.J. Fast inactivation of Nav current in rat adrenal chromaffin cells involves two independent inactivation pathways. *J. Gen. Physiol.* **2021**, *153*, e202012784. [CrossRef]

29. Lingle, C.J.; Martinez-Espinosa, P.L.; Guarina, L.; Carbone, E. Roles of Na^+, Ca^{2+}, and K^+ channels in the generation of repetitive firing and rhythmic bursting in adrenal chromaffin cells. *Pflügers Arch.* **2018**, *470*, 39–52. [CrossRef]

30. Bournaud, R.; Hidalgo, J.; Yu, H.; Jaimovich, E.; Shimahara, T. Low threshold T-type calcium current in rat embryonic chromaffin cells. *J. Physiol.* **2001**, *537*, 35–44. [CrossRef]

31. Favreau, P.; Gilles, N.; Lamthanh, H.; Bournaud, R.; Shimahara, T.; Bouet, F.; Laboute, P.; Letourneux, Y.; Ménez, A.; Molgó, J.; et al. A new omega-conotoxin that targets N-type voltage-sensitive calcium channels with unusual specificity. *Biochemistry* **2001**, *40*, 14567–14575. [CrossRef]

32. García, A.G.; García-De-Diego, A.M.; Gandía, L.; Borges, R.; García-Sancho, J. Calcium signaling and exocytosis in adrenal chromaffin cells. *Physiol. Rev.* **2006**, *86*, 1093–1131. [CrossRef]

33. Padín, J.F.; Fernández-Morales, J.C.; de Diego, A.M.; García, A.G. Calcium Channel Subtypes and Exocytosis in Chromaffin Cells at Early Life. *Curr. Mol. Pharmacol.* **2015**, *8*, 81–86. [CrossRef]

34. Nurse, C.A.; Salman, S.; Scott, A.L. Hypoxia-regulated catecholamine secretion in chromaffin cells. *Cell Tissue Res.* **2018**, *372*, 433–441. [CrossRef] [PubMed]

35. Seidler, F.J.; Slotkin, T.A. Adrenomedullary function in the neonatal rat: Responses to acute hypoxia. *J. Physiol.* **1985**, *358*, 1–16. [CrossRef]

36. Slotkin, T.A.; Seidler, F.J. Adrenomedullary catecholamine release in the fetus and newborn: Secretory mechanisms and their role in stress and survival. *J. Dev. Physiol.* **1988**, *10*, 1–16. [PubMed]

37. Bournaud, R.; Hidalgo, J.; Yu, H.; Girard, E.; Shimahara, T. Catecholamine secretion from rat foetal adrenal chromaffin cells and hypoxia sensitivity. *Pflügers Arch.* **2007**, *454*, 83–92. [CrossRef] [PubMed]

38. Thompson, R.J.; Nurse, C.A. Anoxia differentially modulates multiple K^+ currents and depolarizes neonatal rat adrenal chromaffin cells. *J. Physiol.* **1998**, *512*, 421–434. [CrossRef]

39. Artalejo, A.R.; García, A.G.; Neher, E. Small-conductance Ca^{2+}-activated K^+ channels in bovine chromaffin cells. *Pflug. Arch.* **1993**, *423*, 97–103. [CrossRef]

40. Candia, S.; Garcia, M.L.; Latorre, R. Mode of action of iberiotoxin, a potent blocker of the large conductance Ca^{2+}-activated K^+ channel. *Biophys. J.* **1992**, *63*, 583–590. [CrossRef]

41. Salman, S.; Buttigieg, J.; Zhang, M.; Nurse, C.A. Chronic exposure of neonatal rat adrenomedullary chromaffin cells to opioids in vitro blunts both hypoxia and hypercapnia chemosensitivity. *J. Physiol.* **2013**, *591*, 515–529. [CrossRef]

42. Zhou, Z.; Misler, S. Action potential-induced quantal secretion of catecholamines from rat adrenal chromaffin cells. *J. Biol. Chem.* **1995**, *270*, 3498–3505. [CrossRef]

43. Ghiaroni, V.; Fuwa, H.; Inoue, M.; Sasaki, M.; Miyazaki, K.; Hirama, M.; Yasumoto, T.; Rossini, G.P.; Scalera, G.; Bigiani, A. Effect of ciguatoxin 3C on voltage-gated Na+ and K+ currents in mouse taste cells. *Chem. Senses* **2006**, *31*, 673–680. [CrossRef]

44. Hsu, H.T.; Lo, Y.C.; Huang, Y.M.; Tseng, Y.T.; Wu, S.N. Important modifications by sugammadex, a modified γ-cyclodextrin, of ion currents in differentiated NSC-34 neuronal cells. *BMC Neurosci.* **2017**, *18*, 6. [CrossRef] [PubMed]

45. Kopljar, I.; Labro, A.J.; Cuypers, E.; Johnson, H.W.; Rainier, J.D.; Tytgat, J.; Snyders, D.J. A polyether biotoxin binding site on the lipid-exposed face of the pore domain of Kv channels revealed by the marine toxin gambierol. *Proc. Natl. Acad. Sci. USA* **2009**, *106*, 9896–9901. [CrossRef] [PubMed]

46. Hidalgo, J.; Liberona, J.L.; Molgó, J.; Jaimovich, E. Pacific ciguatoxin-1b effect over Na^+ and K^+ currents, inositol 1,4,5-triphosphate content and intracellular Ca^{2+} signals in cultured rat myotubes. *Br. J. Pharmacol.* **2002**, *137*, 1055–1062. [CrossRef] [PubMed]

47. Birinyi-Strachan, L.C.; Gunning, S.J.; Lewis, R.J.; Nicholson, G.M. Block of voltage-gated potassium channels by Pacific ciguatoxin-1 contributes to increased neuronal excitability in rat sensory neurons. *Toxicol. Appl. Pharmacol.* **2005**, *204*, 175–186. [CrossRef]

48. Schlumberger, S.; Mattei, C.; Molgó, J.; Benoit, E. Dual action of a dinoflagellate-derived precursor of Pacific ciguatoxins (P-CTX-4B) on voltage-dependent K^+ and Na^+ channels of single myelinated axons. *Toxicon* **2010**, *56*, 768–775. [CrossRef]

49. Kopljar, I.; Grottesi, A.; de Block, T.; Rainier, J.D.; Tytgat, J.; Labro, A.J.; Snyders, D.J. Voltage-sensor conformation shapes the intra-membrane drug binding site that determines gambierol affinity in Kv channels. *Neuropharmacology* **2016**, *107*, 160–167. [CrossRef]

50. Hamill, O.P.; Marty, A.; Neher, E.; Sakmann, B.; Sigworth, F.J. Improved patch-clamp techniques for high-resolution current recording from cells and cell-free membrane patches. *Pflug. Arch.* **1981**, *391*, 85–100. [CrossRef]

51. Buttigieg, J.; Brown, S.; Holloway, A.C.; Nurse, C.A. Chronic nicotine blunts hypoxic sensitivity in perinatal rat adrenal chromaffin cells via upregulation of KATP channels: Role of alpha7 nicotinic acetylcholine receptor and hypoxia-inducible factor-2alpha. *J. Neurosci.* **2009**, *29*, 7137–7147. [CrossRef]

52. Wightman, R.M.; Jankowski, J.A.; Kennedy, R.T.; Kawagoe, K.T.; Schroeder, T.J.; Leszczyszyn, D.J.; Near, J.A.; Diliberto, E.J., Jr.; Viveros, O.H. Temporally resolved catecholamine spikes correspond to single vesicle release from individual chromaffin cells. *Proc. Natl. Acad. Sci. USA* **1991**, *88*, 10754–10758. [CrossRef]

Review

Mechanisms of Action of the Peptide Toxins Targeting Human and Rodent Acid-Sensing Ion Channels and Relevance to Their In Vivo Analgesic Effects

Clément Verkest [1,2,†], Miguel Salinas [1,†], Sylvie Diochot [1], Emmanuel Deval [1], Eric Lingueglia [1] and Anne Baron [1,*]

[1] CNRS (Centre National de la Recherche Scientifique), IPMC (Institut de Pharmacologie Moléculaire et Cellulaire), LabEx ICST (Laboratory of Excellence in Ion Channel Science and Therapeutics), FHU InovPain (Fédération Hospitalo-Universitaire "Innovative Solutions in Refractory Chronic Pain"), Université Côte d'Azur, 660 Route des Lucioles, Sophia-Antipolis, 06560 Nice, France

[2] Department of Anesthesiology, University Medical Center Hamburg-Eppendorf, 20251 Hamburg, Germany

* Correspondence: anne.baron@ipmc.cnrs.fr

† These authors contributed equally to this work.

Abstract: Acid-sensing ion channels (ASICs) are voltage-independent H⁺-gated cation channels largely expressed in the nervous system of rodents and humans. At least six isoforms (ASIC1a, 1b, 2a, 2b, 3 and 4) associate into homotrimers or heterotrimers to form functional channels with highly pH-dependent gating properties. This review provides an update on the pharmacological profiles of animal peptide toxins targeting ASICs, including PcTx1 from tarantula and related spider toxins, APETx2 and APETx-like peptides from sea anemone, and mambalgin from snake, as well as the dimeric protein snake toxin MitTx that have all been instrumental to understanding the structure and the pH-dependent gating of rodent and human cloned ASICs and to study the physiological and pathological roles of native ASICs in vitro and in vivo. ASICs are expressed all along the pain pathways and the pharmacological data clearly support a role for these channels in pain. ASIC-targeting peptide toxins interfere with ASIC gating by complex and pH-dependent mechanisms sometimes leading to opposite effects. However, these dual pH-dependent effects of ASIC-inhibiting toxins (PcTx1, mambalgin and APETx2) are fully compatible with, and even support, their analgesic effects in vivo, both in the central and the peripheral nervous system, as well as potential effects in humans.

Keywords: ASIC; sodium channels; toxins; peptide; PcTx1; APETx2; MitTx; mambalgin; pain; nociception

Key Contribution: This review updates the pharmacological and molecular mechanisms of ASIC-targeting animal toxins on pain-related ASICs as well as the data supporting the analgesic effect of ASIC inhibition in vivo in rodents and humans.

Early work by Krishtal et al. in 1980 demonstrated for the first time that application of extracellular acid could evoke inward currents in sensory neurons [1]. Later on, they were attributed to Acid-Sensing Ion Channels (ASICs) [2,3], which are members of the epithelial Na⁺ channel (ENaC) and Degenerin (DEG) ion channel superfamily [4]. Understanding the proton-dependent activation and modulation of ASICs as well as their pathophysiological roles was the subject of active research since their molecular identification and cloning in the late 1990s. Significant efforts were made to discover pharmacological tools to help decipher their function, and peptides isolated from venoms turned out to be a powerful resource for it. Tissue acidosis is associated with pain and ASICs emerged as major pH sensors in sensory neurons where protons also directly affect other receptors such as the Transient Receptor Potential (TRP) vanilloid 1 TRPV1 [5–7], some members of the two-pore domain potassium channel family [8], cyclic nucleotide gated (CGN) channels [9,10] and G-protein coupled receptors [11].

Citation: Verkest, C.; Salinas, M.; Diochot, S.; Deval, E.; Lingueglia, E.; Baron, A. Mechanisms of Action of the Peptide Toxins Targeting Human and Rodent Acid-Sensing Ion Channels and Relevance to Their In Vivo Analgesic Effects. *Toxins* **2022**, *14*, 709. https://doi.org/10.3390/toxins14100709

Received: 7 September 2022
Accepted: 2 October 2022
Published: 17 October 2022

Here, we discuss recent advances in our understanding of ASIC mechanisms of activation by acidic pH and how peptide toxins exert their complex molecular effects on them, with a special emphasis on the in vivo consequences on pain and their use both as pharmacological tools and potential analgesic compounds.

1. Molecular and Functional Properties of ASICs

1.1. Subunits Diversity and Structure

Functional ASICs are formed by the homo- or heterotrimeric association of identical or homologous subunits [12–14] (Figure 1A), each subunit comprising more than 500 amino acids and two transmembrane domains, a large extracellular loop, and intracellular N- and C-termini with a re-entrant N-terminus loop (Figure 1B,C).

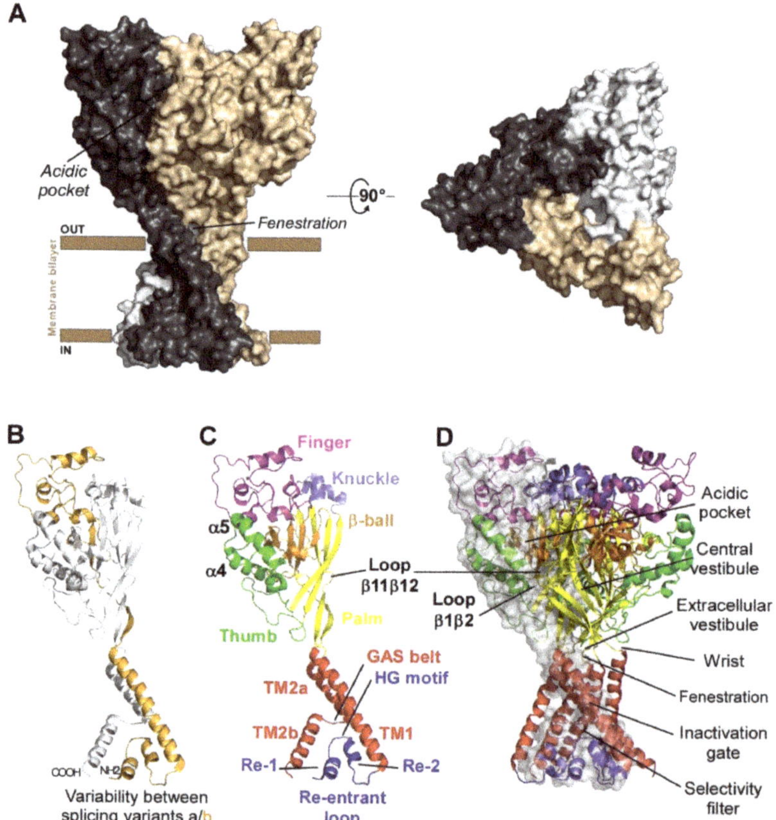

Figure 1. Structure of ASICs. (A) Trimeric organization of ASICs (left panel: side view, right panel: top view). **(B)** Tridimensional skeletal model of a single subunit where variable regions between isoforms "a" and "b" of rat ASIC1 and ASIC2 are highlighted in gold. **(C)** Structure of a single subunit of chicken ASIC1 in resting state (the different sub-domains are shown in specific colors; PDB ID: 6vtl). **(D)** Skeletal 3D representation of a functional channel formed by the assembly of three subunits. A transparent grey surface was added to one subunit to delineate the interface between two adjacent subunits. Same colors as in **(C)** for the different sub-domains, and key structural domains mentioned on the right. Cytoplasmic N- and C-termini, whose structures are unknown, are not shown. Designed with PyMOL software.

Four genes (ACCN1 to ACCN4) encode at least six different ASIC subunits (Table 1) sharing more than 50% amino acid identity: ASIC1a, ASIC1b, ASIC2a, ASIC2b, ASIC3,

ASIC4 (ASIC5, also named BLINaC/BASIC and coded by the ACCN5 gene, only shares 30% amino acid identity and cannot be considered as a genuine ASIC subunit). The difference between the a and b variants of ASIC1 and ASIC2 relies on the first N-terminal third of the subunit (Figure 1B), including the cytoplasmic N-terminal domain, the re-entrant loop (forming part of the pore with the HG motif), the first transmembrane domain TM1 (forearm and wrist domains), and part of the extracellular loop (the palm β1 and β3 sheets, the β-ball β2 sheet and the entire finger domain).

Table 1. Protein sequence comparison of rat and human ASIC subunits.

Isoform	Species	% Identity	Size (aa)	Name in Genbank	Sequence ID
ASIC1a	*Rattus norvegicus*	98.11%	526	ASIC1	NP_077068.1
	Homo sapiens		528	ASIC1 isoform b	NP_001086.2
ASIC1b	*Rattus norvegicus*	93.24%	559	ASIC1 isoform X5	XP_006257440.1
	Homo sapiens		562	ASIC1 isoform c	NP_001243759.1
ASIC2a	*Rattus norvegicus*	99.02%	512	ASIC2 isoform MDEG1	NP_001029186.1
	Homo sapiens		512	ASIC2 isoform MDEG1	NP_001085.2
ASIC2b	*Rattus norvegicus*	98.83%	563	ASIC2 isoform MDEG2	NP_037024.2
	Homo sapiens		563	ASIC2 isoform MDEG2	NP_899233.1
ASIC3	*Rattus norvegicus*	83.68%	533	ASIC3	NP_775158.1
	Homo sapiens		531	ASIC3 isoform a	NP_004760.1
ASIC4	*Rattus norvegicus*	97.22%	539	ASIC4	NP_071570.2
	Homo sapiens		539	ASIC4 isoform 1	NP_061144.4

Percentages of amino acid (aa) identity were calculated using BLAST.

The first crystal structure of an ASIC was solved in 2007 by the group of E. Gouaux from cASIC1 (chicken ASIC1), the chicken ortholog of rat ASIC1a (rASIC1a) [12]. Each subunit was represented as a hand holding a ball and divided into finger, thumb, palm, knuckle, β-ball, wrist, and forearm (transmembrane domains) domains (Figure 1C). An "acidic pocket" containing several pairs of acidic amino acids is present at the interface of each subunit and was proposed to be one of the pH sensors of the channel, whereas cations may access the ion channel by lateral fenestrations, then moving into a broad extracellular vestibule just above the inactivation gate and the selectivity filter (i.e., the structural element in the narrowest part of the pore that determines ionic selectivity) (Figure 1D) [12,15]. The most noticeable structural difference between human (h) ASIC1a (hASIC1a) and cASIC1 is a longer loop that extends down from the α4-helix to the tip of the thumb, due to two extra amino acids (D298 and L299) absent in all other ASIC isoforms [16].

The lowest amino acid identity is 52% between cASIC1 and rASIC3, while rat and human ASIC orthologs show amino acid identities between 83.68% (for ASIC3) and 99.02% (for ASIC2a) (Table 1). Interestingly, hASIC3, but not rASIC3, has three splice variants (a, b, c), resulting in differences in the C-terminal domain, hASIC3a mRNA being the main isoform expressed in human neuronal tissues, although hASIC3c was also significantly detected. A higher level of sequence variability for the same isoform is observed between hASIC3a and rASIC3 or mouse ASIC3 (mASIC3) orthologs. While experimentally determined structures are still lacking for ASIC2 and ASIC3, recent major advances in structure prediction using machine learning have allowed the generation of models for those ASICs, shedding a new light on potential structural variations underlying the functional differences between ASICs [17].

1.2. pH-Dependency

Homo- or heterotrimeric cloned ASICs were found to be voltage-insensitive but highly pH-sensitive upon heterologous expression in *Xenopus* oocytes or in mammalian cell lines. They are sodium selective, with additional low calcium permeability for ASIC1a and hASIC1b [3,18,19]. They are activated by a fast-extracellular acidosis from conditioning physiological pH to acidic test pH and inactivated by sustained extracellular acidosis. Interestingly, rat and human ASIC3 channels can be also activated at neutral (7.4) pH by lipids (arachidonic acid and lysophosphatidylcholine) [20,21] and hASIC3a channels have been shown to be sensitive to both acidic and alkaline pH [22].

The ASIC2b and ASIC4 subunits do not form functional proton-gated channels by themselves, but ASIC2b can associate with other ASIC subunits to confer new properties and regulations to heterotrimeric channels [13,23,24]. ASIC currents are generally transient even if the acidification is maintained, but a sustained phase is associated with expression of ASIC3 or with the presence of the ASIC2b subunit in heterotrimers (Figure 2). A sustained plateau phase is also associated with hASIC1b current, but not with rASIC1b [19]. Two types of sustained currents have been described for ASIC3: a window current at pH around 7.0 (Figure 2) resulting from the overlap of pH-dependent activation and desensitization curves, and a sustained current induced by more acidic test pHs. The TM1 domain modulates the pH-dependent activation, thus contributing to the window current near physiological pH and, combined with the N-terminal domain, the TM1 domain is also the key structural element generating the low pH-evoked sustained current [25].

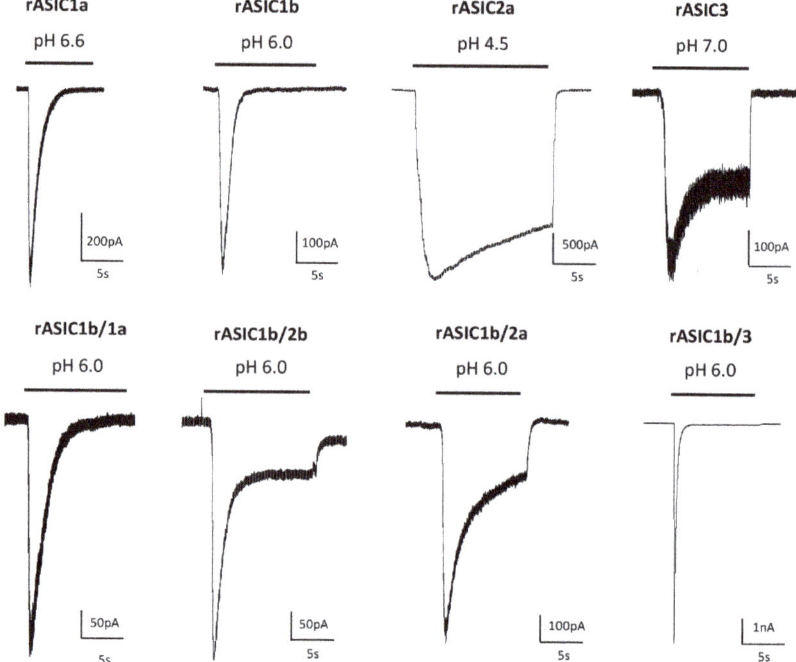

Figure 2. Diversity of currents flowing through homo- and heterotrimeric cloned ASICs. Original current traces of rat heterologously expressed ASIC currents recorded from HEK293 cells depending on the composition in ASIC subunits, activated from pH 7.4 to the indicated test pH, at −60 mV. Homotrimeric channels result from the expression of only one type of ASIC subunit (indicated above each current), whereas heterotrimeric channels result from the co-expression of two different subunits (1:1 ratio in transfection). The corresponding current noted rASIC1b/1a, for example, results from the co-expression of rASIC1b and rASIC1a subunits.

ASICs differ in their biophysical properties depending on their subunit composition, notably in their desensitization time constants. rASIC3, rASIC1a and rASIC1b show significantly faster desensitization kinetics than rASIC2a (Figure 2). Experiments were conducted on heterologously expressed cloned channels by co-expressing two, or more rarely three (or concatemers), different subunits. The stochiometry of association in heterotrimers (i.e., the relative number of each subunit in the channel) is therefore hard to precisely determine in these conditions. The ASIC1a and ASIC2a stoichiometry was at least investigated, showing no preferential association [26].

ASICs differ in their pH sensitivity, and the functional diversity obtained by combining the different subunits in homo- or heterotrimers (Table 2, with references in legend) allows these channels to detect a wide range of pH changes between pH 7.2 and pH 4.0. Sigmoidal curve fit of pH-dependent activation is used to determine the test $pH_{0.5}$, inducing the half-maximal activation that can vary between 6.8–6.3 (rASIC3) and 5.0–3.8 (rASIC2a). ASICs are desensitized depending on the conditioning pH, which is represented by the pH-dependent sigmoidal curve of steady-state desensitization (SSD) and by the conditioning $pH_{0.5}$ of half-maximal SSD that can vary between 7.4–6.8 for rASIC1a and rASIC3 and 6.0–4.7 for hASIC2a (Table 2). These values are generally less acidic than the ones for activation, which means that the SSD mechanism in the presence of a sustained extracellular pH acidification will highly influence the amplitude of an ASIC current triggered by a subsequent rapid drop in pH. The sustained ASIC3 current results from incomplete inactivation and is activated at test pH 6.0 and below for hASIC3a and at pH 6.5 and below for rASIC3.

Table 2. Functional pH ranges of currents flowing through cloned rodent and human ASICs.

Cloned Channel	ACTIVATION		SSD	
	Test pH Threshold/max	$pH_{0.5}$	Conditioning pH Threshold/max	$pH_{0.5}$
rASIC1a	7.0/5.5	6.4–5.8 [chimnqtwxyz]	7.4/6.8	7.3–7.1 [cehimtyz]
rASIC1b	6.4/5.6	6.3–5.7 [fitwxy#]	7.3/6.6	7.0–6.5 [fit#]
m/rASIC2a	6.0/3.0	5.0–3.8 [bnqwxz]	7.0/4.5	6.3–5.6 [mz]
m/rASIC3	7.2/5.5	6.8–6.3 [otwy]	7.4/6.8	7.2–7.0 [sty]
rASIC1a/2a	6.3/4.5	5.6–4.8 [nqrw]		
m/rASIC1a/2b	6.8/6.0	6.4–6.2 [pw]	7.4/7.1	7.3 [p]
rASIC1a/1b		6.3–5.8 [w]		
rASIC1a/3	7.0/5.5	6.7–6.3 [rtw]	7.0/6.8	7.1 [t]
rASIC1b/3	6.6/5.9	6.7–6.2 [tw]	6.9/6.6	6.8 [t]
rASIC1b/2a		4.9 [w]		
rASIC2a/3	7.2/4.5	6.1–5.6 [rw]		
m/rASIC2a/2b		4.8 [bw]		
rASIC2b/3		6.5 [w]		
m/rASIC1a/2a/3		6.4–5.1 [rw]		
rASIC1a/2b/3		6.3 [w]		
rASIC1b/2a/3		4.9 [w]		
hASIC1a	6.8/6.0	6.6–6.3 [dgikov]	7.0/6.7	7.2–6.9 [degiko]
hASIC1b	6.5/5.5	5.9–5.7 [gi]	6.7/6.4	6.5–6.1 [gi]
hASIC2a	6.8/3.5	5.7 [u]	6.0/4.7	5.5 [u]
hASIC3a	7.0/5.5	6.6–6.2 [aj]	7.0/7.9	7.7–7.5 [as]
cASIC1	6.8/6.3	6.6 [l]	7.4/7.1	7.3 [l]

Representative pH ranges (threshold/max) and $pH_{0.5}$ values for pH-dependent activation of cellular ASIC currents activated from conditioning pH 7.4 to variable test pHs, and for pH-dependent steady state desensitization (SSD) of currents maximally activated from variable conditioning pHs with rat (r), mouse (m), chicken (c) and human (h) homotrimeric and heterotrimeric ASICs heterologously expressed in *Xenopus* oocytes or mammalian cell lines. The corresponding current noted rASIC1a/2a, for example, results from the co-expression of rASIC1a and rASIC2a subunits. References: a [27], b [28], c [29], d [30], e [31], f [32], g [19], h [33], i [34], j [22], k [35], l [36], m [37], n [38], o [39], p [40], q [41], r [42], s [43], t [44], u [45], v [46], w [14], x [47], y [48], z [49], # unpublished data.

1.3. pH-Dependent Gating

The molecular mechanism of pH-dependent gating of ASICs was studied through combined approaches including electrophysiology, mutagenesis, molecular dynamic simulations, X-ray crystallography and cryo-electron microscopy (cryo-EM), along with the pharmacological use of ASIC-targeting animal toxins.

The structures of the three conformational states involved in H^+-dependent gating of homotrimeric cASIC1 were solved by X-ray crystallography and Cryo-EM: resting state [50] (Figure 3A), open state [51] (Figure 3B) and desensitized state [12] (Figure 3C). Cryo-EM structures of the hASIC1a in its closed state have also been solved, in complex with the toxin mambalgin-1 [52], and in complex with a specific nanobody Nb.C1 [16]. At the interface of each subunit of the trimeric channel, an acidic pocket is formed by intra-subunit contacts between the thumb, the β-ball and the finger domains, together with residues from the palm domain on the adjacent subunit (Figures 1A,D and 3) [12,50].

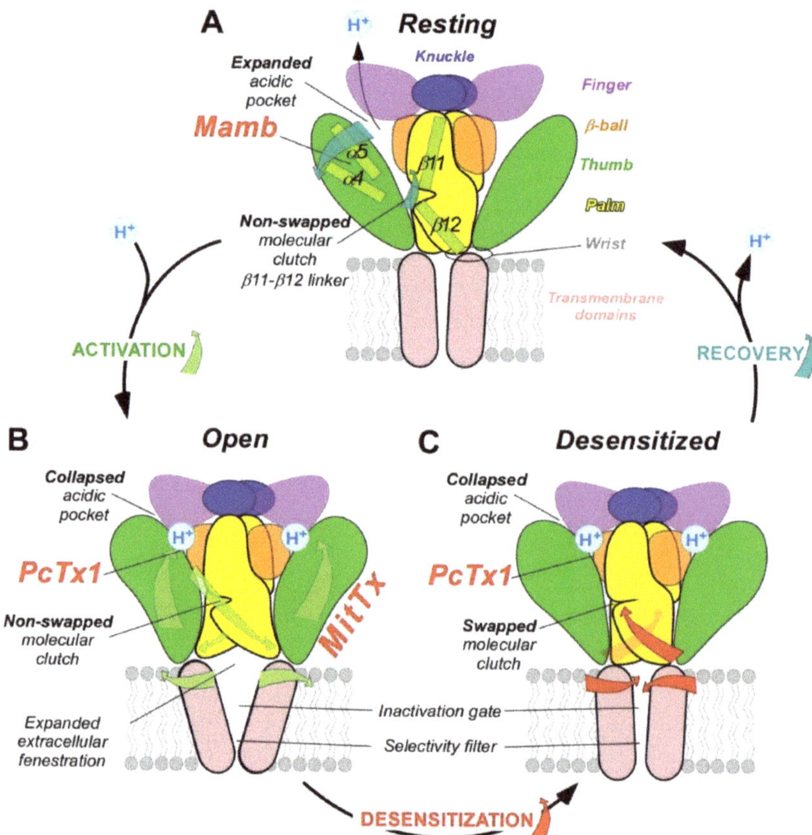

Figure 3. pH-dependent gating mechanisms of ASICs and interaction with toxins. ASIC gating involves three conformational states. (**A**), high pH resting state, which is stabilized by the toxin mambalgin (see Section 2.4) (major domains involved are indicated with the same color code as in Figure 1). (**B**), low pH open state, which is stabilized by the toxin MitTx (see Section 2.3) and also partially by the toxin PcTx1 (see Section 2.2). (**C**), low pH desensitized state also promoted by PcTx1. To illustrate the recovery process in (**A**), the deprotonation mechanism of only one acidic pocket is presented. Blue (**A**), green (**B**) and red (**C**) arrows show critical conformational changes during recovery, activation and desensitization processes, respectively. For clarity, only two subunits are shown.

Upon activation by extracellular acidic pH, protonation of the acidic pocket (Figure 3A,B) leads to its collapsed conformation, which is stabilized by the formation of three pairs of carboxyl–carboxylate interactions between the side chains of aspartate and glutamate residues [12,50]. Cl⁻ ion may play a role in channel gating by stabilizing the collapsed conformation of the acidic pocket at low pH, which seems state dependent since this bound Cl⁻ is absent in the resting state at high pH [53]. The motion of thumb helices α4/α5, resulting from collapse of the acidic pocket, induces a global motion of the thumb domain, which is directly connected to the transmembrane domain through a non-covalent contact forming a part of the wrist region [12,54]. In parallel, anchoring of the α5 helix against the palm of the adjacent subunit induces bending of the lower palm (β1 and β12 strands) toward the transmembrane domains (TM1 and TM2) to which they are covalently connected to form the other part of the wrist region. All together, thumb and lower palm domain motions lead to a rotation of each subunit around the scaffold formed by the knuckle and upper palm domains [50] that induces a translation of TM1 and TM2 leading to the expansion of extracellular fenestrations and to an iris-like opening of the channel gate (Figure 3B).

Ions are then enabled to pass through the selectivity filter of the pore (GAS belt motif, between TM2a and TM2b and HG motif in N-terminal re-entrant loop; Figure 1C) [15,51,55]. The Lys212 of the palm domain is deeply anchored to the thumb domain of the adjacent subunit and seems critical to facilitate the cooperativity between subunits during the global rotation of the extracellular domains of all the subunits [12,15,51,56]. In the lower palm domain, an inter-subunit hydrogen-bond network, close to the wrist region, seems also critical for the correct propagation of conformational changes leading to the expansion of the extracellular fenestration [57,58] (Figure 3B). Several residues at the extracellular side of the transmembrane domain that form contacts within each subunit in desensitized and resting state [12,39] are disrupted after the iris-like opening of the pore [50]. It is interesting to note the arginine in this region that seems also necessary to mediate the potentiation of ASIC currents by lipids [59]. In addition to acid-induced activation, hASIC3 is also sensitive to alkalization, and this property is supported by two arginine residues only present in the human channel and also located close to the boundary between the plasma membrane and the extracellular medium [22].

During prolonged acidification, the β11-β12 linker that demarcates the upper and lower palm domains undergoes a substantial conformational change induced by the switch in sidechain orientations of two residues [60–62]. This plays the role of "molecular clutch" allowing transmembrane domains to relax back into a "resting-like" conformation to permit rapid desensitization by uncoupling the conformational change of the upper extracellular domain from the lower part of the channel leading to the narrowing of the fenestration and the closing of the inactivation gate (Figure 3C) [50,61–65]. Just under the "molecular clutch", a fourth pair of carboxyl–carboxylate interaction between the side chains of glutamate residues was identified [12] that probably influences its stability [66,67]. Moreover, the previously mentioned Lys212, in a loop immediately above the β11-β12 linker, binds a Cl⁻ anion located in the thumb domain of an adjacent subunit [12,50,53] and could explain the mutations in the thumb domain also influencing desensitization [64,68–70].

Finally, when returning to physiological pH, the channel would return to the resting state after deprotonation of acidic residues that drive the expansion of the acidic pocket, allowing the β11-β12 linker to revert back to a non-swapped conformation (Figure 3A) [50,61].

The intra- and inter-subunit network of H-bonds, salt bridges and carboxyl–carboxylate pairs involving several residues with different pKa values, highlights the complexity of the pH-dependent gating of ASICs and explains the different pH-dependent activation and desensitization characteristics of the various homo- or heterotrimeric channels in different species (Table 2), since several domains are directly or indirectly involved in both mechanisms via their intra- and inter-subunit connections.

Among the channels and receptors that respond to acidic pHs, only the proton-activated chloride channel (PAC) is, like ASICs, directly activated by protons through a complex and dedicated mechanism [71,72]. Interestingly, this channel also has a trimeric structure with a large extracellular domain, comprising fenestrations and acidic pockets, whose conformation change is transmitted to transmembrane domains after protonation. This suggests a complex convergent evolutionary process to achieve the pH sensing property of these two unrelated channels, albeit with completely different mechanisms at the molecular level. Very recently, a lysosomal proton (and K⁺) channel [73] with unrelated sequence and structure with ASICs was shown to be also activated by protons via a still unknown molecular mechanism. Proton-mediated gating of the capsaicin receptor TRPV1 is dependent of one key residue of the pore region (Phe660) [7], the inactivation of CNG channels is controlled by extracellular protons leading to the collapse of the pore via the titration of a single glutamate residue within the selectivity filter [10], and the two pore domain potassium channels (K_2P) are modulated by protons through the titration of a key residue in the pore [8].

1.4. Pathophysiological Roles in Pain Sensing

Relying on their pH-dependent gating properties, ASICs were involved in several functions associated with physiological and/or pathological extracellular pH variations [4,74–81]. We will focus here on pain sensing. Besides metabolic disorders producing systemic pH changes, there are other pathophysiological conditions that result in local pH variations, generally associated with increased pain perception. During pathophysiological conditions like inflammation, tissue injury, ischemia or cancer, extracellular pH can drop from physiological values (generally around 7.4) to values around 6.5 or even below. In rodent models, local extracellular pH can for instance decrease to 6.8–6.0 in implanted tumor, to 5.8 in carcinoma, to 6.0–4.0 in bone cancer, to 6.8 upon inflammation, to 5.7 in arthritis, to 6.5 in muscle after incision, to 6.9 during heart angina, to 6.9 in joints in osteoarthritis or rheumatoid arthritis, and to 6.5–6.2 in mouse brain after traumatic injury or stroke [82–85]. In humans, local extracellular pH was found to decrease to 6.0–5.4 in abscess, to 5.7–5.4 in human malignant tumors, to 7.0–6.0 in joint synovial fluid of osteoarthritis patients, to 6.9 in gout, to 6.4 in melanoma, or to 6.7 intracutaneously after muscle exercise [81,84], and localized skin tissue acidification (pH $\geq$ 6.0) causes pain in humans [86,87].

Tissue acidosis occurs therefore in a variety of pathological painful conditions, and ASIC subtypes expressed in nociceptive neurons have all the hallmarks of pain sensors. Several reviews have summarized the role of ASICs in the peripheral nervous system in nociception and also in proprioception [74–76,78–80,88]. ASICs are highly sensitive to moderate acidifications, being for instance 10-fold more sensitive than the heat, capsaicin and proton-sensitive channel TRPV1 also expressed in peripheral sensory neurons. ASICs can generate sustained depolarizing currents upon prolonged tissue acidification compatible with the detection of non-adapting pain. Regulation of their activity by several pain-related mediators beside protons (inflammatory factors, neuropeptides, lipids, etc.) [21,89,90] led to the notion of coincidence detectors, especially for ASIC3, associated with pain detection and peripheral sensitization processes in pathophysiological situations like inflammation or chronic pain [81,91].

It is important to note that ASICs in the pain pathways are expressed not only in sensory neurons but also in dorsal horn neurons of the spinal cord involved in pain processing as well as in the brain, where they could be involved in synaptic transmission and plasticity, activated by the acidification of the synaptic cleft after the co-release of the acidic content of neurotransmitter synaptic vesicles, in particular in the case of chronic pain situations leading to central sensitization processes. Homotrimeric ASIC1a and heterotrimeric ASIC1a/2a were found to be postsynaptic receptors activated in several brain structures in glutamatergic and GABAergic neurons, where they could generate 3–10% of the synaptic current, even 20% at GABAergic synapses, and be involved in diverse forms of synaptic plasticity [77,92]. In the central nervous system, presynaptic or postsynaptic ASICs have thus been proposed to modulate learning and memory and to play a role in epilepsy and mood disorders as well as in neuronal damages associated with stroke and Alzheimer's disease [77,92,93].

2. Dual Effects of Animal Toxins Targeting ASICs

Development of the pharmacology of ASICs was very important for studying their structure, their molecular and cellular functions, and their pathophysiological roles, in combination with knockout or knockdown animals. The pharmacology of ASICs includes poorly selective modulators like synthetic amiloride, GMQ (2-guanidine-4-methylquinazoline), diminazene and non-steroidal anti-inflammatory drugs (NSAIDs), endogenous modulators like lipids, nitric oxide (NO), extracellular cations, polyamines such as agmatine or spermine, neuropeptides (dynorphin A, big dynorphin, RFamide-related peptides), natural compounds among which there are vegetal compounds particularly used in traditional Chinese medicine and finally much more selective animal peptide and protein toxins [94,95].

Four ASIC-targeting peptide toxins have been extensively characterized to date: the peptide toxins PcTx1 (psalmotoxin) from spider [96], APETx2 from sea anemone [97], and mambalgin (Mamb, with isoforms 1, 2 and 3) from mamba snakes [33,98], and the heterodimeric protein MitTx from coral snake [99].

The mechanisms by which toxins modulate ASIC gating are best described for ASIC1a and PcTx1 [29,31,32,47,60,100–103], ASIC1a and MitTx [51,99,104] and ASIC1a and Mamb [33,34,49,52,105–108] through combined approaches including electrophysiology, mutagenesis, molecular dynamics simulations, X-ray crystallography and cryo-EM, which allow to propose a mechanism for the pharmacological effects of these gating modifier toxins, as well as for APETx2 whose binding site on ASIC3 is not yet formally identified, which helps to understand their complex pH-dependent dual effects.

2.1. Preliminary Remarks on the Models and Data Interpretation

Depending on the subunit composition of ASICs, on the animal species and on extracellular pH variations, three different macroscopic effects could be observed on whole-cell currents flowing through heterologously expressed cloned ASICs when toxins are applied at physiological conditioning pH 7.4: inhibition (INH, Table 3 with references included) of the peak H^+-gated current, potentiation (POT, Table 3) of the peak H^+-gated current, or activation (ACT, Table 3) of the current without any change in pH.

These complex and sometimes opposite pH-dependent effects of ASIC-targeting peptide toxins could question the validity and/or extrapolation to humans of some in vivo data especially against pain. We will present evidence on how these dual effects of ASIC-targeting animal toxins on both rodent and human channels are fully compatible with, and even support, the analgesic effects seen in vivo in rodents and potential effects in humans.

The expression system should also be taken into account in interpreting these effects, as opposite results were sometimes observed on cloned ASICs whether they were expressed in *Xenopus* oocytes (O, Table 3) or in mammalian cells (◆, Table 3). For example, potentiating effects of APETx2 on rASIC1b and rASIC2a, as well as on rASIC1b/3 and rASIC2a/3 currents were reported from *Xenopus* oocytes, whereas similar concentrations of toxin were reported to have no effect on the same channels expressed in transfected mammalian cells. These effects observed in *Xenopus* oocytes should thus be considered with caution if they were not confirmed in transfected mammalian cells and/or in native currents recorded from neurons, particularly to extrapolate in vivo effects in mammals.

Furthermore, in structure-activity studies, the impact of a mutation on ASIC pH-dependent gating should be carefully controlled since the effect of mutations strongly affecting the pH-dependent gating can introduce a bias, making it difficult to draw a formal conclusion concerning the direct involvement of a channel residue into the binding site of a toxin.

Table 3. Effects of ASIC-targeting animal toxins on the amplitude of cloned rodent and human ASIC peak whole-cell currents activated from physiological pH 7.4.

Channel	Mamb (1, 2 or 3) IC_{50}/EC_{50}	PcTx1 IC_{50}/EC_{50}	APETx2 IC_{50}/EC_{50}	MitTx EC_{50}
rASIC1a	INH○◆ 3–55 nM [a,p,q,r]	INH○◆ 0.3–3.7 nM [b,c,o,u]	NO○◆ at 10 μM [y,e]	ACT○ 9 nM [i]
rASIC1b	INH○◆ 22–192 nM [a,p,q,r]	POT○ 25–100 nM [d,u]	NO◆ at 3 μM [e]; POT○ at 3–10 μM [y,#]	ACT○ 23 nM [i]
rASIC2a	NO○◆ at 3 μM [a,p]	NO○ at 100 nM [b]	NO○◆ at 3 μM [e,#]; POT○ at 10 μM [y]	POT○ at 75 nM [i]
rASIC3	NO○◆ at 3 μM [a,p]	NO○ at 100 nM [b]	INH○◆ 37–63 nM, test pH 6 [e,f,g]	ACT○ 830 nM [i]
rASIC1a/2a	INH○◆ 152–252 nM [a,p,r]	NO○◆ at 50 nM [b,v,w]		ACT○ at 75 nM [i]
rASIC1a/2b	INH◆ 61 nM [a]	INH○ 3 nM [h]; NO◆ at 300 nM [#]		
rASIC1a/1b	INH◆ 72 nM [a]			
rASIC1a/3	NO◆ at 2 μM [a,p]	NO◆ at 10 nM [b]; INH◆ at 100 nM, test pH 6 [x]	INH○◆ 2 μM [e,#]	NO○ at 75 nM [i]
rASIC1b/3	INH◆ at 1 μM [#]; POT◆ at 1 μM, test pH 6.6 [#]		INH◆ 900 nM, test pH 6 [e]; POT○ at 3 μM, test pH 6 [#]	
rASIC1b/2a	INH◆ at 1 μM [#]			
rASIC1b/2b	INH◆ at 100 nM [#]			
rASIC2a/3			NO◆ at 3 μM [e]; POT○ at 3 μM, test pH 6 [#]	NO○ at 75 nM [i]
rASIC2b/3			INH○◆ 117 nM [e,#]	
mASIC1a/2a/3		NO◆ at 100 nM [x]; NO○◆ at 25–60 nM [m,n,j]		
hASIC1a	INH○◆ 24–203 nM [a,q,s,t]	INH○ 3.3 nM, test pH 6 [u]; POT○ at 60 nM, test pH 6.7 [n]		
hASIC1b	INH○ at 1 μM; POT○◆ 60 nM, test pH6 [q,#]	POT○ at 60 nM [n]		
hASIC2a	NO◆ at 850 nM [a]	NO◆ at 100 nM [j]		
hASIC3a			INH○◆ 175–344 nM [e,z]	
hASIC1a/2a	INH◆ 220 nM [a]	NO◆ at 100 nM [j]; POT○ at 200 nM [h]		
hASIC1a/1b	INH○ 256 nM [q]			
hASIC1a/3	INH○ 462 nM [q]			
hASIC1b/3	INH○ at 1 μM [q]; POT○ at 1 μM, test pH 6 [q]			
cASIC1	INH◆ 124 nM, test pH 6 [t]	ACT○◆ 189 nM [l,k,p]; POT○◆ at 20 nM [k,m]		

Rat (r), mouse (m), chicken (c) and human (h) homotrimeric and heterotrimeric ASICs were heterologously expressed in *Xenopus* oocytes (○) or mammalian cell lines (◆) and activated from pH 7.4 by an acidic maximal stimulation to pH 5.0 or below (unless another test pH is mentioned). The corresponding current noted rASIC1a/2a, for example, results from the co-expression of rASIC1a and rASIC2a subunits. Inhibition (INH) or potentiation (POT) of the H^+-activated ASIC peak current, or activation (ACT) in the absence of acid stimulation is shown in red, blue and purple, respectively. The IC_{50} or EC_{50} values are indicated when available. If not, the highest dose tested is noted; NO: no effect (with the highest dose tested indicated). Toxins were applied at physiological conditioning pH 7.4. References: a [33], b [96], c [29], d [32], e [97], f [109], g [110], h [40], i [99], j [111], k [112], l [60], m [30], n [19], o [100,101], p [98], q [34], r [105], s [113], t [108], u [31], v [114], w [115]; x [116], y [117], z [118], # unpublished data.

2.2. PcTx1 and Related Toxins

2.2.1. Pharmacological Profile

This peptide of forty amino acids was isolated from the South American tarantula *Psalmopoeus cambridgei* venom [96]. It folds according to the inhibitor cystine knot (ICK) motif [100,119]. Discovered as a potent inhibitor of cloned rASIC1a (IC_{50} = 0.3–3.7 nM),

PcTx1 applied at conditioning physiological pH 7.4 was also shown to inhibit rASIC1a/3 at higher concentrations (25–100 nM) (Table 3, references included) and to inhibit hASIC1a at test pH 6 with an IC_{50} of 13 nM [31]. Inhibition of rASIC1a/2b was only reported once in the *Xenopus* oocytes expression system [40] but was not confirmed from mammalian cell experiments performed with up to 300 nM of PcTx1 (unpublished data). Depending on the pH conditions, PcTx1 also shows potentiating effects on rASIC1b, hASIC1a, hASIC1b, hASIC1a/2a and on cASIC1 currents (Table 3) [19,31,32] and can also directly activate cASIC1a (EC_{50}~189 nM) [98,112].

PcTx1 was reported to exert complex state-dependent effects on ASIC1a- and ASIC1b-containing channels, which would depend on its concentration but also on the animal species, on the pH-dependent properties of the channels and on the pH at which the toxin is applied. For example, whereas exerting almost no effect on the hASIC1a current activated from conditioning pH 7.4 to test pH 5.0 [19,30,111], PcTx1 was reported to exert an inhibitory effect on the current maximally activated from conditioning pH 7.2, or on the current activated from conditioning pH 7.4 to test pH 6.0 at high concentration [30,31], and a potentiation of the current submaximally activated from conditioning pH 7.4 to test pH 7.2–6.2 [19] (Figure 4B). However, if inhibitory effects were obtained both from *Xenopus* oocytes and mammalian cells data, potentiating effects of PcTx1 on the rASIC1b, hASIC1a, hASIC1b and hASIC1a/2a need to be considered with caution because they are only obtained from *Xenopus* oocytes experiments.

PcTx1 has no effect on the other ASIC channels, nor on a variety of Kv, Nav and Cav channels [96]. The best specificity of the toxin for ASICs is supported by binding experiments with an iodinated form of the peptide, which shows similar binding properties on rat brain membranes and heterologously expressed cloned ASIC1a channels [47].

Despite these dual effects described on heterologously expressed cloned ASICs, the effects of PcTx1 on native neuronal rodent ASIC currents were mostly inhibitory [37,120,121], and generally used to support the participation of homotrimeric ASIC1a in the whole-cell ASIC current. The observed inhibition of native neuronal ASIC currents is in good agreement with in vivo analgesic effects confirmed by genetic invalidations, particularly after central (intrathecal, i.t., or intracerebroventricular, i.c.v.) injections of the peptide in different animal pain models (chemical, inflammatory and neuropathic pain) [122,123].

The spider peptide Hm3a (π-TRTX-Hm3a, from *Heteroscodra maculata*) shares high sequence similarity with PcTx1 (37 amino acids, with 5 different amino acids and three residues missing at the C-terminus) but is more resistant to enzymatic, chemical and thermal degradation although in vivo studies are yet to confirm this efficacy. It inhibits the rASIC1a current with an affinity (IC_{50} = 1.3–2.6 nM) similar to PcTx1, as well as hASIC1a (IC_{50} = 0.52 nM), and potentiates rASIC1b at higher concentration (EC_{50} = 46.5 nM) as well as rASIC1a/1b (EC_{50} = 17.4 nM) and hASIC1b (EC_{50} = 178 nM), all recorded from *Xenopus* oocytes [124].

Another spider toxin, Hi1a (from *Hadronyche infensa*), is a double knot peptide composed of 75 amino acids that looks like two PcTx1 in tandem (each with 62% and 50% identity with PcTx1) [125,126]. It inhibits rASIC1a (IC_{50} = 0.4 nM), as well as hASIC1a (IC_{50} = 0.5 nM) currents with an affinity similar to PcTx1, and potentiates rASIC1b at higher concentration (EC_{50} = 46 nM), with a more durable effect than PcTx1 on currents recorded from *Xenopus* oocytes [125]. Hi1a does not exert a major effect on a panel of human ion channels involved in cellular excitability (hNav1.5, hKv4.3/hKChIP2, hCav1.2, hKv11.1/hERG, hKv7.1/hKCNQ1 or hKir2.1 currents) [127].

2.2.2. PcTx1, a Gating Modifier Stabilizing Open and Desensitized States
Biophysical Mechanisms

When co-applied at physiological conditioning pH 7.4, PcTx1 is able to bind to the closed state of ASIC1a and to induce a conformational change that is, however, not sufficient to directly open the channel in most cases [103]. By mimicking a local protonation of the acidic pocket, PcTx1 triggers its H^+-dependent collapsed conformation thus promoting both open and desensitized states [51], as evidenced by the apparent greater affinity of protons (leftward shift) for both the activation and SSD curves of rASIC1a and hASIC1a (Figure 4A,B) [19,29–31,128]. The amplitude of this shift is dependent on the concentration of PcTx1 [31]. Except in the case of cASIC1 [98] and of chimeric channels on which PcTx1 behaves as a direct agonist at pH 7.4 [32], PcTx1 does not directly open ASIC1a, probably because other key residues must be protonated by a pH value lower than 7.4 to cause the pore to open completely.

PcTx1 inhibitory effect on rASIC1a current from physiological conditioning pH 7.4 to every test pH is mostly due to its pH-dependent SSD promoting effect, whereas no more inhibition was observed from pH 8 instead revealing a potentiation of the current at test pH values in the activation curve pH range (7.2–6.2), due to the opposite potentiating effect by a leftward shift of the activation curve (Figure 4A).

Rat and human ASIC1a differ by five residues in the thumb domain, which render the pH-dependent SSD of hASIC1a less sensitive to pH and thus prevent the potent inhibitory effect of PcTx1 at conditioning pH 7.4, revealing instead a current potentiation of hASIC1a through the shift of the pH-dependent activation that can be seen at test pH 7.2–6.2, i.e., in the activation pH range only [19] (Figure 4A,B). Inhibition of hASIC1a can only be restored if PcTx1 is applied with slightly acidic conditioning pH (pH 7.2 in the presence of 1 nM PcTx1 [31]), i.e., in the range of the effect of PcTx1 on the pH-dependent SSD (Figure 4B), or by increasing the PcTx1 concentration to 3–10 nM to further shift the SSD curve above pH 7.4 [31]. The PcTx1 effect on currents submaximally activated from conditioning pH < 7.4 will result from the sum of inhibiting and potentiating effects.

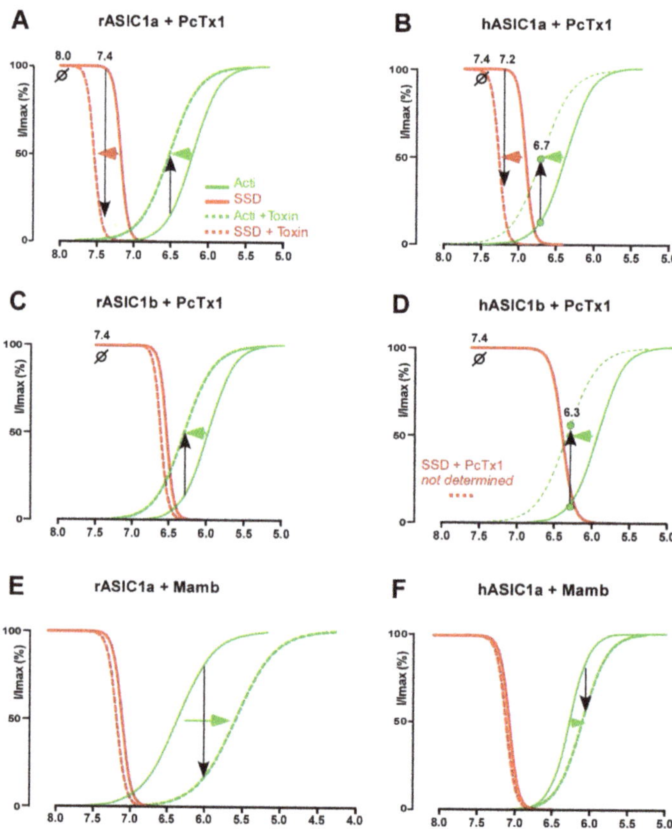

Figure 4. Toxins interfere with the pH-dependent gating of rodent and human ASICs. Schematic

pH-dependent curves of normalized (I/I$_{max}$%) activation (Acti, green) and steady state desensitization (SSD, red) of heterologously expressed cloned rat and human ASIC1a and ASIC1b in the absence (solid line) and in the presence of PcTx1 (dashed line) (**A–D**) and of rat and human ASIC1a in the presence and in the absence of mambalgin (Mamb) (**E,F**). All curves were adapted from published data. (**A,B**), **rASIC1a (A) and hASIC1a (B) gating modulation by PcTx1.** The toxin increases the apparent H$^+$ affinity of rASIC1a current thus inducing a leftward shift of both the activation and SSD curves towards more alkaline pH values. (**A**), PcTx1 inhibitory effect on rASIC1a current from physiological conditioning pH 7.4 (black downward arrow) to every test pH is mostly due to its pH-dependent SSD promoting effect, whereas no more inhibition was observed from pH 8.0 (⊘) instead revealing a potentiation of the current at test pH in the activation curve pH range (7.2–6.2), due to the opposite potentiating effect by a leftward shift of the activation curve (black upward arrow) [29] (PcTx1 10 nM). (**B**), PcTx1 exerts almost no effect on the hASIC1a current maximally activated from conditioning pH 7.4 (⊘), an inhibitory effect on the hASIC1a current maximally activated from conditioning pH 7.2 (black downward arrow), and a potentiation on the hASIC1a current submaximally activated from conditioning pH 7.4 (black upward arrow). Curves adapted from [30,31] (PcTx1 1 nM), with the shift of activation curve deduced from the PcTx1-induced current potentiation at test pH 6.7 (green points, PcTx1 60 nM) [19]. (**C,D**), **rASIC1b (C) and hASIC1b (D) gating modulation by PcTx1.** The toxin promotes opening of rASIC1b and hASIC1b through a leftward shift of the activation curve towards less acidic pH with almost no effect on the SSD curve. Consequently, PcTx1 does no inhibit the current maximally activated from conditioning pH 7.4 (⊘), and potentiates the current submaximally activated from pH 7.4 to test pH 6.8–5.8 (black upward arrow). Curves adapted from [32] (PcTx1 100 nM), and the shift of hASIC1b activation curve is deduced from the PcTx1-induced potentiation of the current at test pH 6.3 (green points, PcTx1 60 nM) [19]. The effect of PcTx1 on hASIC1b SSD curve is not yet known. (**E,F**), **rASIC1a (E) and hASIC1a (F) gating modulation by Mamb**. Mamb inhibits rASIC1a and hASIC1a currents mainly by a rightward shift of the pH-dependent activation curve towards more acidic pH values. Curves for rASIC1a current adapted from [33] (Mamb-1, 200 nM), and for hASIC1a from [34] (Mamb-3, 10 nM; note that this concentration is below IC$_{50}$ value of Mamb on hASIC1a (see Table 3) and that a higher shift could thus be expected with a higher Mamb concentration). Red and green arrows illustrate shifts (acidic rightward, alkaline leftward) in the pH-dependent curves of activation and/or SSD by toxins. Data on the gating modulation of hASIC1b and rASIC1b by mambalgin-1 are shown in another following figure.

PcTx1 promotes opening of rASIC1b and hASIC1b by acidic drop from physiological resting pH 7.4 (EC$_{50}$~100 nM) through a leftward shift of their activation curve towards less acidic pH (Figure 4C,D), with almost no effect on the SSD curve. Consequently, PcTx1 does not inhibit the current maximally activated from conditioning pH 7.4, and potentiates the current submaximally activated from pH 7.4 to test pH 6.8–5.8, in the activation curve pH range. The absence of PcTx1 effect on the pH-dependent SSD of rASIC1b comes from an alteration of the contact between the toxin and the divergent adjacent upper part of the palm domain of ASIC1b (Figure 5A) [32,47]. PcTx1 is also able to constitutively activate cASIC1 at resting pH 7.4 [112] and to potentiate the H$^+$-activated cASIC1 current [98], presumably stabilizing only the open state of the channel like for ASIC1b.

Structural Mechanisms

PcTx1 primarily binds to the thumb domain through Trp7 and Trp24 that anchor the toxin to Phe351 of cASIC1, crucial for the specificity of PcTx1 [51,103] but also inserts an arginine-rich hairpin into the acidic pocket, making polar interactions mimicking local protonation of the acidic pocket (Figure 5A). Together, these polar and non-polar interactions link the finger, β-ball, and thumb domains of one subunit and the palm domain of the adjacent subunit [47,51] and lead to the collapsed conformation of the acidic pocket, which characterizes both the open and the desensitized states (Figure 3B,C). Three PcTx1 molecules thus bind at three equivalent sites on one homotrimeric ASIC1.

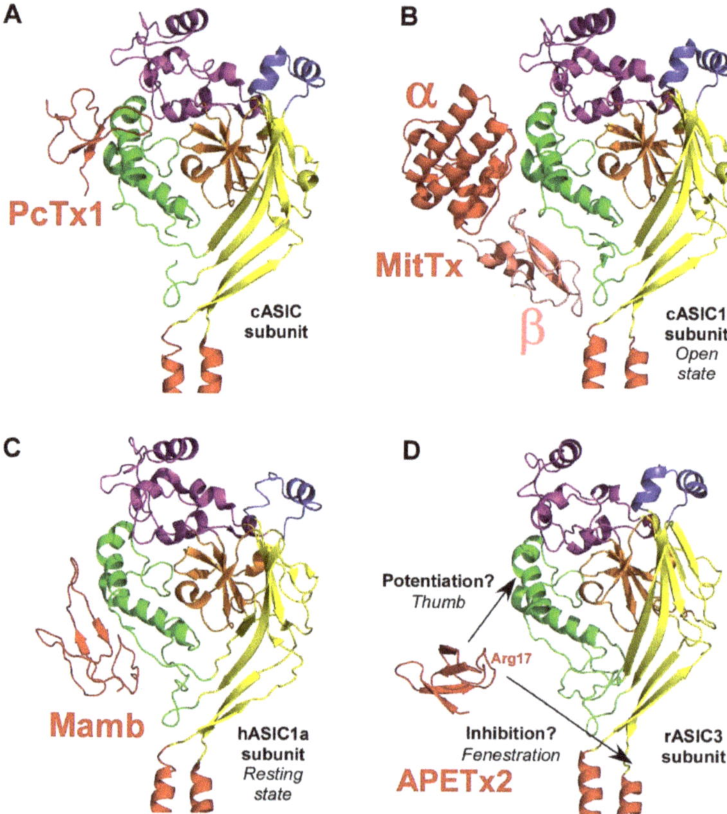

Figure 5. Toxin binding sites on one ASIC subunit. (**A**) Structure of a single cASIC1 subunit (rotated view of the skeletal 3D representation shown in Figure 1C) in complex with PcTx1 (PDB ID: 3s3x) [51]. (**B**), Structure of a single cASIC1 subunit in complex with MitTx (heterodimeric complex of MitTx-α and MitTx-β (PDB ID: 4NTY) [60]. (**C**), Cryo-EM structure of a single hASIC1a subunit in complex with mambalgin-1 (Mamb) at pH 8.0 (PDB ID: 7CFT) [52]. (**D**), Model of a single rASIC3 subunit extrapolated from cASIC1 structure, along with APETx2 (PDB ID: 2MUB) at the same scale, with two potential binding sites (black arrows) [129]. Designed with PyMOL software.

A series of chimeras realized between rASIC1a and rASIC1b or between rASIC1a and rASIC2a showed that the thumb, the β-ball and the palm domains of the adjacent subunit are crucial in explaining the difference in sensitivity of PcTx1 between the different ASIC isoforms. When ASIC1a residues of the β-ball were exchanged with the ones belonging to the PcTx1-insensitive ASIC2a (mutant 1a-RDQ190,258,259KQE) [49]), PcTx1 cannot inhibit the channel and rather induces a strong potentiation, in good agreement with the fact that Arg190 was involved in the interface with PcTx1 [51]. A similar effect is also observed when a part of the β-ball domain of ASIC1a was exchanged for the one of ASIC2a [47], showing the crucial role of the β-ball domain in the mechanism of PcTx1 inhibition of ASIC1a via the leftward shift of its pH-dependent SSD (Figure 4A,B).

When part of the palm domain of ASIC1a was exchanged with that of ASIC2a [49], PcTx1 also exerted a potentiating effect that cannot be attributed to an indirect effect via a change of the pH-dependence of SSD and can only be interpreted as a loss of contact between PcTx1 and the palm domain [47,49] preventing the modulation of the pH-dependent desensitization by PcTx1. The palm domains are different between rASIC1a and rASIC1b subunits, and interestingly, PcTx1 is also not able to inhibit the rASIC1b current by shifting its pH-dependent SSD curve [32] (Figure 4C), but rather induces a potentiating effect through the shift of the pH-dependent curve of activation. Accordingly, the introduction of part of the palm domain of rASIC1a into rASIC1b restored the inhibition by PcTx1 [47]. A similar mechanism could also take place in the case of hASIC1b [19] and cASIC1 [60,64,130], which all have palm domains divergent from ASIC1a and are all potentiated by PcTx1, and also explains why PcTx1 is able to open cASIC1 at pH 7.4 [98]. Moreover, an overlap between the effects of PcTx1 and GMQ was shown [131], which is known to act through the $\beta11$–$\beta12$ linker of the palm responsible for the desensitization (molecular clutch), highlighting the role of the palm domain to mediate the effect of PcTx1 on the desensitization process.

PcTx1 was reported to exert no inhibition (applied at pH 7.4, Table 3) on heterotrimeric ASIC1a/2a highly expressed in central neurons along with homotrimeric ASIC1a, and even a small potentiation on mASIC1a/2a when applied at conditioning pH 7.9 [40]. This can also be explained by the pH-dependent gating properties of these channels that show a half-maximal SSD around pH 6.8 [114,115], i.e., more acidic than rASIC1a, rendering the PcTx1-induced shift of the pH-dependent SSD curve not sufficient to desensitize ASIC1a/2a at conditioning pH 7.9, whereas the potentiation induced by the shift of the pH-dependent activation takes place. Accordingly, PcTx1 was able to inhibit heterotrimeric ASIC1a/2a when applied at the slightly acidic conditioning pH 7.0 [114,115].

PcTx1-Related Compounds

The peptide toxin **Hm3a** has five amino acid substitutions compared to PcTx1 and is three residues shorter at the C-terminus [124]. Only the R28K substitution is found in the active site, but has little apparent effect on its potency. Similar to PcTx1, Hm3a produces different effects on rASIC1a and rASIC1b that also depend on the palm domain of the adjacent subunit diverging between the two isoforms (especially Arg175 and Glu177 of rASIC1a corresponding to Cys and Gly residues in rASIC1b, respectively) [32,47].

The toxin **Hi1a** with two PcTx1-like peptides in tandem [125,126] inhibits rASIC1a and hASIC1a currents with a similar affinity and shifts (at 5 nM) the $pH_{0.5}$ of activation from 6.13 to 6.01 for rASIC1a, and from 6.22 to 6.04 for hASIC1a, suggesting a stabilization of the closed state [125]. However, the $pH_{0.5}$ of SSD is also shifted from 7.33 to 7.47 for rASIC1a and from 6.96 to 7.37 for hASIC1a, suggesting that Hi1a also promotes the desensitized state similarly to PcTx1. It seems that the Hi1a two-fold structure allows stabilization of the ASIC1a closed state as in the case of Mamb, in addition to promoting SSD like PcTx1 but weaker. This double effect, promoting desensitization and stabilization of the closed state, explains the inhibitory potency of Hi1a, which is not completely independent of conditioning pH as proposed by Chassagnon et al. [125]. When an alkaline conditioning pH is chosen to avoid the inhibitory effect taking place through the shift of the pH-dependent SSD curve, it reveals the effect generated by the shift of the pH-dependent activation curve, i.e., inhibition for Hi1a and potentiation for PcTx1.

The **C5b compound** was developed from the molecular knowledge of the binding of PcTx1 to ASIC1a. It binds in the acidic pocket, thus inhibiting hASIC1a and mASIC1a currents in a pH-dependent manner with an affinity decreasing with the acidification of the test pH value, as expected for a competitive proton inhibitor ($IC_{50} = 22$ nM, 100 nM or 7 µM when hASIC1a current is elicited from a conditioning pH 7.4 to test pH 6.7, 6.0 or 5.0, respectively) [132,133]. It is therefore more potent at mild than at extreme test pHs. C5b shifts the pH-dependent activation of hASIC1a towards lower pH, with a $pH_{0.5}$ value shifted from 6.57 to 6.4 by 100 nM C5b, and it reduces the maximal (pH 5.0-evoked) current, in good agreement with a mechanism of competition between C5b and protons, further suggesting that C5b prevents the collapse of the acidic pocket, leading to a stabilization of the closed state. On mouse brain slices, C5b (100 nM) is able to inhibit the ASIC part of EPSCs recorded in the anterior cingulate cortex as well as the LTP induction in the hippocampal CA3–CA1 pathway [132,133], and when it is i.v. injected in mice, C5b appears to cross the blood–brain barrier [133]. However, the C5b compound appears to be less specific than PcTx1, also inhibiting rASIC3 and heterotrimeric mASIC1a/2, but with a reduced affinity (no effect on ASIC2a and ASIC2a/2b, ASIC1b not tested).

2.3. MitTx, a Painful Toxin

2.3.1. Pharmacological Profile

MitTx was identified from the venom of Texas coral snake *Micrurus tener tener* as an α-bungarotoxin-like structure with two noncovalent subunits, a MitTx-α subunit consisting of a Kunitz type peptide of 60 amino acids, and a MitTx-β subunit, which is a 120 amino-acid phospholipase A2-like protein [99].

Independently of extracellular pH variations, MitTx was shown to constitutively activate several recombinant rodent homotrimeric and heterotrimeric ASICs [99,104], particularly rASIC1a and rASIC1b (EC_{50} = 9 and 23 nM, respectively), with a much lower effect on rASIC3 (EC_{50} = 830 nM) and on heterotrimeric rASIC1a/2a. At neutral pH, rASIC2a is not sensitive to MitTx but its proton-evoked activation is massively potentiated under more acidic conditions (pH 6.5) (Table 3, references included). Interestingly, the effect of PcTx1 and MitTx on rASIC1a were not additive [99], suggesting common binding sites. The effect of MitTx on native mouse ASIC channels in sensory trigeminal ganglion neurons, as well as the painful sensation induced by injection into the mouse hind paw, seems to mainly depend on ASIC1a-containing channels because these effects disappear in ASIC1a-KO mice.

MitTx, which locks ASICs in the open state, was co-crystallized with the cASIC1a channel to solve the first physiologically relevant open structure of these channels and to address the structure of the selectivity filter [51]. It nicely illustrates how toxins targeting ASICs are important tools not only to decode the physiological roles of these channels, but also to decrypt their structural and functional features.

2.3.2. MitTx, a Gating Modifier Stabilizing the Open State

Unlike PcTx1, which locally mimics protons by targeting the acidic pocket at the interface of two ASIC subunits, MitTx interacts with a single subunit by forming extensive interactions with the wrist, palm and thumb domains of ASIC1 and acts like a "churchkey" bottle opener [51] (Figures 3B and 5B). Several key contacts are necessary to set up this mechanism. The MitTx-α subunit insinuates the aromatic ring of its Phe14 at the interface of two channel subunits and splays them apart by forming extensive interactions with the β1-β2 linker and with the thumb domain of the adjacent subunit, both critical for the gating [12,134–139]. On the other hand, the MitTx-α subunit insinuates an ammonium group provided by its Lys16 into the wrist region, thereby coupling the base of the thumb to the TM1 domain (Figure 5B). It is interesting to note that the ammonium group occupies the same position as Cs^+ ions in the ASIC open state [60], underscoring the role of thumb/TM1 contact in the stabilization of the open conformation of the pore. All these contacts, together with the interaction of the MitTx-β subunit with the upper part of the thumb domain, stabilize the open state of the channel where the acidic pocket is collapsed, the molecular clutch formed by linker β11–β12 is not switched, and the extracellular vestibule is extended, leading to a stabilized symmetric open pore (Figure 3B), which does not evolve towards a desensitized state as in the case when the channel is activated by protons, or in the presence of PcTx1. During the pore opening induced by MitTx, the extensive TM2-mediated intersubunit contacts, that define the occlusion of the desensitized and closed ion channel, are disrupted [51]. This stabilization of the open state can be shown on the potentiation of the ASIC2a current by MitTx, involving a drastic shifting of its pH-dependent activation curve towards less acidic pH ($pH_{0.5}$ shifted from 3.5 to 6.0) [99]. The overlap of the binding between MitTx-β and PcTx1 on the thumb domain explains why the binding and biological activity of MitTx and PcTx1 are mutually exclusive [99,140].

2.4. Mambalgin

2.4.1. Pharmacological Profile

Mambalgin (Mamb) is a three finger peptide toxin of fifty-seven amino acids with 3 isoforms (each differing by only one amino acid) identified from the venom of the African black mamba *Dendroaspis polylepis* (mambalgin-1 and mambalgin-2) and from the venom of the green mamba *Dendroaspis angusticeps* (mambalgin-3) [33,98]. The three Mamb isoforms display the same pharmacological properties, and will thus not be distinguished in this review.

Mamb applied at conditioning pH 7.4 was shown to inhibit cloned rodent and human ASIC1a, rASIC1b as well as other rASIC1a-containing and rASIC1b-containing heterotrimeric channels with IC_{50} ranging from 11 to 252 nM, as well as cASIC1, without any effect on rASIC2a and rASIC3 (up to 3 µM) nor hASIC2a [98] (Table 3, references included), generally expressed in both *Xenopus* oocytes and mammalian cells, neither on a variety of ligand/voltage-gated ion channels [33].

Potentiating pH-dependent effects of Mamb were also observed on hASIC1b (from *Xenopus* oocytes and mammalian cells) and rat and human ASIC1b/3 (from mammalian cells for rASIC1b/3 and *Xenopus* oocytes for hASIC1b/3), but only for sub-maximal test pH values (6.6–6.0).

Mamb drastically inhibits ASIC currents of spinal cord and hippocampal neurons [33,37,141] in good correlation with the fact that Mamb inhibits different combinations of homo- and heteromeric ASICs thought to be expressed in central neurons (i.e., ASIC1a, ASIC1a/2a and/or ASIC1a/2b). In rat sensory neurons, Mamb inhibits about 60% of ASIC mean current amplitude and PcTx1 about 40%. The difference was attributed to the additional inhibition by Mamb of ASIC1b-containing channels in addition to homotrimeric ASIC1a also inhibited by PcTx1 [33,120,142]. Accordingly, in vivo analgesic effects in rodents are in good agreement with an inhibition of ASICs, confirmed by experiments with genetic invalidation of either ASIC1a or ASIC1b [33,142,143]. Mamb also potently inhibits (by 90%) hASIC currents recorded from human stem cell-derived sensory neurons [144].

2.4.2. Mambalgin, a Gating Modifier Stabilizing the Closed State
Biophysical Mechanisms and Relevance to In Vivo Analgesic Effects

On rASIC1a and hASIC1a, Mamb acts by a rightward shift of the pH-dependent activation curve towards more acidic pH values, thus stabilizing the channel closed state (Figure 4E,F) [33,34], without significant effect on the shift of the SSD curve [131]. Contrary to what was observed on rASIC1a and hASIC1a, Mamb is able to shift the pH-dependent SSD curve of rASIC1b and hASIC1b (Figure 6A,D) [34] towards more alkaline pHs, as observed with PcTx1 on rASIC1a and hASIC1a (Figure 4A,B).

Consequently, Mamb actually shows dual effects on hASIC1b and hASIC1b/3, either potentiation or inhibition, depending on both conditioning and test pH for channel activation, as illustrated by original data shown in Figure 6. Because of the shift of the activation curve towards less acidic pH (Figure 6A,B), the hASIC1b current activated from pH 7.4 to 6.0 is potentiated by Mamb. A partial inhibition is observed when the current is activated from pH 7.4 to 5.0 [34] that can be further increased when starting from a conditioning pH 6.6, because of the shift of the SSD curve towards less acidic pH (Figure 6A,C). The inhibitory effect of Mamb on rASIC1b is also pH-dependent, being stronger when the conditioning pH is slightly acidified. When the current is activated from pH 7.4 to 6.0, Mamb produces a partial inhibition (Figure 6E), whereas the current is fully inhibited upon activation by a pH drop from 6.6 to 5.0 (Figure 6F). This would support a higher potency of Mamb on both rodent and human ASIC1b in pathological situations where the extracellular pH is thought to be slightly acidified [84,145]. Like human channels [34], heterotrimeric rASIC1b/ASIC3 are weakly inhibited when the current is activated from 7.4 to 5.0 (Figure 6G,I) and can be potentiated upon activation by moderate acidosis (pH 6.6, Figure 6G,H).

These pH-dependent dual effects do not hinder the peripheral analgesic effects of Mamb in vivo against inflammatory and neuropathic pain described in rats and mice nor the participation of rASIC1b in these effects [33,142,146]. As it is clearly not the case in rodents despite similar pH-dependent effects, the dual pH-dependent effects of Mamb on hASIC1b-containing channels is therefore not expected to compromise possible analgesic effects in humans, as questioned recently [34]. In addition, analgesic effects of Mamb are not only supported by the inhibition of peripheral ASIC1b-containing channels, but also by the inhibition of ASIC1a-containing channels, as described in the central nervous system [33,142], and the rodent and human homotrimeric ASIC1a can be potently blocked by the peptide as well as heterotrimeric hASIC1a-containing channels (ASIC1a/2a, ASIC1a/1b, ASIC1a/3) [33,34].

Structural Mechanisms

Mamb interacts with rASIC1a directly through the thumb domain, but its inhibitory effect probably requires indirect influence of the palm and the β-ball domains [49]. In the structure of the hASIC1a/Mamb complex obtained by cryo-electron microscopy [52], Mamb preferentially binds to a channel conformation similar to the closed state [50,52]. In the cASIC1/Mamb complex, fingers I and II of Mamb bind to the α4 and α5 helices of the thumb domain, delimiting a part of the acidic pocket [108]. At the rat ASIC1a/Mamb interface, it was shown that the binding site is composed of four key residues forming a hinge between the α4 and α5 helices [106] (Figures 3A and 5C). Mamb locks it, preventing its motion during proton activation and leading to stabilization of the expanded shape of the acidic pocket and thus to stabilization of a channel conformation similar to the closed state [50,106]. Contrary to PcTx1 and similarly to MitTx, Mamb would not interfere directly with the acidic pocket [49,105] but only on the thumb domain [106,108]. However, contrary to MitTx, the core of Mamb would have no contact with the lower part of the thumb domain [106], which could explain their different effects. Contrary to PcTx1 and GMQ effects on rASIC1a, that showed an overlap of their mechanisms, no overlap is observed with Mamb, in good agreement with the fact that this toxin does not significantly affect the desensitization process [131].

By homology with the PcTx1 mechanism which modulates the SSD, the difference in Mamb behavior between the ASIC1a and ASIC1b isoforms could be supported by the palm domain, i.e., Mamb could modify ASIC1b SSD by interfering with the palm domain of ASIC1b but not with the one of ASIC1a. This hypothesis would be interesting to explore.

Although structural models deduced from the studies of the hASIC1a/Mamb [52] and of the rASIC1a/Mamb complexes [106] are very similar, there are differences in the pharmacological behaviors depending on the species or the subunit subtypes. In the rASIC1a/Mamb interaction, the Arg28 side chain of Mamb is freely exposed to the solvent and cannot be assigned to Glu342, Asp345 or Asp349 into the acidic pocket contrary to what is suggested for the hASIC1a/Mamb interaction [52]. The direct interaction of Arg28 of Mamb with Asp351 of hASIC1a in the acidic pocket would hinder its interaction with Arg190 located in the β-ball which is mandatory for collapse of the acidic pocket [12,52]. However, Mamb was shown to still be able to partially inhibit a rASIC1a channel where Arg190 was mutated, thus disrupting the inhibitory effect of PcTx1 targeting the acidic pocket [49], which does not support a central involvement of Arg190 in the Mamb inhibition, at least in rASIC1a.

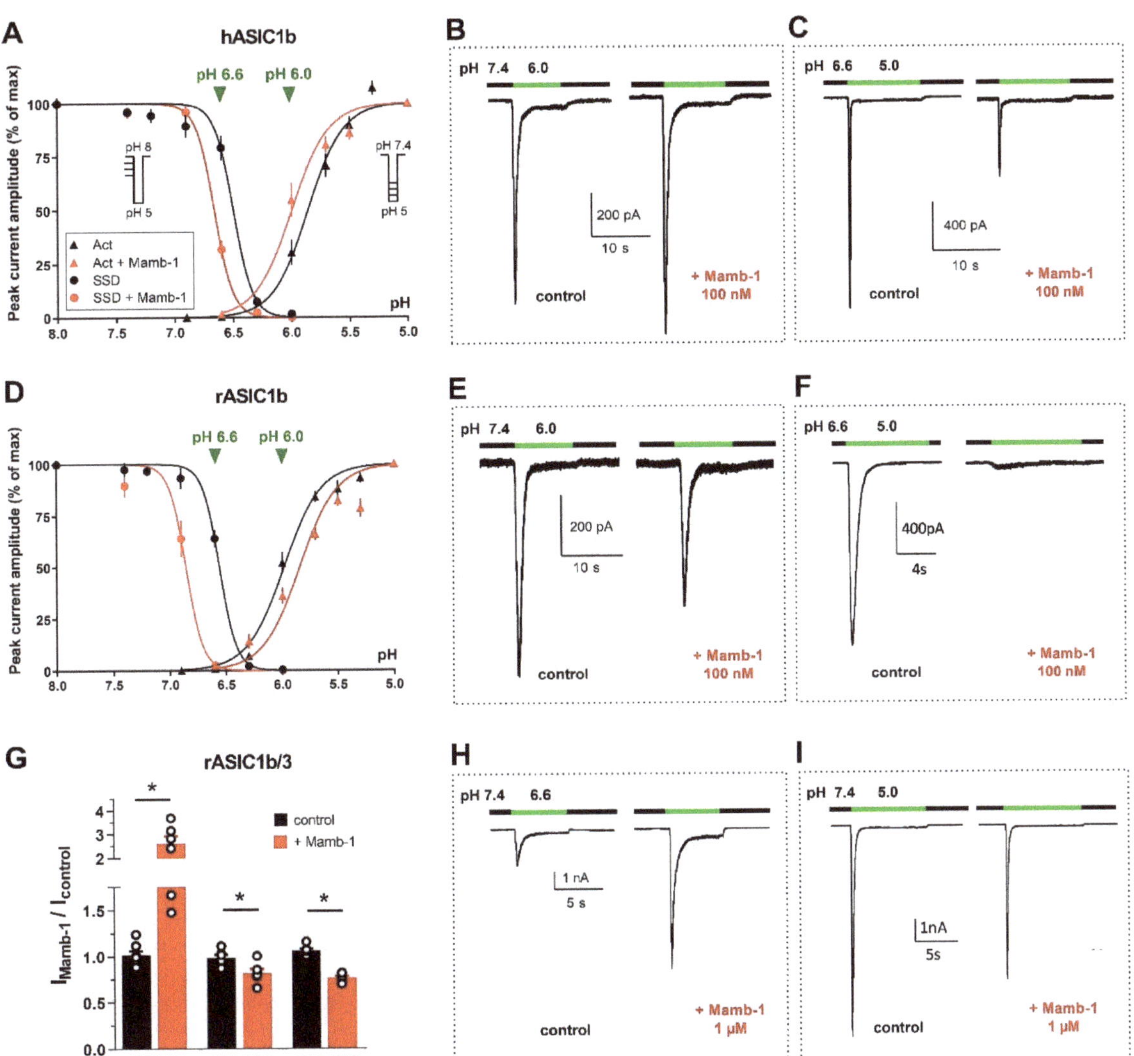

Figure 6. Dual pH-dependent effects of mambalgin-1 on whole-cell currents flowing through hASIC1b (**A–C**), rASIC1b (**D–F**) and rASIC1b/3 (**G–I**) heterologously expressed in HEK293 cells (unpublished data). (**A**) on hASIC1b current, mambalgin-1 (Mamb-1, 100 nM) induced a leftward shift of both the pH-dependent activation (Act) curve ($pH_{0.5}$ shifted from 5.86 to 6.0, $p = 0.02$, Mann–Whitney test) supporting a potentiating effect, and of the pH-dependent SSD curve ($pH_{0.5}$ shifted from 6.50 to 6.66, $p = 0.004$) supporting an inhibitory effect. The protocols used for activation and SSD are shown in inset in A (mean $\pm$ SEM; n = 4–12 cells per point). (**B,C**), original hASIC1b whole-cell current traces recorded at -60 mV illustrating the dual effects of Mamb-1 (100 nM, applied 30 s before the pH drop) on a current activated by a pH drop from 7.4 to 6.0 (potentiation, B), and on a current activated by a pH drop from 6.6 to 5.0 (inhibition, C). (**D**) on rASIC1b current, Mamb-1 (100 nM) induced a rightward shift of the pH-dependent activation curve ($pH_{0.5}$ shifted from 5.98 to 5.85, $p = 0.002$) and a leftward shift of the pH-dependent SSD curve ($pH_{0.5}$ shifted from 6.56 to 6.86, $p = 0.0002$), both supporting an inhibitory effect (same protocols and curve labels as in A; mean $\pm$ SEM; n = 4–12 cells per point). (**E,F**), original rASIC1b whole-cell current traces recorded in the same conditions as in

B-C and illustrating the partial inhibition by Mamb-1 of a current activated by a pH drop from 7.4 to 6.0 (**E**), and the full inhibition of a current activated by a pH drop from 6.6 to 5.0 (**F**). (**G**) bar graph quantification with individual data points of the effects of Mamb-1 (1 µM) on rASIC1b/3 heteromeric current. Mamb-1 induced a potentiation of the current when activated by a pH drop from 7.4 to 6.6 ($p = 0.02$, Wilcoxon paired test), but a partial inhibition when activated by a pH drop from 7.4 to 6.0 ($p = 0.03$) or 5.0 ($p = 0.03$). Mean ± SEM; n = 6–7 cells per condition. * $p < 0.05$. (**H,I**), original rASIC1b/3 whole-cell current traces recorded in the same conditions as in (**B,C**) and illustrating the potentiating effect (**H**) and the partial inhibition (**I**) by Mamb-1 depending on the test pH value.

Mamb is more efficient on rASIC1a (IC_{50} = 3–55 nM, Table 3) than on hASIC1a (IC_{50} = 24–203 nM, see Table 3 for references) despite a high degree of sequence identity (98.11%, Table 1). There are two interesting differences between rat and human sequences (also true for ASIC1b), near the wrist region in the thumb domain: an insertion of two residues (Asp298 and Leu299) is found in the human sequence, as well as a Lysine at position 291 instead of Asn291 in the rat sequence. Both positions have been tested [128] and it was revealed that most of the difference of Mamb affinity between hASIC1a and rASIC1a comes from the N291K variation. Another difference could be due to the fact that, in the rASIC1a/Mamb interaction, the Lys-8 of Mamb cannot be assigned to Asp298 or Asp296, as proposed in the hASIC1a/Mamb model [52,106].

2.5. APETx2 and APETx-like Peptides

2.5.1. Pharmacological Profile

APETx2 is a peptide of forty-two amino acids isolated from the venom of the sea anemone *Anthopleura elegantissima* [97], belonging to the disulfide-rich all-β structural family [147]. It was shown to inhibit rASIC3 (IC_{50} of 37–63 nM), hASIC3 (IC_{50} of 175–344 nM), rASIC2b/3 (IC_{50} of 117 nM) and, at higher concentrations, rASIC1a/3 (IC_{50} of 2 µM) and rASIC1b/3 (IC_{50} of 900 nM) (Table 3, references included). APETx2 rapidly and reversibly inhibits the transient ASIC3 peak current and the sustained window current evoked at test pH 7.0 [148], as well as the alkali-induced hASIC3 current [22], but the toxin does not affect the sustained component evoked at acidic test pHs [97].

Potentiating effects of APETx2 in the micromolar range were also reported on rASIC1b, rASIC2a, rASIC1b/3 and rASIC2a/3 heterologously expressed in *Xenopus* oocytes ([75] and unpublished data). However, no effect of similar concentrations was reported on the same channels expressed in a mammalian cell line [97], highlighting differences that can occur depending of the expression system.

On native ASIC currents of rodent neurons, APETx2 was only described to exert inhibitory effects interpreted as the inhibition of ASIC3-containing channels involved in the total current [97,120,143]. APETx2 injections in rodents pain models induced anti-hyperalgesic or analgesic effects against inflammatory pain [75,91,118,149–151], chemical pain [152], migraine [153], joint pain [154,155], ocular pain [156], muscular pain [148,157,158] and bone pain [159–161]. However, inter-animal variability was reported once in a rat model of inflammatory hyperalgesia using outbred animals, with 50% of non-responsive animals proposed to be linked to the potentiating effects of APETx2 at high dose [117] on ASIC currents that could counteract the analgesic effects.

At generally higher concentrations, APETx2 was shown to inhibit recombinant and native Nav1.8 voltage-dependent channel, a sensory neuron-specific Nav channel (IC_{50} of 2.6 µM for native channel in rat sensory neurons, 55 nM for recombinant rat channel and 6.6–18.7 µM for recombinant human Nav1.8 channel [162,163]), as well as Nav1.2 (IC_{50} = 114 nM [163]), Nav1.6 currents and the cardiac hERG channel in the micromolar range [164].

Hcr 1b-1, 2, 3, 4 are APETx-like peptides from the sea anemone *Heteractis crispa* that have different effects. Hcr 1b-1 and Hcr 1b-2 inhibit hASIC3 like APETx2, but with a lower affinity (IC_{50} of 5.5, and 15.9 µM, respectively), while Hcr 1b-2 also inhibits rASIC1a with IC$_{50}$ 4.8 $\pm$ 0.3 μM [165,166], and Hcr 1b-3 inhibits rASIC1a and rASIC3 with IC$_{50}$ of 5 and 17 μM, respectively. Hcr 1b-4 was found to be the first potentiator of ASIC3, simultaneously inhibiting rASIC1a at similar concentrations, with an EC$_{50}$ of 1.53 μM and an IC$_{50}$ of 1.25 μM, respectively. Hcr1b-2 showed an analgesic activity in vivo, significantly reducing the number of writhings in an acetic acid-induced writhing test, but promiscuous effects were also reported for Hcr1b-2 (1 μM), mostly inhibiting but also potentiating Kv, Nav and Cav (T type) channels [167].

2.5.2. APETx2, a Pore Blocker?

Even if APETx2 was identified in 2004 [97], its binding site and mechanism of action were poorly characterized. Only one study based entirely on docking and clustering proposes two potential binding sites on ASIC3, one on the thumb domain and the other one in the lower part of the extracellular domain (Figure 5D) [129]. Both sites remain plausible but the Arg17 of APETx2, which appears to be a critical determinant of ASIC3 inhibition, would only interact at the interface in the putative site near the bottom of the extracellular domain [110].

Inhibition by APETx2 is pH-independent (pH-dependent curves of activation and SSD are not shifted by APETx2 [128]), and this is what is expected by the APETx2 hypothetical binding site in the lower part of the palm domain, allowing occlusion of the extracellular fenestration, and suggesting a pore blocking behavior and not a gating modulation like PcTx1, MitTx and Mamb. The potentiation of rASIC1b and rASIC2a currents, as well as rASIC1b/3 and rASIC2a/3 expressed in *Xenopus* oocytes induced by concentrations of APETx2 30- to 100-fold higher than the concentration inhibiting rASIC3 (Table 3), could suggest that the second binding site located on the thumb domain might be involved in regulating its gating but with a much lower affinity.

2.6. Other Animal Toxins Targeting ASICs

Sea anemone toxins were shown to inhibit ASICs besides APETx2. The pi-AnmTx Ugr 9a-1 (**Ugr 9-1**) peptide, with a "twisted β–hairpin" structure, from *Urticina grebelnyi* inhibits hASIC3 expressed in *Xenopus* oocytes. It completely blocks the transient peak current with an IC$_{50}$ around 10 μM and partially inhibits the sustained current with an IC$_{50}$ of 1.4 μM [168]. Ugr 9-1 showed analgesic effects in two rodent models of inflammatory pain [118,168]. **PhcrTx1** from *Phymanthus crucifer* was also shown to partially inhibit native ASIC currents from rat sensory neurons (IC$_{50}$~100 nM) but also voltage-gated K$^+$ currents in the μM range. It presents an ICK scaffold and is the first member of a new structural group of sea anemone toxins [169]. The peptide **RPRFamide** from the marine cone snail *Conus textile*, related to the snail FMRFamide peptide, potentiates rASIC3 expressed in *Xenopus* oocytes, particularly the sustained component of the current (10–100 μM), therefore potentiating muscular acid-induced pain in mice in an ASIC3-dependent manner [170]. **α-Dendrotoxin** from the green mamba *Dendroaspis angusticeps* reversibly inhibits the transient ASIC currents in rat DRG neurons with an IC$_{50}$ of 0.8 μM, also inhibiting the sustained current at 3 μM [171]. It is a peptide of fifty-nine amino acids with a single Kunitz domain fold [172] similar to the α-subunit of MitTx (32% identity, 55% homology [173]), and is a well-known low nanomolar Kv1.x channel blocker. Finally, the short peptide **Sa12b** (10 amino acids, no cysteine) from the wasp *Sphex argentatus* was described to inhibit ASIC currents from rat sensory neurons (IC$_{50}$ = 81 nM) when applied in the conditioning as well as in the acidic test pH. Sa12b activity would not be pH-dependent, not interacting with the proton-gating mechanism [174]. An inhibitory effect of Sa12b (1 μM) was observed on cloned rASIC1a ($-$38.3 $\pm$ 6.5%, n = 8, p = 0.019 with paired t-test, synthetic Sa12b applied in conditioning pH 7.4 and test pH 5.5, unpublished data) expressed in *Xenopus* oocytes.

3. Expression of ASICs in the Nervous System and Peptide Toxin Effects on Native Currents

3.1. Expression of ASICs in Neurons

In vivo, the effects of ASIC-targeting compounds on pathophysiological processes are thought to depend on the mixture of functional ASICs involved in native ASIC currents, depending on the expression pattern of ASIC subunits in each cell type. Although expressed in various tissues in rodents like in humans [76,81], ASICs are largely found in neurons of both the central and the peripheral nervous system (Figure 7).

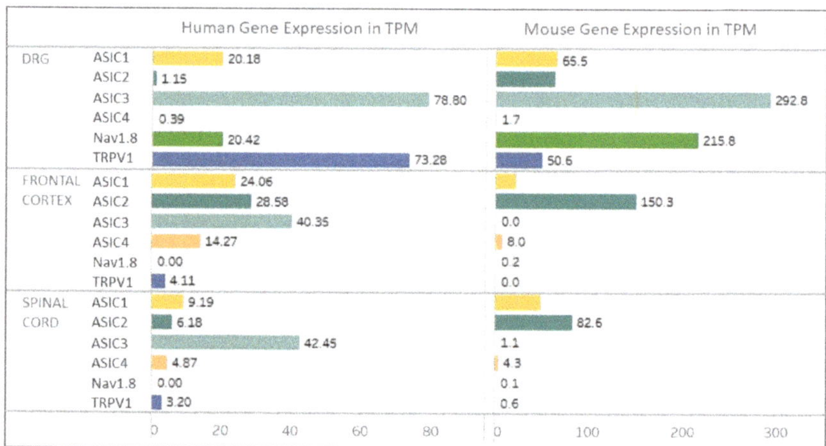

		Human Gene Expression in TPM	Mouse Gene Expression in TPM
DRG	ASIC1	20.18	65.5
	ASIC2	1.15	
	ASIC3	78.80	292.8
	ASIC4	0.39	1.7
	Nav1.8	20.42	215.8
	TRPV1	73.28	50.6
FRONTAL CORTEX	ASIC1	24.06	
	ASIC2	28.58	150.3
	ASIC3	40.35	0.0
	ASIC4	14.27	8.0
	Nav1.8	0.00	0.2
	TRPV1	4.11	0.0
SPINAL CORD	ASIC1	9.19	
	ASIC2	6.18	82.6
	ASIC3	42.45	1.1
	ASIC4	4.87	4.3
	Nav1.8	0.00	0.1
	TRPV1	3.20	0.6

Figure 7. Expression of ASICs in rodent and human nervous tissues. Expression of ASIC genes in central and peripheral nervous system from RNAseq data for human and mouse ASIC1-4, Nav 1.8 and TRPV1 genes from dorsal root ganglia (DRG), frontal cortex and spinal cord. Nav 1.8 and TRPV1 channels were shown as typically expressed in nociceptor sensory neurons. X axis represents TPM: Transcripts Per Million. Adapted from website: https://sensoryomics.shinyapps.io/RNA-Data/ [175] (accessed on June 2022).

3.1.1. Expression in Peripheral Sensory Neurons

In rodents, almost all ASIC genes (except ASIC4) are expressed in sensory neurons of the peripheral nervous system [76,79] in the dorsal root ganglia DRG (Figure 7) [42,176,177], the trigeminal ganglion TG [178,179], and the nodose and jugular ganglia from the vagus nerve [180–182]. The mesencephalic trigeminal nucleus only expresses ASIC1b, ASIC2a and ASIC3 [179,183].

Among the 17 subtypes of rodent DRG sensory neurons classified by large-scale single-cell RNA sequencing (scRNAseq), ASIC genes were found to be expressed in 14 subtypes [184]. ASIC1 (without discriminating between ASIC1a and ASIC1b splice variants) is expressed in most peptidergic neurons (four subtypes over eight, including myelinated Aδ-nociceptors), in neurofilament-expressing myelinated low-threshold mechanoreceptor (LTMR) and proprioceptor neurons for which ASIC1 mRNA is a characteristic marker, and to a lesser extent in some unmyelinated non-peptidergic neurons C-LTMRs (low threshold mechanoreceptors). ASIC2 is particularly highly expressed in all unmyelinated non-peptidergic neurons (six subtypes including C-LTMR and C-nociceptors), but also in peptidergic (six PEP subtypes over eight including unmyelinated C-nociceptors and myelinated Aδ nociceptors) and neurofilament-expressing myelinated neurons (LTMR). ASIC3 is highly expressed in unmyelinated peptidergic C-nociceptors and myelinated Aδ nociceptors (for which it is a marker), but also in neurofilament-expressing myelinated neurons (including LTMRs and proprioceptors) (data from the Mouse Brain Atlas of the Linnarsson Lab., http://mousebrain.org/) [177,184–186] (accessed on 1 June 2022). Classification of mice TG sensory neurons based on scRNAseq identified 13 neuronal types showing great similarities with the DRG subtypes, despite the lack of proprioceptors. Among TG sensory neurons, ASIC1 is expressed in cold nociceptors, large mechanosensory touch neurons (LTMRs) and peptidergic nociceptors, ASIC2 is present in touch C-fibers, non-peptidergic and peptidergic heat nociceptors and mechanonociceptors, and ASIC3 is expressed in large mechanosensory touch neurons, large nociceptors and peptidergic heat nociceptors [187,188].

An in situ hybridization study (RNAscope) on mouse lumbar DRG neurons showed that none of the five ASIC (except ASIC4) mRNAs showed a similar distribution [189]. ASIC mRNAs, including the splice variants, were expressed in myelinated neurons (including LTMRs, proprioceptors and Aδ-nociceptors). In non-myelinated C-nociceptors, ASIC2b was expressed in almost all neurons, ASIC1a, ASIC1b and ASIC3 were only expressed in peptidergic neurons, and ASIC2a was mostly expressed in non-peptidergic neurons [143,189]. Non-peptidergic nociceptors thus showed low expression of ASIC3 and the highest levels of ASIC2a and ASIC2b expression, with ASIC1a and ASIC1b not detected, in agreement with electrophysiological, immunohistochemical, and other in situ hybridization studies [190]. This is contrasting with the peptidergic subpopulation, in which more than 60% of neurons express the ASIC3 mRNA and approximately 25% express ASIC1a and ASIC1b mRNAs [189]. An increased expression of ASICs is associated with an increase in evoked and spontaneous excitability of small size nociceptor neurons, which may contribute to hyperalgesia and chronic inflammatory pain [191–193].

In humans, recent scRNAseq experiments using lumbar DRGs [194] showed that nociceptors represented ~60 to 70% of all sensory neurons. Humans also have Aβ-fiber nociceptors but non-peptidergic neurons do not exist (i.e., all sensory neurons are peptidergic). Like in rodents, ASIC1 and ASIC3 are expressed in human DRG (Figure 7) and TG neurons, ASIC2 being expressed at a lower level [175,195], with little or no expression of ASIC4. ASIC3 is part of the 30 most selectively expressed ion channels in the human TG and DRG compared to the brain or other non-nervous tissues, and ASIC1 and ASIC2 were found to be enriched in neurons compared to non-neuronal cells in rat and human TG and DRG [196]. Like in mice, ASIC1 is a marker of human proprioceptors [185]. ASIC1 and ASIC3 are notably expressed in putative "silent" nociceptors in humans [194], which correspond to a subset of C-fibers specifically expressing the cholinergic receptor nicotinic alpha 3 subunit (*CHRNA3*) that innervate joints, viscera and skin and are often referred to mechano-insensitive C-fibers [197]. They are unresponsive to noxious mechanical stimuli under normal conditions, but they can be sensitized after inflammatory stimulation as they express a wide array of receptors amongst which there is ASIC3, which might be important in certain pain disorders. Comparison with gene expression in mice [184,185] shows that ASIC1 is more widely expressed among DRG subpopulations in humans than in mice although it remains enriched in Aβ-LTMRs and proprioceptors. ASIC2, in contrast, is less widely expressed in human DRGs than in mice, mainly in Aδ-nociceptors and proprioceptors, and ASIC3 is largely expressed and to a higher level than in mice, principally in pruritogen receptor enriched nociceptors, in Aδ- and Aβ-nociceptors [194].

3.1.2. Expression in Central Neurons

In the rodent central nervous system, ASIC1a, ASIC2a and ASIC2b isoforms are widely expressed (Figure 7). scRNAseq experiments from mice cortex and hippocampus show that ASIC1 and ASIC2 genes are largely expressed in inhibitory and excitatory neurons in all neuronal subtypes, while ASIC3 (shown to be expressed in some brain areas) and ASIC4 are more expressed in inhibitory (GABAergic) neurons and some excitatory neurons. ASIC1 and ASIC4 were found to be characteristic markers of two distinct subclasses of inhibitory neurons in the midbrain [184,198]. The expression of ASIC1 and ASIC2 genes was shown in most inhibitory and excitatory neuronal subpopulations of the spinal cord [184,199,200] (Figure 7), in good agreement with functional electrophysiological studies on cultured neurons or dorsal spinal cord slices showing that ASIC currents were flowing through a mixture of homotrimeric ASIC1a and heterotrimeric ASIC1a/2 [37,122,123,201]. Although mainly expressed in sensory neurons, ASIC3 was also shown to be expressed in some areas of the rodent brain, particularly in pathologic states. In the hypothalamus and trigeminal nucleus caudalis, its expression is up regulated in a dural inflammatory mediated preclinical model of migraine [202], and in neuropathic mouse, ASIC3 was shown to be expressed in three brain regions (nucleus accumbens, medial prefrontal cortex and periacqueductal grey) of the pain brain network [203].

In humans, scRNAseq experiments from multiple cortical areas show that ASIC2 is widely expressed in inhibitory and excitatory neurons in all neuronal subtypes, but ASIC1 is mainly present in inhibitory neurons, whereas it is less widely expressed than in mice, and ASIC4 is likewise expressed in inhibitory (GABAergic) neurons [198]. However, contrary to rodents, the ASIC3 gene appears to be expressed in human spinal cord neurons (Sensoryomics website, https://paincenter.utdallas.edu/sensoryomics/ (accessed on 1 June 2022)) [22,175] (Figure 7).

3.1.3. Expression in Glial Cells

At lower expression levels than in neurons, ASICs (ASIC1a, ASIC2a and to a lesser degree ASIC3 or ASIC4) were shown to also be expressed in rodent and human glial cells that are involved in synaptic transmission and inflammatory responses in the nervous system [204–208]. ASIC1a expression seems prevalent, but the LPS (lipopolysaccharide) stimulation of cultured rat microglia induces an up regulation of ASIC1a, ASIC2a and ASIC3 expression and an increase in ASIC current, leading to a subsequent increase in intracellular calcium and expression of inflammatory cytokines that could be partly inhibited by PcTx1 [205]. Similar results were obtained from mouse astrocytes along with an increase in ASIC1a expression by pharmacological induction of epilepsy that also appears in hippocampal astrocytes from epileptic patients [207]. This suggests that substantial expression of ASICs in glial cells could also possibly be related to neuronal pathological states.

3.2. Effects of ASIC-Targeting Peptide Toxins on Native Currents

In rodent sensory neurons, Mamb was shown to inhibit about 60% of ASIC currents and for PcTx1 about 40%, which was attributed to the supplemental inhibition by Mamb of ASIC1b-containing channels in addition to ASIC1a-containing channels, while APETx2 inhibits Mamb-insensitive ASIC3-containing channels [33,97,120,143]. In central neurons, Mamb was accordingly found to drastically inhibit the ASIC currents of spinal cord and hippocampal neurons, whereas PcTx1 was found to only inhibit about 30% of their amplitude [33,37,121], supporting a combined expression of homotrimeric ASIC1a and heterotrimeric ASIC1a/ASIC2.

Data obtained on native human ASIC currents are in good agreement with the results from rodent neurons. Two studies on cultured human DRG neurons (after therapeutic ganglionectomy on patients suffering from chronic intractable pain) show that every recorded neuron was able to generate a transient ASIC-like current upon extracellular acidification, sometimes associated with a sustained TRPV1-like current [209,210]. Another study shows that ASIC currents recorded from human stem cell-derived sensory neurons [144] were inhibited by PcTx1, APETx2 and Mamb, with the order of efficacy Mamb > APETx2 > PcTx1, suggesting the involvement in human sensory neurons of ASIC1a- and/or ASIC1b-containing heterotrimeric channels, and also of ASIC3-containing channels. Another study combining transcriptomic RNAseq and electrophysiology on DRG neurons showed a sexual dimorphism in neuropathic patients, ASIC1 and ASIC3 genes being more expressed in males than in females [211].

In cultured central cortical neurons from patients undergoing craniotomies for the removal of brain tumor, 10 nM PcTx1 was found to inhibit by 70% the amplitude of the native ASIC current [212], suggesting the major involvement of homomeric ASIC1a in human central neurons, although the ASIC2a subunit was found co-expressed with ASIC1a with a similar ratio as in mice (with ASIC2b less expressed) [213]. A higher membrane targeting of the ASIC1a subunit was observed in acutely resected human cortical tissue (from patients undergoing surgical treatment of intractable epilepsy) compared to mice, possibly linked to a more efficient trafficking due to an amino acid difference at position 285 between mASIC1a and hASIC1a [213]. A dominant expression of ASIC1 compared to ASIC2 was also reported from human central neuronal cell line cultures [214], associated with a dominant ASIC1a current (strongly inhibited by PcTx1) and a native hASIC1a current was also recorded from neuroblastoma differentiated into neuronal-like phenotype, associated with the co-expression of ASIC2 but not ASIC3, leading to some Mamb-sensitive heterotrimeric ASIC1a/2a [215].

4. Pathophysiological Relevance of ASICs and in Vivo Effects of ASIC-Targeting Peptide Toxins

4.1. Relevance in Pain

Clinical data support the involvement of ASICs in **cutaneous pain** in humans. A decrease in pH in the skin of human volunteers was associated with non-adapting pain [216] and this cutaneous acid-induced pain is blocked by amiloride and/or NSAIDs [86,87,217], with a prominent effect for 7.4 < pH < 6.0 while pain associated with more acidic pH is also sensitive to capsazepine (an inhibitor of TRPV1) [87]. The respective role of ASIC and TRPV1 channels in human cutaneous acidic pain may be complex, as a recent clinical psychophysical study on 32 healthy volunteers suggests that TRPV1 would be the predominant sensor of pH 6.0-induced pain in skin [218]. The extreme pain evoked by the Texas coral snake's bite in humans [219] has also been attributed to the constitutive activation of human ASICs by MitTx present in the snake's venom [99], that would be an advantage on an evolutionary point of view against mammals that represent a threat (the snake living in urbanized territories). In rodents, APETx2 has shown potent analgesic effects after local application in cutaneous acidic and inflammatory pain supporting the role of peripheral ASIC3 in primary thermal and mechanical hyperalgesia [91,117,149,152,168], and the toxin Ugr 9-1, which also inhibits ASIC3 channels, reversed inflammatory and acid-induced pain after i.v. injection [168]. Intraplantar analgesic effects of Mamb support the involvement of DRG-specific ASIC1b-containing channels in rodent pain sensing [33]. ASIC1a channels could also be involved, particularly in the orofacial region (innervated by TG neurons), where PcTx1 was shown to reduce chemically induced sub-cutaneous pain [178], whereas it was generally without effect when applied elsewhere on the body skin, perhaps due to a higher ASIC1a expression in TG neurons than in DRG neurons [220] in rodents. From an ethologic point of view, it is interesting to note that the activation of cASIC1a channels by PcTx1 would represent an evolutionary advantage for spiders to induce pain in avian species that pose a threat in their environment, whereas inducing analgesic effects in prey like rodents could be useful to prevent them for fleeing or reacting to their bite. Similarly, analgesic properties of Mamb could constitute an advantage for mamba snakes against prey (rodents).

In rodent **migraine** models, amiloride [221] and APETx2 [153,222] show analgesic effects, suggesting a role of peripheral ASIC3 in dural afferents in migraine-related behavior, whereas Mamb effects [146] also suggested a role for peripheral ASIC1, most probably ASIC1b-containing channels, particularly in the chronification of cutaneous allodynia. In a small open clinical study on migraine, amiloride showed some efficacy for the reduction of aura and headache symptoms [223,224].

Blockade of ASIC3-containing channels exerts analgesic effects in several animal models of **bone pain** [159–161]. Inhibiting ASIC3 attenuates pain behaviors in animal models of osteoporosis, bone cancer and osteoarthritis, and bone pathologies in which inflammation is a major component [154,159,225]. The in vivo effects of APETx2 and electrophysiological recordings of bone afferent neurons on a bone-nerve rat preparation both suggest a role for ASIC3-containing channels in Aδ and C-fiber bone afferent neurons in the pathogenesis of inflammatory bone pain [226].

Regarding **joint pain**, ASIC3 was found to be expressed in more than 30% of DRG neurons innervating the knee joint in mice [227], and ASIC expression in DRG is increased in mice models of acute arthritis or rheumatoid arthritis [227,228]. LPC and arachidonic acid (AA) were shown to induce a slow constitutive activation of ASIC3 including the human isoform [21], and high levels of lysophosphatidylcholine (LPC) were measured in synovial fluids of two independent cohorts of patients with rheumatic diseases, correlated with pain outcomes in the cohort of osteoarthritis (OA) patients [20]. LPC also evokes a robust depolarizing current in DRG neurons at physiological pH 7.4, increases the firing of spinal nociceptive neuron innervated by nociceptive C-fiber, and induces pain behavior in rats and mice after subcutaneous co-injection with arachidonic acid, effects that are significantly reduced by ASIC3 blockers, including APETx2, or in ASIC3 knockout mice [20,21,229]. In a pathology-derived mouse model, intra-articular injections of LPC trigger a chronic pain state associated with anxiety-like behaviors that involves ASIC3-containing channels and is significantly reduced by intra-articular APETx2 [20]. APETx2 was also shown to reduce pain progression when injected in the early phase of an OA rat model [154]. This suggests a role for ASIC3 in triggering chronic joint pain, with potential implications of its inhibition for pain management in OA and possibly across other rheumatic diseases.

The role of ASICs was documented in **gastrointestinal pain** [230]. Increased hASIC3 expression in inflamed Crohn's disease intestine and small diameter sensory neurons of the neuronal plexus suggests a role in pain or dysmotility [231]. Considering oesophageal heartburn pain associated with gastroesophageal reflux disease (GERD), ASIC1 and ASIC3 expression were recently found to be increased in biopsies of patients compared to healthy subjects, which positively correlates with symptom severity of heartburn and regurgitation [232]. In the same study, injections of PcTx1 or APETx2 were found to normalize pain response to oesophageal distension in a rat model of reflux oesophagitis, leading to visceral mechanical hypersensitivity [232]. A clinical study was conducted in 2015 in a model of GERD showing that a small molecule (PPC-5650) with a weak selectivity for ASIC1a, reduced the sensitization to mechanical stimulation of the oesophagus [233].

ASICs were also proposed to be involved in **fibromyalgia and muscle pain** [234]. ASIC3 is expressed in more than 50% of small muscle sensory afferents in rat [148,186], and ASIC expression in DRG is increased in mouse models of muscle inflammation [235]. Fibromyalgia is commonly considered as a stress-related chronic pain disorder. The involvement of ASIC3 channels in this pathology was proposed based on their potentiation by LPC [21], as excessive oxidative stress and LPC (the LPC16:0 species) levels were reported in patients with fibromyalgia [236]. Moreover, LPC level was correlated with pain symptoms in patients with high oxidative stress and disease severity, and an increase in LPC was also observed in a mouse model of stress-induced chronic hyperalgesia, in which pharmacological or genetic inhibition of ASIC3 impeded the development of chronic hyperalgesia [236]. By combining ASIC3-knockout mice and APETx2, ASIC3 was also involved in mechanical hyperalgesia in a mouse model of fibromyalgia induced by repeated intramuscular acid injections [143,157], as well as in a rat reserpine-induced pain model [237], associated with an increased expression of ASIC3 in DRG [237,238], spinal cord and thalamus [238]. Activation of ASIC1b-containing channels was involved in the mouse model of fibromyalgia induced by intramuscular acid injections, with the development of hyperalgesia absent in ASIC1b-knockout mice and blocked by Mamb but not by PcTx1 [143]. PcTx1 was, however, shown to prevent activity-induced muscular hyperalgesia in mice [116], suggesting that ASIC1a could be important in the generation of muscle inflammation [235]. APETx2 was also able to relieve pain in a rat plantar incision model of **postoperative pain** [148].

In rodents, ASIC3 was shown to contribute to **orthodontic pain** [239–241], and recent data show that periodontal acidification (around pH 7.0) induced by tooth movement results in mechanical tooth pain hypersensitivity that was partially reversed by an injection of APETx2 in the periodontal tissue in rats [242]. A genome-wide association study suggests that a genetic variation in the ASIC2 gene could be significantly associated with severe gingival inflammation, linked to periodontitis [243].

In addition to peripheral sensory mechanisms, ASICs were shown to be involved in **spinal and supra-spinal pain processing**. In mice, i.t. and i.c.v. injections of PcTx1 induce potent analgesic effects in inflammatory and neuropathic pain models [123], as well as in a rat model of irritable bowel syndrome [244], involving the blockade of ASIC1a which causes a release of endogenous Met-enkephalin in the cerebrospinal fluid. Central injections (i.t. or i.c.v.) of Mamb in mice also induce analgesic effects on acute, inflammatory and neuropathic pain [33,142].

4.2. Relevance in Other Pathological Situations

ASICs are widely expressed in neurons outside the pain pathways, as well as in various non-neuronal tissues [76,81], thus supporting their involvement in others pathophysiological processes.

Interestingly, **neuroprotective effects** could be expected from ASIC inhibition, which would be complementary to the analgesic effects, particularly when neuronal damages are associated with pain, like in neuropathic or ischemic-related pain. In rodent CNS, ASICs, and particularly ASIC1a (whose opening could induce direct and indirect Ca^{2+} entry in neurons), were shown to participate in acidotoxicity and neuronal death associated with **ischemia** or **traumatic injury** [93,245,246]. Surviving neurons after ischemia/reperfusion protocol had increased levels of ASIC2a expression, whereas ASIC1a and ASIC2b levels remained unchanged, suggesting a potential protective role of ASIC2a-containing channels [247,248]. The i.c.v. injection of amiloride or PcTx1 protects against severe focal ischemia by reducing the infarct volume by more than 50% [246,249], i.c.v. administration of Hi1a up to eight hours after stroke shows neuroprotective potency in a rat focal ischemia model [125,250], and, in rat spinal cord, i.t. injection of PcTx1 reduces the lesion volume induced by traumatic injury [251]. When i.v. injected in mice, the C5b compound, developed from molecular interaction of PcTx1 with ASIC1a, appears to cross the blood–brain barrier and shows neuroprotective ASIC1a-dependent action, rescuing cerebral ischemia damages [133]. Amiloride showed neuroprotective as well as **myeloprotective effects** in animal models of multiple sclerosis [252,253]. Accordingly, chronic brain lesions of patients with progressive multiple sclerosis show an increased expression of ASIC1 in axons, and a pilot clinical study showed that orally given amiloride could exert neuroprotective effects [254]. Three **neurodegenerative diseases** are also suggested to have ASICs involved in their etiologies, including Parkinson's, Huntington's and Alzheimer's disease [255,256]. Both amiloride and PcTx1 were found to be neuroprotective in a mouse model of Parkinson's disease [257], and the neuroprotective effect of paeoniflorin, the principal active ingredient of an anti-Parkinson's disease traditional Chinese medicine, may involve inhibition of ASIC1a [258,259].

In **neonatal hyperbilirubinemia**, accumulation of bilirubin in the CNS results in neurotoxicity in various brain regions. A recent study showed that bilirubin potentiated the currents mediated by ASIC1a in an acidic environment and increased neuronal excitability, Ca^{2+} overload, spike firings and cell death [260]. Consistent with these results, neonatal conditioning with concurrent hyperbilirubinemia and acidosis primed long-term impairment of sensory and cognitive deficits in vivo in mice, suggesting potential benefits of ASIC inhibition [77].

ASIC inhibitors could also be useful to **relieve intervertebral disc degeneration and arthritis**. An up-regulation of ASIC1, ASIC2 and ASIC3 expression was described in the rodent and human nucleus pulposus in intervertebral disc degeneration [261,262]. Accordingly, PcTx1 reduces the acid-induced apoptosis and Ca^{2+} levels in apoptosis of endplate chondrocytes, supporting the involvement of ASIC1a [263], and PcTx1 as well as APETx2 were also shown to block the expression of acid-induced senescence-related markers in rat and human articular chondrocytes and chondrocyte cell lines [262,264]. In degenerated intervertebral discs and **osteoporosis**, extracellular acidosis induces osteoclastogenesis through intracellular Ca^{2+} rise, and both PcTx1 and specific ASIC1a siRNA significantly inhibit these events associated with bone resorption [265]. The increase in osteoclast activity not only leads to bone remodeling but is also a source of pronociceptive factors that sensitize the bone-innervating nociceptors. Recently, in a mouse model of **rheumatoid arthritis** a link between bone erosion and pain was found in a state of subclinical inflammation that could be relieved by APETx2 and ASIC3 genetic invalidation [228]. Cartilage and bone protective effect of ASIC-targeting compounds could be of therapeutic interest associated with analgesic effects against bone and joint pain. The ASIC-targeting wasp toxin Sa12b was recently found to improve the biological activity of cultured human nucleus pulposus mesenchymal stem cells isolated from patients who underwent lumbar disc herniation surgery [266] and the design of a new hydrogel containing Sa12b is now proposed for tissue engineering clinical trials to regenerate damaged nucleus pulposus in intervertebral disc [267].

Regarding **cardiovascular homeostasis**, the role of ASICs in the local vascular control is supported by the expression of ASIC3 in muscle metaboreceptors, the sensory nerves that innervate muscle arterioles and detect changes in muscle metabolism [186,268]. Inhibition of ASIC3-containing channels in sensory neurons by APETx2, amiloride and the NSAID diclofenac block the skin vasodilation response to direct pressure (pressure-induced vasodilation or PIV) in both humans and rodents [269], and a greater protein expression of ASIC3 was measured by immunoblotting in hypertensive SHR rats [270]. Genetic and pharmacological data using PcTx1 also demonstrated a role for ASIC1a in neurons in the regulation of microvascular tone and response to CO_2 via nitric oxide production and vasodilation [271]. In addition, ASIC1a was recently identified as a critical determinant in **heart ischemia reperfusion injury** through human genetics studies, human stem cell-derived cardiomyocytes and mouse models, confirming its potency in multiple ischemia and stroke injury models [125,127,246]. The treatment with Hi1a or PcTx1 reduces human stem cell-derived cardiomyocytes death by half after in vitro ischemia-reperfusion, and in mice, the genetic ablation or the pharmacological blockade of ASIC1a improves cardiomyocyte recovery after acute ischemia-reperfusion injury without affecting heart functional homeostasis [127].

ASIC1a channel inhibitors (PcTx1 and amiloride) cause a significant reduction of **tumor growth** and tumor load in mice [272]. Expression of ASIC1a was reported to be high in cancer patients, in vitro experiments revealed that PcTx1 or ASIC1a siRNA could weaken the migration, proliferation and invasion of tumor cells, and PcTx1 (i.v.) could inhibit breast tumor growth in mice [273]. Mamb was shown to also be efficient against growth and migration of glioma, leukemia and melanoma cells [274–276].

Anxiolytic-like effects of PcTx1 have also been described, which would be again complementary to the analgesic effects. A genetic variation of the human gene coding for ASIC1a was associated with panic disorder and with anxiety phenotypes linked to amygdala dysfunction [277,278]. Genetic disruption of ASIC1a in neurons or i.c.v. injection of PcTx1 was shown to have antidepressant effects in mice [279,280]. However, other data show that activation of ASIC1a channels in the rat basolateral amygdala decreases anxiety-like behavior, while inhibition by PcTx1 would increase the level of anxiety in rats [281]. These discrepancies could rely on activity of ASICs in other brain regions than the amygdala (like hippocampus) or on differences in ASIC1a contribution in innate and acquired fear [77].

ASIC involvement in **epilepsy** is also still unclear, with some in vivo rodent data suggesting that ASIC inhibition (particularly by i.c.v. PcTx1) can reduce epileptic symptoms [77], while other data indicated that ASICs, and especially ASIC1a, could play a role in seizure termination through activation of inhibitory interneurons when brain pH decreases [282]. In humans, genetic study suggested an association between single nucleotide polymorphisms in ASIC1 and temporal lobe epilepsy [283]. A down regulation of the expression of ASIC1a was reported in cortical lesions of patients with focal cortical dysplasia, a recognized cause of medically intractable epilepsy [284], which suggests that ASIC1a loss may contribute to epileptogenesis in these patients. An anti-epileptic role has also been suggested for ASIC3-containing channels expressed in interneurons, associated with elevated expression in the brains of temporal lobe epilepsy patients and rats, and inhibition of ASIC3-containing channels with APETx2 in rat models of epilepsy increases seizures susceptibility [285]. In addition to neuronal ASICs, an increased ASIC1a expression also occurs in mouse astrocytes after induction of epilepsy, as well as in hippocampal astrocytes from epileptic patients [207].

5. Conclusions

Toxins have not only been instrumental in the study of ASICs to understand the molecular features of these channels and their pH-dependent gating, but also to study the physiological and pathological roles of native channels both in vitro and in vivo in rodent pain models, supporting ASICs as therapeutic targets in pain and beyond. Furthermore, the molecular knowledge of ASIC gating and interaction with ASIC-targeting toxin inhibitors now allows to design new molecules like C5b and to predict their pharmacological potential and possibly their therapeutic relevance. Other putative applications of these peptides are also emerging, such as the design of a fusion protein incorporating an alpaca-derived nanobody targeting hASIC1a and the peptide toxin PcTx1 to achieve potent and, contrary to PcTx1 alone, more stable inhibition of ASIC1a (~84–87% current inhibition), improving the potency of PcTx1 and its potential applications [16].

ASICs exhibit complex and highly pH-dependent gating properties, and it is therefore not surprising that ASIC-targeting peptide toxins interact with channel gating also by a complex and pH-dependent mechanism. However, despite sometimes complex behaviors with pH-dependent activating, potentiating or inhibitory effects on rodent and human cloned channels recorded in vitro, the effects of ASIC-targeting toxins described on neuronal native ASIC currents in rodent and, when possible, in human neurons were mostly inhibitory for PcTx1, APETx2 and Mamb, or stimulatory for MitTx, in good agreement with in vivo effects. Insights from rodent and human studies on various pain-related processes [78] show that the ASIC-inhibiting toxins PcTx1, Mamb and APETx2 always induce a reduction of acute or pathological pain, whereas MitTx increased pain-related behaviors, which is consistent with the effects of other ASIC inhibitors like amiloride or NSAIDs, and with the effects of ASIC genetic invalidation or knockdown. Dual pH-dependent effects of the ASIC-targeting compounds do not therefore compromise their analgesic relevance both in the central and peripheral nervous system. Furthermore, complementary effects of some ASIC-targeting analgesic toxins like neuroprotective and anxiolytic effects could even be beneficial.

ASICs are thus interesting potential drug targets regarding the need to develop new and more effective analgesics with limited adverse side effects, notably in the context of the opioid crisis [286]. In addition, several administration pathways could be used to alternatively target local, peripheral or central pain mechanisms, depending on the ASIC subtype targeted, the physicochemical properties of the compounds and their blood–brain barrier permeability. Some of the ASIC peptide blockers described here could be interesting potential leads for pain relief, including in the context of chronic and inflammatory pain, with relevance in migraine, bone, joint and muscle pain, fibromyalgia, and postoperative, gastrointestinal or tooth pain, thus deserving further characterization of their effect on native ASICs in human neurons.

Author Contributions: All authors contributed to drafting the manuscript and have read and agreed to its published version.

Funding: Work in the authors' laboratory was supported by grants from the Centre National de la Recherche Scientifique and the Institut National de la Santé et de la Recherche Médicale, and by grants from the Agence Nationale de la Recherche (ANR-17-CE18-0019, ANR-17-CE16-0018, and ANR-11-LABX-0015-01) and from the Association Française contre les Myopathies (AFM grants #19618, #23731).

Institutional Review Board Statement: Data involving animal experimentation were obtained with the agreement of the Institutional Local Ethical Committee and authorized by the French Ministry of Research according to the European Union regulations and Directive 2010/63/EU (Agreements E061525 and APAFIS#14012-2018030214144339).

Acknowledgments: The authors thank members of the Ion Channels and Pain team for discussions, and T. Price for authorization to use graphical data from Sensoryomics website.

Conflicts of Interest: The authors declare no conflict of interest.

References

1. Krishtal, O.A.; Pidoplichko, V.I. A receptor for protons in the nerve cell membrane. *Neuroscience* **1980**, *5*, 2325–2327. [CrossRef]
2. Waldmann, R.; Bassilana, F.; de Weille, J.; Champigny, G.; Heurteaux, C.; Lazdunski, M. Molecular cloning of a non-inactivating proton-gated Na$^+$ channel specific for sensory neurons. *J. Biol. Chem.* **1997**, *272*, 20975–20978. [CrossRef] [PubMed]
3. Waldmann, R.; Champigny, G.; Bassilana, F.; Heurteaux, C.; Lazdunski, M. A proton-gated cation channel involved in acid-sensing. *Nature* **1997**, *386*, 173–177. [CrossRef] [PubMed]
4. Kellenberger, S.; Schild, L.; Ohlstein, E.H. International Union of Basic and Clinical Pharmacology. XCI. Structure, Function, and Pharmacology of Acid-Sensing Ion Channels and the Epithelial Na$^+$ Channel. *Pharmacol. Rev.* **2015**, *67*, 1–35. [CrossRef] [PubMed]
5. Rosenbaum, T.; Morales-Lazaro, S.L.; Islas, L.D. TRP channels: A journey towards a molecular understanding of pain. *Nat. Rev. Neurosci.* **2022**, *23*, 596–610. [CrossRef]
6. Comes, N.; Gasull, X.; Callejo, G. Proton Sensing on the Ocular Surface: Implications in Eye Pain. *Front. Pharm.* **2021**, *12*, 773871. [CrossRef]
7. Aneiros, E.; Cao, L.; Papakosta, M.; Stevens, E.B.; Phillips, S.; Grimm, C. The biophysical and molecular basis of TRPV1 proton gating. *EMBO J.* **2011**, *30*, 994–1002. [CrossRef]
8. Feliciangeli, S.; Chatelain, F.C.; Bichet, D.; Lesage, F. The family of K2P channels: Salient structural and functional properties. *J. Physiol.* **2015**, *593*, 2587–2603. [CrossRef]
9. Napolitano, L.M.R.; Torre, V.; Marchesi, A. CNG channel structure, function, and gating: A tale of conformational flexibility. *Pflug. Arch.* **2021**, *473*, 1423–1435. [CrossRef]
10. Marchesi, A.; Arcangeletti, M.; Mazzolini, M.; Torre, V. Proton transfer unlocks inactivation in cyclic nucleotide-gated A1 channels. *J. Physiol.* **2015**, *593*, 857–870. [CrossRef]
11. Zha, X.M.; Xiong, Z.G.; Simon, R.P. pH and proton-sensitive receptors in brain ischemia. *J. Cereb. Blood Flow Metab.* **2022**, *42*, 1349–1363. [CrossRef] [PubMed]
12. Jasti, J.; Furukawa, H.; Gonzales, E.B.; Gouaux, E. Structure of acid-sensing ion channel 1 at 1.9 A resolution and low pH. *Nature* **2007**, *449*, 316–323. [CrossRef] [PubMed]
13. Lingueglia, E.; de Weille, J.R.; Bassilana, F.; Heurteaux, C.; Sakai, H.; Waldmann, R.; Lazdunski, M. A modulatory subunit of acid sensing ion channels in brain and dorsal root ganglion cells. *J. Biol. Chem.* **1997**, *272*, 29778–29783. [CrossRef]
14. Hesselager, M.; Timmermann, D.B.; Ahring, P.K. pH Dependency and Desensitization Kinetics of Heterologously Expressed Combinations of Acid-sensing Ion Channel Subunits. *J. Biol. Chem.* **2004**, *279*, 11006–11015. [CrossRef] [PubMed]
15. Gonzales, E.B.; Kawate, T.; Gouaux, E. Pore architecture and ion sites in acid-sensing ion channels and P2X receptors. *Nature* **2009**, *460*, 599–604. [CrossRef] [PubMed]
16. Wu, Y.; Chen, Z.; Sigworth, F.J.; Canessa, C.M. Structure and analysis of nanobody binding to the human ASIC1a ion channel. *eLife* **2021**, *10*, e67115. [CrossRef]
17. Jumper, J.; Evans, R.; Pritzel, A.; Green, T.; Figurnov, M.; Ronneberger, O.; Tunyasuvunakool, K.; Bates, R.; Žídek, A.; Potapenko, A.; et al. Highly accurate protein structure prediction with AlphaFold. *Nature* **2021**, *596*, 583–589. [CrossRef]
18. Bassler, E.L.; Ngo-Anh, T.J.; Geisler, H.S.; Ruppersberg, J.P.; Grunder, S. Molecular and functional characterization of acid-sensing ion channel (ASIC) 1b. *J. Biol. Chem.* **2001**, *276*, 33782–33787. [CrossRef]
19. Hoagland, E.N.; Sherwood, T.W.; Lee, K.G.; Walker, C.J.; Askwith, C.C. Identification of a Calcium Permeable Human Acid-sensing Ion Channel 1 Transcript Variant. *J. Biol. Chem.* **2010**, *285*, 41852–41862. [CrossRef]
20. Jacquot, F.; Khoury, S.; Labrum, B.; Delanoe, K.; Pidoux, L.; Barbier, J.; Delay, L.; Bayle, A.; Aissouni, Y.; Barriere, D.A.; et al. Lysophosphatidylcholine 16: 0 mediates chronic joint pain associated to rheumatic diseases through acid-sensing ion channel 3. *Pain* **2022**, *163*, 1999–2013. [CrossRef]
21. Marra, S.; Ferru-Clément, R.; Breuil, V.; Delaunay, A.; Christin, M.; Friend, V.; Sebille, S.; Cognard, C.; Ferreira, T.; Roux, C.; et al. Non-acidic activation of pain-related Acid-Sensing Ion Channel 3 by lipids. *EMBO J.* **2016**, *35*, 414–428. [CrossRef] [PubMed]
22. Delaunay, A.; Gasull, X.; Salinas, M.; Noël, J.; Friend, V.; Lingueglia, E.; Deval, E. Human ASIC3 channel dynamically adapts its activity to sense the extracellular pH in both acidic and alkaline directions. *Proc. Natl. Acad. Sci. USA* **2012**, *109*, 13124–13129. [CrossRef]
23. Deval, E.; Salinas, M.; Baron, A.; Lingueglia, E.; Lazdunski, M. ASIC2b-dependent Regulation of ASIC3, an Essential Acid-sensing Ion Channel Subunit in Sensory Neurons via the Partner Protein PICK-1. *J. Biol. Chem.* **2004**, *279*, 19531–19539. [CrossRef]
24. Sivils, A.; Yang, F.; Wang, J.Q.; Chu, X.-P. Acid-Sensing Ion Channel 2: Function and Modulation. *Membranes* **2022**, *12*, 113. [CrossRef] [PubMed]
25. Salinas, M.; Lazdunski, M.; Lingueglia, E. Structural Elements for the Generation of Sustained Currents by the Acid Pain Sensor ASIC3. *J. Biol. Chem.* **2009**, *284*, 31851–31859. [CrossRef] [PubMed]

26. Bartoi, T.; Augustinowski, K.; Polleichtner, G.; Gründer, S.; Ulbrich, M.H. Acid-sensing ion channel (ASIC) 1a/2a heteromers have a flexible 2:1/1:2 stoichiometry. *Proc. Natl. Acad. Sci. USA* **2014**, *111*, 8281–8286. [CrossRef]

27. de Weille, J.R.; Bassilana, F.; Lazdunski, M.; Waldmann, R. Identification, functional expression and chromosomal localisation of a sustained human proton-gated cation channel. *FEBS Lett.* **1998**, *433*, 257–260. [CrossRef]

28. Askwith, C.C.; Wemmie, J.A.; Price, M.P.; Rokhlina, T.; Welsh, M.J. Acid-sensing Ion Channel 2 (ASIC2) Modulates ASIC1 H⁺-activated Currents in Hippocampal Neurons. *J. Biol. Chem.* **2004**, *279*, 18296–18305. [CrossRef]

29. Chen, X.; Kalbacher, H.; Grunder, S. The tarantula toxin psalmotoxin 1 inhibits acid-sensing ion channel (ASIC) 1a by increasing its apparent H⁺ affinity. *J. Gen. Physiol.* **2005**, *126*, 71–79. [CrossRef]

30. Sherwood, T.W.; Askwith, C.C. Endogenous Arginine-Phenylalanine-Amide-related Peptides Alter Steady-state Desensitization of ASIC1a. *J. Biol. Chem.* **2008**, *283*, 1818–1830. [CrossRef]

31. Cristofori-Armstrong, B.; Saez, N.J.; Chassagnon, I.R.; King, G.F.; Rash, L.D. The modulation of acid-sensing ion channel 1 by PcTx1 is pH-, subtype- and species-dependent: Importance of interactions at the channel subunit interface and potential for engineering selective analogues. *Biochem. Pharmacol.* **2019**, *163*, 381–390. [CrossRef] [PubMed]

32. Chen, X.; Kalbacher, H.; Grunder, S. Interaction of acid-sensing ion channel (ASIC) 1 with the tarantula toxin psalmotoxin 1 is state dependent. *J. Gen. Physiol.* **2006**, *127*, 267–276. [CrossRef] [PubMed]

33. Diochot, S.; Baron, A.; Salinas, M.; Douguet, D.; Scarzello, S.; Dabert-Gay, A.-S.; Debayle, D.; Friend, V.; Alloui, A.; Lazdunski, M.; et al. Black mamba venom peptides target acid-sensing ion channels to abolish pain. *Nature* **2012**, *490*, 552–555. [CrossRef]

34. Cristofori-Armstrong, B.; Budusan, E.; Rash, L.D. Mambalgin-3 potentiates human acid-sensing ion channel 1b under mild to moderate acidosis: Implications as an analgesic lead. *Proc. Natl. Acad. Sci. USA* **2021**, *118*, e2021581118. [CrossRef] [PubMed]

35. Vaithia, A.; Vullo, S.; Peng, Z.; Alijevic, O.; Kellenberger, S. Accelerated Current Decay Kinetics of a Rare Human Acid-Sensing ion Channel 1a Variant That Is Used in Many Studies as Wild Type. *Front. Mol. Neurosci.* **2019**, *12*, 133. [CrossRef]

36. Rook, M.L.; Miaro, M.; Couch, T.; Kneisley, D.L.; Musgaard, M.; MacLean, D.M. Mutation of a conserved glutamine residue does not abolish desensitization of acid-sensing ion channel 1. *J. Gen. Physiol.* **2021**, *153*, e202012855. [CrossRef]

37. Baron, A.; Voilley, N.; Lazdunski, M.; Lingueglia, E. Acid sensing ion channels in dorsal spinal cord neurons. *J. Neurosci.* **2008**, *28*, 1498–1508. [CrossRef]

38. Baron, A.; Schaefer, L.; Lingueglia, E.; Champigny, G.; Lazdunski, M. Zn²⁺ and H⁺ are coactivators of acid-sensing ion channels. *J. Biol. Chem.* **2001**, *276*, 35361–35367. [CrossRef]

39. Chen, Z.; Kuenze, G.; Meiler, J.; Canessa, C.M. An arginine residue in the outer segment of hASIC1a TM1 affects both proton affinity and channel desensitization. *J. Gen. Physiol.* **2021**, *153*, e202012802. [CrossRef]

40. Sherwood, T.W.; Lee, K.G.; Gormley, M.G.; Askwith, C.C. Heteromeric Acid-Sensing Ion Channels (ASICs) Composed of ASIC2b and ASIC1a Display Novel Channel Properties and Contribute to Acidosis-Induced Neuronal Death. *J. Neurosci.* **2011**, *31*, 9723–9734. [CrossRef]

41. Bassilana, F.; Champigny, G.; Waldmann, R.; de Weille, J.R.; Heurteaux, C.; Lazdunski, M. The acid-sensitive ionic channel subunit ASIC and the mammalian degenerin MDEG form a heteromultimeric H⁺-gated Na⁺ channel with novel properties. *J. Biol. Chem.* **1997**, *272*, 28819–28822. [CrossRef] [PubMed]

42. Hattori, T.; Chen, J.; Harding, A.M.S.; Price, M.P.; Lu, Y.; Abboud, F.M.; Benson, C.J. ASIC2a and ASIC3 Heteromultimerize to Form pH-Sensitive Channels in Mouse Cardiac Dorsal Root Ganglia Neurons. *Circ. Res.* **2009**, *105*, 279–286. [CrossRef] [PubMed]

43. Osmakov, D.I.; Koshelev, S.G.; Andreev, Y.A.; Kozlov, S.A. Endogenous Isoquinoline Alkaloids Agonists of Acid-Sensing Ion Channel Type 3. *Front. Mol. Neurosci.* **2017**, *10*, 282. [CrossRef] [PubMed]

44. Chen, X.; Paukert, M.; Kadurin, I.; Pusch, M.; Gründer, S. Strong modulation by RFamide neuropeptides of the ASIC1b/3 heteromer in competition with extracellular calcium. *Neuropharmacology* **2006**, *50*, 964–974. [CrossRef]

45. Malysz, J.; Scott, V.E.; Faltynek, C.; Gopalakrishnan, M. Characterization of human ASIC2a homomeric channels stably expressed in murine Ltk− cells. *Life Sci.* **2008**, *82*, 30–40. [CrossRef]

46. Cho, J.-H.; Askwith, C.C. Potentiation of acid-sensing ion channels by sulfhydryl compounds. *Am. J. Physiol.-Cell. Physiol.* **2007**, *292*, C2161–C2174. [CrossRef]

47. Salinas, M.; Rash, L.D.; Baron, A.; Lambeau, G.; Escoubas, P.; Lazdunski, M. The receptor site of the spider toxin PcTx1 on the proton-gated cation channel ASIC1a. *J. Physiol.* **2006**, *570*, 339–354. [CrossRef]

48. Sutherland, S.P.; Benson, C.J.; Adelman, J.P.; McCleskey, E.W. Acid-sensing ion channel 3 matches the acid-gated current in cardiac ischemia-sensing neurons. *Proc. Natl. Acad. Sci. USA* **2001**, *98*, 711–716. [CrossRef]

49. Salinas, M.; Besson, T.; Delettre, Q.; Diochot, S.; Boulakirba, S.; Douguet, D.; Lingueglia, E. Binding Site and Inhibitory Mechanism of the Mambalgin-2 Pain-relieving Peptide on Acid-sensing Ion Channel 1a. *J. Biol. Chem.* **2014**, *289*, 13363–13373. [CrossRef]

50. Yoder, N.; Yoshioka, C.; Gouaux, E. Gating mechanisms of acid-sensing ion channels. *Nature* **2018**, *555*, 397–401. [CrossRef]

51. Baconguis, I.; Bohlen, C.J.; Goehring, A.; Julius, D.; Gouaux, E. X-Ray Structure of Acid-Sensing Ion Channel 1–Snake Toxin Complex Reveals Open State of a Na⁺-Selective Channel. *Cell* **2014**, *156*, 717–729. [CrossRef] [PubMed]

52. Sun, D.; Liu, S.; Li, S.; Zhang, M.; Yang, F.; Wen, M.; Shi, P.; Wang, T.; Pan, M.; Chang, S.; et al. Structural insights into human acid-sensing ion channel 1a inhibition by snake toxin mambalgin1. *eLife* **2020**, *9*, e57096. [CrossRef] [PubMed]

53. Yoder, N.; Gouaux, E. Divalent cation and chloride ion sites of chicken acid sensing ion channel 1a elucidated by x-ray crystallography. *PLoS ONE* **2018**, *13*, e0202134. [CrossRef]

54. Li, T.; Yang, Y.; Canessa, C.M. Interaction of the Aromatics Tyr-72/Trp-288 in the Interface of the Extracellular and Transmembrane Domains Is Essential for Proton Gating of Acid-sensing Ion Channels. *J. Biol. Chem.* **2009**, *284*, 4689–4694. [CrossRef]

55. Yoder, N.; Gouaux, E. The His-Gly motif of acid-sensing ion channels resides in a reentrant 'loop' implicated in gating and ion selectivity. *eLife* **2020**, *9*, e56527. [CrossRef]

56. Bargeton, B.; Kellenberger, S. The Contact Region between Three Domains of the Extracellular Loop of ASIC1a Is Critical for Channel Function. *J. Biol. Chem.* **2010**, *285*, 13816–13826. [CrossRef]

57. Paukert, M.; Babini, E.; Pusch, M.; Gründer, S. Identification of the Ca^{2+} Blocking Site of Acid-sensing Ion Channel (ASIC) 1. *J. Gen. Physiol.* **2004**, *124*, 383–394. [CrossRef]

58. Paukert, M.; Chen, X.; Polleichtner, G.; Schindelin, H.; Gründer, S. Candidate Amino Acids Involved in H^+ Gating of Acid-sensing Ion Channel 1a. *J. Biol. Chem.* **2008**, *283*, 572–581. [CrossRef]

59. Klipp, R.C.; Bankston, J.R. Structural determinants of acid-sensing ion channel potentiation by single chain lipids. *J. Gen. Physiol.* **2022**, *154*, e202213156. [CrossRef]

60. Baconguis, I.; Gouaux, E. Structural plasticity and dynamic selectivity of acid-sensing ion channel–spider toxin complexes. *Nature* **2012**, *489*, 400–405. [CrossRef]

61. Wu, Y.; Chen, Z.; Canessa, C.M. A valve-like mechanism controls desensitization of functional mammalian isoforms of acid-sensing ion channels. *eLife* **2019**, *8*, e45851. [CrossRef] [PubMed]

62. Li, T.; Yang, Y.; Canessa, C.M. Asn415 in the β11-β12 Linker Decreases Proton-dependent Desensitization of ASIC1. *J. Biol. Chem.* **2010**, *285*, 31285–31291. [CrossRef] [PubMed]

63. Rook, M.L.; Williamson, A.; Lueck, J.D.; Musgaard, M.; Maclean, D.M. β11-12 linker isomerization governs acid-sensing ion channel desensitization and recovery. *eLife* **2020**, *9*, e51111. [CrossRef] [PubMed]

64. Rook, M.L.; Ananchenko, A.; Musgaard, M.; MacLean, D.M. Molecular Investigation of Chicken Acid-Sensing Ion Channel 1 β11-12 Linker Isomerization and Channel Kinetics. *Front. Cell. Neurosci.* **2021**, *15*, 761813. [CrossRef] [PubMed]

65. Rook, M.L.; Musgaard, M.; MacLean, D.M. Coupling structure with function in acid-sensing ion channels: Challenges in pursuit of proton sensors. *J. Physiol.* **2020**, *599*, 417–430. [CrossRef] [PubMed]

66. Cushman, K.A.; Marsh-Haffner, J.; Adelman, J.P.; McCleskey, E.W. A Conformation Change in the Extracellular Domain that Accompanies Desensitization of Acid-sensing Ion Channel (ASIC) 3. *J. Gen. Physiol.* **2007**, *129*, 345–350. [CrossRef]

67. Della Vecchia, M.C.; Rued, A.C.; Carattino, M.D. Gating Transitions in the Palm Domain of ASIC1a*. *J. Biol. Chem.* **2013**, *288*, 5487–5495. [CrossRef]

68. Krauson, A.J.; Carattino, M.D. The Thumb Domain Mediates Acid-sensing Ion Channel Desensitization. *J. Biol. Chem.* **2016**, *291*, 11407–11419. [CrossRef] [PubMed]

69. Kusama, N.; Gautam, M.; Harding, A.M.S.; Snyder, P.M.; Benson, C.J. Acid-sensing ion channels (ASICs) are differentially modulated by anions dependent on their subunit composition. *Am. J. Physiol.-Cell. Physiol.* **2013**, *304*, C89–C101. [CrossRef]

70. Kusama, N.; Harding, A.M.S.; Benson, C.J. Extracellular Chloride Modulates the Desensitization Kinetics of Acid-sensing Ion Channel 1a (ASIC1a). *J. Biol. Chem.* **2010**, *285*, 17425–17431. [CrossRef]

71. Ruan, Z.; Osei-Owusu, J.; Du, J.; Qiu, Z.; Lu, W. Structures and pH-sensing mechanism of the proton-activated chloride channel. *Nature* **2020**, *588*, 350–354. [CrossRef]

72. Osei-Owusu, J.; Kots, E.; Ruan, Z.; Mihaljevic, L.; Chen, K.H.; Tamhaney, A.; Ye, X.; Lu, W.; Weinstein, H.; Qiu, Z. Molecular determinants of pH sensing in the proton-activated chloride channel. *Proc. Natl. Acad. Sci. USA* **2022**, *119*, e2200727119. [CrossRef] [PubMed]

73. Hu, M.; Li, P.; Wang, C.; Feng, X.; Geng, Q.; Chen, W.; Marthi, M.; Zhang, W.; Gao, C.; Reid, W.; et al. Parkinson's disease-risk protein TMEM175 is a proton-activated proton channel in lysosomes. *Cell* **2022**, *185*, 2292–2308.e2220. [CrossRef] [PubMed]

74. Karsan, N.; Gonzales, E.B.; Dussor, G. Targeted Acid-Sensing Ion Channel Therapies for Migraine. *Neurotherapeutics* **2018**, *15*, 402–414. [CrossRef] [PubMed]

75. Lee, C.H.; Chen, C.C. Roles of ASICs in Nociception and Proprioception. *Adv. Exp. Med. Biol.* **2018**, *1099*, 37–47. [CrossRef] [PubMed]

76. Deval, E.; Lingueglia, E. Acid-Sensing Ion Channels and nociception in the peripheral and central nervous systems. *Neuropharmacology* **2015**, *94*, 49–57. [CrossRef]

77. Storozhuk, M.; Cherninskyi, A.; Maximyuk, O.; Isaev, D.; Krishtal, O. Acid-Sensing Ion Channels: Focus on Physiological and Some Pathological Roles in the Brain. *Curr. Neuropharmacol.* **2021**, *19*, 1570–1589. [CrossRef]

78. Heusser, S.A.; Pless, S.A. Acid-sensing ion channels as potential therapeutic targets. *Trends Pharmacol. Sci.* **2021**, *42*, 1035–1050. [CrossRef]

79. Lin, S.-H.; Sun, W.-H.; Chen, C.-C. Genetic exploration of the role of acid-sensing ion channels. *Neuropharmacology* **2015**, *94*, 99–118. [CrossRef]

80. Lin, J.-H.; Hung, C.-H.; Han, D.-S.; Chen, S.-T.; Lee, C.-H.; Sun, W.-Z.; Chen, C.-C. Sensing acidosis: Nociception or sngception? *J. Biomed. Sci.* **2018**, *25*, 85. [CrossRef]

81. Dulai, J.S.; Smith, E.S.J.; Rahman, T. Acid-sensing ion channel 3: An analgesic target. *Channels* **2021**, *15*, 94–127. [CrossRef] [PubMed]

82. Ritzel, R.M.; He, J.; Li, Y.; Cao, T.; Khan, N.; Shim, B.; Sabirzhanov, B.; Aubrecht, T.; Stoica, B.A.; Faden, A.I.; et al. Proton extrusion during oxidative burst in microglia exacerbates pathological acidosis following traumatic brain injury. *Glia* **2020**, *69*, 746–764. [CrossRef] [PubMed]

83. Lin, L.-H.; Jones, S.; Talman, W.T. Cellular Localization of Acid-Sensing Ion Channel 1 in Rat Nucleus Tractus Solitarii. *Cell. Mol. Neurobiol.* **2017**, *38*, 219–232. [CrossRef] [PubMed]

84. Wood, J.N.; Stein, C.; Gaveriaux-Ruff, C. Opioids and Pain. In *The Oxford Handbook of the Neurobiology of Pain*; Oxford University Press: Oxford, UK, 2020; pp. 727–769.

85. Chesler, M. Regulation and Modulation of pH in the Brain. *Physiol. Rev.* **2003**, *83*, 1183–1221. [CrossRef]

86. Ugawa, S.; Ueda, T.; Ishida, Y.; Nishigaki, M.; Shibata, Y.; Shimada, S. Amiloride-blockable acid-sensing ion channels are leading acid sensors expressed in human nociceptors. *J. Clin. Invest.* **2002**, *110*, 1185–1190. [CrossRef] [PubMed]

87. Jones, N.G.; Slater, R.; Cadiou, H.; McNaughton, P.; McMahon, S.B. Acid-induced pain and its modulation in humans. *J. Neurosci.* **2004**, *24*, 10974–10979. [CrossRef] [PubMed]

88. Ruan, N.; Tribble, J.; Peterson, A.M.; Jiang, Q.; Wang, J.Q.; Chu, X.-P. Acid-Sensing Ion Channels and Mechanosensation. *Int. J. Mol. Sci.* **2021**, *22*, 4810. [CrossRef]

89. Wang, Y.; O'Bryant, Z.; Wang, H.; Huang, Y. Regulating Factors in Acid-Sensing Ion Channel 1a Function. *Neurochem. Res.* **2015**, *41*, 631–645. [CrossRef] [PubMed]

90. Cullinan, M.M.; Klipp, R.C.; Bankston, J.R. Regulation of acid-sensing ion channels by protein binding partners. *Channels* **2021**, *15*, 635–647. [CrossRef]

91. Deval, E.; Noël, J.; Lay, N.; Alloui, A.; Diochot, S.; Friend, V.; Jodar, M.; Lazdunski, M.; Lingueglia, E. ASIC3, a sensor of acidic and primary inflammatory pain. *EMBO J.* **2008**, *27*, 3047–3055. [CrossRef]

92. Du, J.; Reznikov, L.R.; Price, M.P.; Zha, X.-m.; Lu, Y.; Moninger, T.O.; Wemmie, J.A.; Welsh, M.J. Protons are a neurotransmitter that regulates synaptic plasticity in the lateral amygdala. *Proc. Natl. Acad. Sci. USA* **2014**, *111*, 8961–8966. [CrossRef] [PubMed]

93. Huang, Y.; Jiang, N.; Li, J.; Ji, Y.-H.; Xiong, Z.-G.; Zha, X.-m. Two aspects of ASIC function: Synaptic plasticity and neuronal injury. *Neuropharmacology* **2015**, *94*, 42–48. [CrossRef] [PubMed]

94. Baron, A.; Lingueglia, E. Pharmacology of acid-sensing ion channels—Physiological and therapeutical perspectives. *Neuropharmacology* **2015**, *94*, 19–35. [CrossRef]

95. Rash, L.D. Acid-Sensing Ion Channel Pharmacology, Past, Present, and Future. *Adv. Pharm.* **2017**, *79*, 35–66. [CrossRef]

96. Escoubas, P.; De Weille, J.R.; Lecoq, A.; Diochot, S.; Waldmann, R.; Champigny, G.; Moinier, D.; Menez, A.; Lazdunski, M. Isolation of a tarantula toxin specific for a class of proton-gated Na$^+$ channels. *J. Biol. Chem.* **2000**, *275*, 25116–25121. [CrossRef] [PubMed]

97. Diochot, S.; Baron, A.; Rash, L.D.; Deval, E.; Escoubas, P.; Scarzello, S.; Salinas, M.; Lazdunski, M. A new sea anemone peptide, APETx2, inhibits ASIC3, a major acid-sensitive channel in sensory neurons. *EMBO J.* **2004**, *23*, 1516–1525. [CrossRef] [PubMed]

98. Baron, A.; Diochot, S.; Salinas, M.; Deval, E.; Noël, J.; Lingueglia, E. Venom toxins in the exploration of molecular, physiological and pathophysiological functions of acid-sensing ion channels. *Toxicon* **2013**, *75*, 187–204. [CrossRef]

99. Bohlen, C.J.; Chesler, A.T.; Sharif-Naeini, R.; Medzihradszky, K.F.; Zhou, S.; King, D.; Sánchez, E.E.; Burlingame, A.L.; Basbaum, A.I.; Julius, D. A heteromeric Texas coral snake toxin targets acid-sensing ion channels to produce pain. *Nature* **2011**, *479*, 410–414. [CrossRef]

100. Saez, N.J.; Mobli, M.; Bieri, M.; Chassagnon, I.R.; Malde, A.K.; Gamsjaeger, R.; Mark, A.E.; Gooley, P.R.; Rash, L.D.; King, G.F. A Dynamic Pharmacophore Drives the Interaction between Psalmotoxin-1 and the Putative Drug Target Acid-Sensing Ion Channel 1a. *Mol. Pharmacol.* **2011**, *80*, 796–808. [CrossRef]

101. Saez, N.J.; Deplazes, E.; Cristofori-Armstrong, B.; Chassagnon, I.R.; Lin, X.; Mobli, M.; Mark, A.E.; Rash, L.D.; King, G.F. Molecular dynamics and functional studies define a hot spot of crystal contacts essential for PcTx1 inhibition of acid-sensing ion channel 1a. *Br. J. Pharmacol.* **2015**, *172*, 4985–4995. [CrossRef]

102. Dawson, R.J.P.; Benz, J.; Stohler, P.; Tetaz, T.; Joseph, C.; Huber, S.; Schmid, G.; Hügin, D.; Pflimlin, P.; Trube, G.; et al. Structure of the Acid-sensing ion channel 1 in complex with the gating modifier Psalmotoxin 1. *Nat. Commun.* **2012**, *3*, 936. [CrossRef] [PubMed]

103. Borg, C.B.; Heusser, S.A.; Colding, J.M.; Pless, S.A. Conformational decoupling in acid-sensing ion channels uncovers mechanism and stoichiometry of PcTx1-mediated inhibition. *eLife* **2022**, *11*, e73384. [CrossRef]

104. Bohlen, C.J.; Julius, D. Receptor-targeting mechanisms of pain-causing toxins: How ow? *Toxicon* **2012**, *60*, 254–264. [CrossRef] [PubMed]

105. Mourier, G.; Salinas, M.; Kessler, P.; Stura, E.A.; Leblanc, M.; Tepshi, L.; Besson, T.; Diochot, S.; Baron, A.; Douguet, D.; et al. Mambalgin-1 Pain-relieving Peptide, Stepwise Solid-phase Synthesis, Crystal Structure, and Functional Domain for Acid-sensing Ion Channel 1a Inhibition. *J. Biol. Chem.* **2016**, *291*, 2616–2629. [CrossRef] [PubMed]

106. Salinas, M.; Kessler, P.; Douguet, D.; Sarraf, D.; Tonali, N.; Thai, R.; Servent, D.; Lingueglia, E. Mambalgin-1 pain-relieving peptide locks the hinge between α4 and α5 helices to inhibit rat acid-sensing ion channel 1a. *Neuropharmacology* **2021**, *185*, 108453. [CrossRef]

107. Schroeder, C.I.; Rash, L.D.; Vila-Farrés, X.; Rosengren, K.J.; Mobli, M.; King, G.F.; Alewood, P.F.; Craik, D.J.; Durek, T. Chemical Synthesis, 3D Structure, and ASIC Binding Site of the Toxin Mambalgin-2. *Angew. Chem. Int. Ed.* **2014**, *53*, 1017–1020. [CrossRef]

108. Sun, D.; Yu, Y.; Xue, X.; Pan, M.; Wen, M.; Li, S.; Qu, Q.; Li, X.; Zhang, L.; Li, X.; et al. Cryo-EM structure of the ASIC1a–mambalgin-1 complex reveals that the peptide toxin mambalgin-1 inhibits acid-sensing ion channels through an unusual allosteric effect. *Cell Discov.* **2018**, *4*, 27. [CrossRef]

109. Jensen, J.E.; Durek, T.; Alewood, P.F.; Adams, D.J.; King, G.F.; Rash, L.D. Chemical synthesis and folding of APETx2, a potent and selective inhibitor of acid sensing ion channel 3. *Toxicon* **2009**, *54*, 56–61. [CrossRef]

110. Anangi, R.; Chen, C.-C.; Lin, Y.-W.; Cheng, Y.-R.; Cheng, C.-H.; Chen, Y.-C.; Chu, Y.-P.; Chuang, W.-J. Expression in *Pichia pastoris* and characterization of APETx2, a specific inhibitor of acid sensing ion channel 3. *Toxicon* **2010**, *56*, 1388–1397. [CrossRef]

111. Qadri, Y.J.; Berdiev, B.K.; Song, Y.; Lippton, H.L.; Fuller, C.M.; Benos, D.J. Psalmotoxin-1 Docking to Human Acid-sensing Ion Channel-1. *J. Biol. Chem.* **2009**, *284*, 17625–17633. [CrossRef]

112. Samways, D.S.K.; Harkins, A.B.; Egan, T.M. Native and recombinant ASIC1a receptors conduct negligible Ca^{2+} entry. *Cell Calcium* **2009**, *45*, 319–325. [CrossRef] [PubMed]

113. Pan, C.-x.; Wu, F.-r.; Wang, X.-y.; Tang, J.; Gao, W.-f.; Ge, J.-f.; Chen, F.-h. Inhibition of ASICs reduces rat hepatic stellate cells activity and liver fibrosis: An in vitro and in vivo study. *Cell Biol. Int.* **2014**, *38*, 1003–1012. [CrossRef] [PubMed]

114. Joeres, N.; Augustinowski, K.; Neuhof, A.; Assmann, M.; Gründer, S. Functional and pharmacological characterization of two different ASIC1a/2a heteromers reveals their sensitivity to the spider toxin PcTx1. *Sci. Rep.* **2016**, *6*, 27647. [CrossRef] [PubMed]

115. Liu, Y.; Hagan, R.; Schoellerman, J. Dual actions of Psalmotoxin at ASIC1a and ASIC2a heteromeric channels (ASIC1a/2a). *Sci. Rep.* **2018**, *8*, 7179. [CrossRef]

116. Gregory, N.S.; Gautam, M.; Benson, C.J.; Sluka, K.A. Acid Sensing Ion Channel 1a (ASIC1a) Mediates Activity-induced Pain by Modulation of Heteromeric ASIC Channel Kinetics. *Neuroscience* **2018**, *386*, 166–174. [CrossRef]

117. Lee, J.Y.P.; Saez, N.J.; Cristofori-Armstrong, B.; Anangi, R.; King, G.F.; Smith, M.T.; Rash, L.D. Inhibition of acid-sensing ion channels by diminazene and APETx2 evoke partial and highly variable antihyperalgesia in a rat model of inflammatory pain. *Br. J. Pharmacol.* **2018**, *175*, 2204–2218. [CrossRef]

118. Andreev, Y.; Osmakov, D.; Koshelev, S.; Maleeva, E.; Logashina, Y.; Palikov, V.; Palikova, Y.; Dyachenko, I.; Kozlov, S. Analgesic Activity of Acid-Sensing Ion Channel 3 (ASIC3) Inhibitors: Sea Anemones Peptides Ugr9-1 and APETx2 versus Low Molecular Weight Compounds. *Mar. Drugs* **2018**, *16*, 500. [CrossRef]

119. Escoubas, P.; Bernard, C.; Lambeau, G.; Lazdunski, M.; Darbon, H. Recombinant production and solution structure of PcTx1, the specific peptide inhibitor of ASIC1a proton-gated cation channels. *Protein Sci.* **2003**, *12*, 1332–1343. [CrossRef]

120. Verkest, C.; Diochot, S.; Lingueglia, E.; Baron, A. C-Jun N-Terminal Kinase Post-Translational Regulation of Pain-Related Acid-Sensing Ion Channels 1b and 3. *J. Neurosci.* **2021**, *41*, 8673–8685. [CrossRef]

121. Baron, A.; Waldmann, R.; Lazdunski, M. ASIC-like, proton-activated currents in rat hippocampal neurons. *J. Physiol.* **2002**, *539*, 485–494. [CrossRef]

122. Duan, B.; Wu, L.J.; Yu, Y.Q.; Ding, Y.; Jing, L.; Xu, L.; Chen, J.; Xu, T.L. Upregulation of acid-sensing ion channel ASIC1a in spinal dorsal horn neurons contributes to inflammatory pain hypersensitivity. *J. Neurosci.* **2007**, *27*, 11139–11148. [CrossRef] [PubMed]

123. Mazzuca, M.; Heurteaux, C.; Alloui, A.; Diochot, S.; Baron, A.; Voilley, N.; Blondeau, N.; Escoubas, P.; Gelot, A.; Cupo, A.; et al. A tarantula peptide against pain via ASIC1a channels and opioid mechanisms. *Nat. Neurosci.* **2007**, *10*, 943–945. [CrossRef] [PubMed]

124. Er, S.Y.; Cristofori-Armstrong, B.; Escoubas, P.; Rash, L.D. Discovery and molecular interaction studies of a highly stable, tarantula peptide modulator of acid-sensing ion channel 1. *Neuropharmacology* **2017**, *127*, 185–195. [CrossRef]

125. Chassagnon, I.R.; McCarthy, C.A.; Chin, Y.K.Y.; Pineda, S.S.; Keramidas, A.; Mobli, M.; Pham, V.; De Silva, T.M.; Lynch, J.W.; Widdop, R.E.; et al. Potent neuroprotection after stroke afforded by a double-knot spider-venom peptide that inhibits acid-sensing ion channel 1a. *Proc. Natl. Acad. Sci. USA* **2017**, *114*, 3750–3755. [CrossRef]

126. Duggan, N.M.; Saez, N.J.; Clayton, D.; Budusan, E.; Watson, E.E.; Tucker, I.J.; Rash, L.D.; King, G.F.; Payne, R.J. Total Synthesis of the Spider-Venom Peptide Hi1a. *Org. Lett.* **2021**, *23*, 8375–8379. [CrossRef] [PubMed]

127. Redd, M.A.; Scheuer, S.E.; Saez, N.J.; Yoshikawa, Y.; Chiu, H.S.; Gao, L.; Hicks, M.; Villanueva, J.E.; Joshi, Y.; Chow, C.Y.; et al. Therapeutic Inhibition of Acid-Sensing Ion Channel 1a Recovers Heart Function After Ischemia–Reperfusion Injury. *Circulation* **2021**, *144*, 947–960. [CrossRef]

128. Cristofori-Armstrong, B. Discovery and modulation of acid-sensing ion channel modulating venom peptides. Ph.D. Thesis, The University of Queensland, Brisbane, Australia, 29 March 2019.

129. Rahman, T.; Smith, E.S.J. In silico assessment of interaction of sea anemone toxin APETx2 and acid sensing ion channel 3. *Biochem. Biophys. Res. Commun.* **2014**, *450*, 384–389. [CrossRef]

130. Gründer, S.; Augustinowski, K. Toxin binding reveals two open state structures for one acid-sensing ion channel. *Channels* **2014**, *6*, 409–413. [CrossRef]

131. Besson, T.; Lingueglia, E.; Salinas, M. Pharmacological modulation of Acid-Sensing Ion Channels 1a and 3 by amiloride and 2-guanidine-4-methylquinazoline (GMQ). *Neuropharmacology* **2017**, *125*, 429–440. [CrossRef] [PubMed]

132. Buta, A.; Maximyuk, O.; Kovalskyy, D.; Sukach, V.; Vovk, M.; Ievglevskyi, O.; Isaeva, E.; Isaev, D.; Savotchenko, A.; Krishtal, O. Novel Potent Orthosteric Antagonist of ASIC1a Prevents NMDAR-Dependent LTP Induction. *J. Med. Chem.* **2015**, *58*, 4449–4461. [CrossRef] [PubMed]

133. Qi, X.; Lu, J.-F.; Huang, Z.-Y.; Liu, Y.-J.; Cai, L.-B.; Wen, X.-L.; Song, X.-L.; Xiong, J.; Sun, P.-Y.; Zhang, H.; et al. Pharmacological Validation of ASIC1a as a Druggable Target for Neuroprotection in Cerebral Ischemia Using an Intravenously Available Small Molecule Inhibitor. *Front. Pharmacol.* **2022**, *13*, 849498. [CrossRef] [PubMed]

134. Roy, S.; Boiteux, C.; Alijevic, O.; Liang, C.; Bernèche, S.; Kellenberger, S. Molecular determinants of desensitization in an ENaC/degenerin channel. *FASEB J.* **2013**, *27*, 5034–5045. [CrossRef] [PubMed]

135. Springauf, A.; Bresenitz, P.; Gründer, S. The Interaction between Two Extracellular Linker Regions Controls Sustained Opening of Acid-sensing Ion Channel 1. *J. Biol. Chem.* **2011**, *286*, 24374–24384. [CrossRef]

136. Aldrich, R.W.; Yang, H.; Yu, Y.; Li, W.-G.; Yu, F.; Cao, H.; Xu, T.-L.; Jiang, H. Inherent Dynamics of the Acid-Sensing Ion Channel 1 Correlates with the Gating Mechanism. *PLoS Biol.* **2009**, *7*, e1000151. [CrossRef]

137. Swain, S.M.; Bera, A.K. Coupling of Proton Binding in Extracellular Domain to Channel Gating in Acid-Sensing Ion Channel. *J. Mol. Neurosci.* **2013**, *51*, 199–207. [CrossRef] [PubMed]

138. Li, T.; Yang, Y.; Canessa, C.M. Leu85 in the β1-β2 Linker of ASIC1 Slows Activation and Decreases the Apparent Proton Affinity by Stabilizing a Closed Conformation. *J. Biol. Chem.* **2010**, *285*, 22706–22712. [CrossRef]

139. Li, T.; Yang, Y.; Canessa, C.M. Two residues in the extracellular domain convert a nonfunctional ASIC1 into a proton-activated channel. *Am. J. Physiol.-Cell. Physiol.* **2010**, *299*, C66–C73. [CrossRef]

140. Salinas Castellanos, L.C.; Uchitel, O.D.; Weissmann, C. Signaling Pathways in Proton and Non-proton ASIC1a Activation. *Front. Cell. Neurosci.* **2021**, *15*, 735414. [CrossRef]

141. Baron, A.; Deval, E.; Salinas, M.; Lingueglia, E.; Voilley, N.; Lazdunski, M. Protein kinase C stimulates the acid-sensing ion channel ASIC2a via the PDZ domain-containing protein PICK1. *J. Biol. Chem.* **2002**, *277*, 50463–50468. [CrossRef]

142. Diochot, S.; Alloui, A.; Rodrigues, P.; Dauvois, M.; Friend, V.; Aissouni, Y.; Eschalier, A.; Lingueglia, E.; Baron, A. Analgesic effects of mambalgin peptide inhibitors of acid-sensing ion channels in inflammatory and neuropathic pain. *Pain* **2016**, *157*, 552–559. [CrossRef]

143. Chang, C.-T.; Fong, S.W.; Lee, C.-H.; Chuang, Y.-C.; Lin, S.-H.; Chen, C.-C. Involvement of Acid-Sensing Ion Channel 1b in the Development of Acid-Induced Chronic Muscle Pain. *Front. Neurosci.* **2019**, *13*, 1247. [CrossRef] [PubMed]

144. Young, G.T.; Gutteridge, A.; Fox, H.D.E.; Wilbrey, A.L.; Cao, L.; Cho, L.T.; Brown, A.R.; Benn, C.L.; Kammonen, L.R.; Friedman, J.H.; et al. Characterizing Human Stem Cell–derived Sensory Neurons at the Single-cell Level Reveals Their Ion Channel Expression and Utility in Pain Research. *Mol. Ther.* **2014**, *22*, 1530–1543. [CrossRef] [PubMed]

145. Woo, Y.C.; Park, S.S.; Subieta, A.R.; Brennan, T.J. Changes in tissue pH and temperature after incision indicate acidosis may contribute to postoperative pain. *Anesthesiology* **2004**, *101*, 468–475. [CrossRef]

146. Verkest, C.; Piquet, E.; Diochot, S.; Dauvois, M.; Lanteri-Minet, M.; Lingueglia, E.; Baron, A. Effects of systemic inhibitors of acid-sensing ion channels 1 (ASIC1) against acute and chronic mechanical allodynia in a rodent model of migraine. *Br. J. Pharmacol.* **2018**, *175*, 4154–4166. [CrossRef]

147. Chagot, B.; Diochot, S.; Pimentel, C.; Lazdunski, M.; Darbon, H. Solution structure of APETx1 from the sea anemone Anthopleura elegantissima: A new fold for an HERG toxin. *Proteins* **2005**, *59*, 380–386. [CrossRef] [PubMed]

148. Deval, E.; Noel, J.; Gasull, X.; Delaunay, A.; Alloui, A.; Friend, V.; Eschalier, A.; Lazdunski, M.; Lingueglia, E. Acid-Sensing Ion Channels in Postoperative Pain. *J. Neurosci.* **2011**, *31*, 6059–6066. [CrossRef]

149. Karczewski, J.; Spencer, R.H.; Garsky, V.M.; Liang, A.; Leitl, M.D.; Cato, M.J.; Cook, S.P.; Kane, S.; Urban, M.O. Reversal of acid-induced and inflammatory pain by the selective ASIC3 inhibitor, APETx2. *Br. J. Pharmacol.* **2010**, *161*, 950–960. [CrossRef]

150. Gilbert, H.T.J.; Hodson, N.; Baird, P.; Richardson, S.M.; Hoyland, J.A. Acidic pH promotes intervertebral disc degeneration: Acid-sensing ion channel -3 as a potential therapeutic target. *Sci. Rep.* **2016**, *6*, 37360. [CrossRef]

151. Wang, X.Y.; Yan, W.W.; Zhang, X.L.; Liu, H.; Zhang, L.C. ASIC3 in the cerebrospinal fluid-contacting nucleus of brain parenchyma contributes to inflammatory pain in rats. *Neurol. Res.* **2013**, *36*, 270–275. [CrossRef]

152. Martínez-Rojas, V.A.; Barragán-Iglesias, P.; Rocha-González, H.I.; Murbartián, J.; Granados-Soto, V. Role of TRPV1 and ASIC3 in formalin-induced secondary allodynia and hyperalgesia. *Pharmacol. Rep.* **2014**, *66*, 964–971. [CrossRef]

153. Yan, J.; Wei, X.; Bischoff, C.; Edelmayer, R.M.; Dussor, G. pH-Evoked Dural Afferent Signaling Is Mediated by ASIC3 and Is Sensitized by Mast Cell Mediators. *Headache J. Head Face Pain* **2013**, *53*, 1250–1261. [CrossRef] [PubMed]

154. Izumi, M.; Ikeuchi, M.; Ji, Q.; Tani, T. Local ASIC3 modulates pain and disease progression in a rat model of osteoarthritis. *J. Biomed. Sci.* **2012**, *19*, 77. [CrossRef] [PubMed]

155. Sugimura, N.; Ikeuchi, M.; Izumi, M.; Kawano, T.; Aso, K.; Kato, T.; Ushida, T.; Yokoyama, M.; Tani, T. Repeated intra-articular injections of acidic saline produce long-lasting joint pain and widespread hyperalgesia. *Eur. J. Pain* **2015**, *19*, 629–638. [CrossRef] [PubMed]

156. Callejo, G.; Castellanos, A.; Castany, M.; Gual, A.; Luna, C.; Acosta, M.C.; Gallar, J.; Giblin, J.P.; Gasull, X. Acid-sensing ion channels detect moderate acidifications to induce ocular pain. *Pain* **2015**, *156*, 483–495. [CrossRef] [PubMed]

157. Chen, W.-N.; Lee, C.-H.; Lin, S.-H.; Wong, C.-W.; Sun, W.-H.; Wood, J.N.; Chen, C.-C. Roles of ASIC3, TRPV1, and NaV1.8 in the Transition from Acute to Chronic Pain in a Mouse Model of Fibromyalgia. *Mol. Pain* **2014**, *10*, 40. [CrossRef] [PubMed]

158. Gregory, J.; McGowan, L. An examination of the prevalence of acute pain for hospitalised adult patients: A systematic review. *J. Clin. Nurs.* **2016**, *25*, 583–598. [CrossRef]

159. Hiasa, M.; Okui, T.; Allette, Y.M.; Ripsch, M.S.; Sun-Wada, G.-H.; Wakabayashi, H.; Roodman, G.D.; White, F.A.; Yoneda, T. Bone Pain Induced by Multiple Myeloma Is Reduced by Targeting V-ATPase and ASIC3. *Cancer Res.* **2017**, *77*, 1283–1295. [CrossRef]

160. Hanaka, M.; Iba, K.; Dohke, T.; Kanaya, K.; Okazaki, S.; Yamashita, T. Antagonists to TRPV1, ASICs and P2X have a potential role to prevent the triggering of regional bone metabolic disorder and pain-like behavior in tail-suspended mice. *Bone* **2018**, *110*, 284–294. [CrossRef]

161. Kanaya, K.; Iba, K.; Abe, Y.; Dohke, T.; Okazaki, S.; Matsumura, T.; Yamashita, T. Acid-sensing ion channel 3 or P2X2/3 is involved in the pain-like behavior under a high bone turnover state in ovariectomized mice. *J. Orthop. Res.* **2016**, *34*, 566–573. [CrossRef]

162. Blanchard, M.G.; Rash, L.D.; Kellenberger, S. Inhibition of voltage-gated Na$^+$ currents in sensory neurones by the sea anemone toxin APETx2. *Br. J. Pharmacol.* **2012**, *165*, 2167–2177. [CrossRef]

163. Peigneur, S.; Béress, L.; Möller, C.; Marí, F.; Forssmann, W.G.; Tytgat, J. A natural point mutation changes both target selectivity and mechanism of action of sea anemone toxins. *FASEB J.* **2012**, *26*, 5141–5151. [CrossRef] [PubMed]

164. Jensen, J.E.; Cristofori-Armstrong, B.; Anangi, R.; Rosengren, K.J.; Lau, C.H.Y.; Mobli, M.; Brust, A.; Alewood, P.F.; King, G.F.; Rash, L.D. Understanding the Molecular Basis of Toxin Promiscuity: The Analgesic Sea Anemone Peptide APETx2 Interacts with Acid-Sensing Ion Channel 3 and hERG Channels via Overlapping Pharmacophores. *J. Med. Chem.* **2014**, *57*, 9195–9203. [CrossRef]

165. Kalina, R.; Gladkikh, I.; Dmitrenok, P.; Chernikov, O.; Koshelev, S.; Kvetkina, A.; Kozlov, S.; Kozlovskaya, E.; Monastyrnaya, M. New APETx-like peptides from sea anemone Heteractis crispa modulate ASIC1a channels. *Peptides* **2018**, *104*, 41–49. [CrossRef]

166. Kalina, R.S.; Koshelev, S.G.; Zelepuga, E.A.; Kim, N.Y.; Kozlov, S.A.; Kozlovskaya, E.P.; Monastyrnaya, M.M.; Gladkikh, I.N. APETx-Like Peptides from the Sea Anemone Heteractis crispa, Diverse in Their Effect on ASIC1a and ASIC3 Ion Channels. *Toxins* **2020**, *12*, 266. [CrossRef] [PubMed]

167. Pinheiro-Junior, E.L.; Kalina, R.; Gladkikh, I.; Leychenko, E.; Tytgat, J.; Peigneur, S. A Tale of Toxin Promiscuity: The Versatile Pharmacological Effects of Hcr 1b-2 Sea Anemone Peptide on Voltage-Gated Ion Channels. *Mar. Drugs* **2022**, *20*, 147. [CrossRef]

168. Osmakov, D.I.; Kozlov, S.A.; Andreev, Y.A.; Koshelev, S.G.; Sanamyan, N.P.; Sanamyan, K.E.; Dyachenko, I.A.; Bondarenko, D.A.; Murashev, A.N.; Mineev, K.S.; et al. Sea Anemone Peptide with Uncommon β-Hairpin Structure Inhibits Acid-sensing Ion Channel 3 (ASIC3) and Reveals Analgesic Activity. *J. Biol. Chem.* **2013**, *288*, 23116–23127. [CrossRef] [PubMed]

169. Rodríguez, A.A.; Salceda, E.; Garateix, A.G.; Zaharenko, A.J.; Peigneur, S.; López, O.; Pons, T.; Richardson, M.; Díaz, M.; Hernández, Y.; et al. A novel sea anemone peptide that inhibits acid-sensing ion channels. *Peptides* **2014**, *53*, 3–12. [CrossRef]

170. Reimers, C.; Lee, C.-H.; Kalbacher, H.; Tian, Y.; Hung, C.-H.; Schmidt, A.; Prokop, L.; Kauferstein, S.; Mebs, D.; Chen, C.-C.; et al. Identification of a cono-RFamide from the venom of Conus textile that targets ASIC3 and enhances muscle pain. *Proc. Natl. Acad. Sci. USA* **2017**, *114*, E3507–E3515. [CrossRef]

171. Báez, A.; Salceda, E.; Fló, M.; Graña, M.; Fernández, C.; Vega, R.; Soto, E. α-Dendrotoxin inhibits the ASIC current in dorsal root ganglion neurons from rat. *Neurosci. Lett.* **2015**, *606*, 42–47. [CrossRef]

172. Skarżyński, T. Crystal structure of α-dendrotoxin from the green mamba venom and its comparison with the structure of bovine pancreatic trypsin inhibitor. *J. Mol. Biol.* **1992**, *224*, 671–683. [CrossRef]

173. Cristofori-Armstrong, B.; Rash, L.D. Acid-sensing ion channel (ASIC) structure and function: Insights from spider, snake and sea anemone venoms. *Neuropharmacology* **2017**, *127*, 173–184. [CrossRef] [PubMed]

174. Hernández, C.; Konno, K.; Salceda, E.; Vega, R.; Zaharenko, A.J.; Soto, E. Sa12b Peptide from Solitary Wasp Inhibits ASIC Currents in Rat Dorsal Root Ganglion Neurons. *Toxins* **2019**, *11*, 585. [CrossRef] [PubMed]

175. Ray, P.; Torck, A.; Quigley, L.; Wangzhou, A.; Neiman, M.; Rao, C.; Lam, T.; Kim, J.-Y.; Kim, T.H.; Zhang, M.Q.; et al. Comparative transcriptome profiling of the human and mouse dorsal root ganglia: An RNA-seq–based resource for pain and sensory neuroscience research. *Pain* **2018**, *159*, 1325–1345. [CrossRef] [PubMed]

176. Chen, C.C.; Zimmer, A.; Sun, W.H.; Hall, J.; Brownstein, M.J. A role for ASIC3 in the modulation of high-intensity pain stimuli. *Proc. Natl. Acad. Sci. USA* **2002**, *99*, 8992–8997. [CrossRef]

177. Price, M.P.; McIlwrath, S.L.; Xie, J.; Cheng, C.; Qiao, J.; Tarr, D.E.; Sluka, K.A.; Brennan, T.J.; Lewin, G.R.; Welsh, M.J. The DRASIC cation channel contributes to the detection of cutaneous touch and acid stimuli in mice. *Neuron* **2001**, *32*, 1071–1083. [CrossRef]

178. Fu, H.; Fang, P.; Zhou, H.-Y.; Zhou, J.; Yu, X.-W.; Ni, M.; Zheng, J.-Y.; Jin, Y.; Chen, J.-G.; Wang, F.; et al. Acid-sensing ion channels in trigeminal ganglion neurons innervating the orofacial region contribute to orofacial inflammatory pain. *Clin. Exp. Pharmacol. Physiol.* **2016**, *43*, 193–202. [CrossRef]

179. Nakamura, M.; Jang, I.-S. Characterization of proton-induced currents in rat trigeminal mesencephalic nucleus neurons. *Brain Res.* **2014**, *1583*, 12–22. [CrossRef]

180. Lu, Y.; Ma, X.; Sabharwal, R.; Snitsarev, V.; Morgan, D.; Rahmouni, K.; Drummond, H.A.; Whiteis, C.A.; Costa, V.; Price, M.; et al. The Ion Channel ASIC2 Is Required for Baroreceptor and Autonomic Control of the Circulation. *Neuron* **2009**, *64*, 885–897. [CrossRef]

181. Dusenkova, S.; Ru, F.; Surdenikova, L.; Nassenstein, C.; Hatok, J.; Dusenka, R.; Banovcin, P.; Kliment, J.; Tatar, M.; Kollarik, M. The expression profile of acid-sensing ion channel (ASIC) subunits ASIC1a, ASIC1b, ASIC2a, ASIC2b, and ASIC3 in the esophageal vagal afferent nerve subtypes. *Am. J. Physiol. -Gastrointest. Liver Physiol.* **2014**, *307*, G922–G930. [CrossRef]

182. Kupari, J.; Häring, M.; Agirre, E.; Castelo-Branco, G.; Ernfors, P. An Atlas of Vagal Sensory Neurons and Their Molecular Specialization. *Cell Rep.* **2019**, *27*, 2508–2523.e2504. [CrossRef]

183. Wu, W.L.; Lin, Y.W.; Min, M.Y.; Chen, C.C. Mice lacking Asic3 show reduced anxiety-like behavior on the elevated plus maze and reduced aggression. *Genes Brain Behav.* **2010**, *9*, 603–614. [CrossRef] [PubMed]

184. Zeisel, A.; Hochgerner, H.; Lönnerberg, P.; Johnsson, A.; Memic, F.; van der Zwan, J.; Häring, M.; Braun, E.; Borm, L.E.; La Manno, G.; et al. Molecular Architecture of the Mouse Nervous System. *Cell* **2018**, *174*, 999–1014.e1022. [CrossRef] [PubMed]

185. Usoskin, D.; Furlan, A.; Islam, S.; Abdo, H.; Lönnerberg, P.; Lou, D.; Hjerling-Leffler, J.; Haeggström, J.; Kharchenko, O.; Kharchenko, P.V.; et al. Unbiased classification of sensory neuron types by large-scale single-cell RNA sequencing. *Nat. Neurosci.* **2014**, *18*, 145–153. [CrossRef] [PubMed]

186. Molliver, D.C.; Immke, D.C.; Fierro, L.; Pare, M.; Rice, F.L.; McCleskey, E.W. ASIC3, an acid-sensing ion channel, is expressed in metaboreceptive sensory neurons. *Mol. Pain* **2005**, *1*, 35. [CrossRef]

187. Nguyen, M.Q.; Le Pichon, C.E.; Ryba, N. Stereotyped transcriptomic transformation of somatosensory neurons in response to injury. *eLife* **2019**, *8*, e49679. [CrossRef]

188. Obukhov, A.G.; Nguyen, M.Q.; Wu, Y.; Bonilla, L.S.; von Buchholtz, L.J.; Ryba, N.J.P. Diversity amongst trigeminal neurons revealed by high throughput single cell sequencing. *PLoS ONE* **2017**, *12*, e0185543. [CrossRef]

189. Papalampropoulou-Tsiridou, M.; Labrecque, S.; Godin, A.G.; De Koninck, Y.; Wang, F. Differential Expression of Acid—Sensing Ion Channels in Mouse Primary Afferents in Naïve and Injured Conditions. *Front. Cell. Neurosci.* **2020**, *14*, 103. [CrossRef]

190. Páez, O.; Segura-Chama, P.; Almanza, A.; Pellicer, F.; Mercado, F. Properties and Differential Expression of H^+ Receptors in Dorsal Root Ganglia: Is a Labeled-Line Coding for Acid Nociception Possible? *Front. Physiol.* **2021**, *12*, 733267. [CrossRef]

191. Duzhyy, D.E.; Voitenko, N.V.; Belan, P.V. Peripheral Inflammation Results in Increased Excitability of Capsaicin-Insensitive Nociceptive DRG Neurons Mediated by Upregulation of ASICs and Voltage-Gated Ion Channels. *Front. Cell. Neurosci.* **2021**, *15*, 723295. [CrossRef]

192. Mamet, J.; Baron, A.; Lazdunski, M.; Voilley, N. Proinflammatory mediators, stimulators of sensory neuron excitability via the expression of acid-sensing ion channels. *J. Neurosci.* **2002**, *22*, 10662–10670. [CrossRef]

193. Voilley, N.; de Weille, J.; Mamet, J.; Lazdunski, M. Nonsteroid anti-inflammatory drugs inhibit both the activity and the inflammation-induced expression of acid-sensing ion channels in nociceptors. *J. Neurosci.* **2001**, *21*, 8026–8033. [CrossRef] [PubMed]

194. Tavares-Ferreira, D.; Shiers, S.; Ray, P.R.; Wangzhou, A.; Jeevakumar, V.; Sankaranarayanan, I.; Cervantes, A.M.; Reese, J.C.; Chamessian, A.; Copits, B.A.; et al. Spatial transcriptomics of dorsal root ganglia identifies molecular signatures of human nociceptors. *Sci. Transl. Med.* **2022**, *14*, eabj8186. [CrossRef] [PubMed]

195. McKemy, D.D.; Flegel, C.; Schöbel, N.; Altmüller, J.; Becker, C.; Tannapfel, A.; Hatt, H.; Gisselmann, G. RNA-Seq Analysis of Human Trigeminal and Dorsal Root Ganglia with a Focus on Chemoreceptors. *PLoS ONE* **2015**, *10*, e0128951. [CrossRef]

196. LaPaglia, D.M.; Sapio, M.R.; Burbelo, P.D.; Thierry-Mieg, J.; Thierry-Mieg, D.; Raithel, S.J.; Ramsden, C.E.; Iadarola, M.J.; Mannes, A.J. RNA-Seq investigations of human post-mortem trigeminal ganglia. *Cephalalgia* **2017**, *38*, 912–932. [CrossRef]

197. Prato, V.; Taberner, F.J.; Hockley, J.R.F.; Callejo, G.; Arcourt, A.; Tazir, B.; Hammer, L.; Schad, P.; Heppenstall, P.A.; Smith, E.S.; et al. Functional and Molecular Characterization of Mechanoinsensitive "Silent" Nociceptors. *Cell Rep.* **2017**, *21*, 3102–3115. [CrossRef]

198. Bakken, T.E.; Jorstad, N.L.; Hu, Q.; Lake, B.B.; Tian, W.; Kalmbach, B.E.; Crow, M.; Hodge, R.D.; Krienen, F.M.; Sorensen, S.A.; et al. Comparative cellular analysis of motor cortex in human, marmoset and mouse. *Nature* **2021**, *598*, 111–119. [CrossRef]

199. Häring, M.; Zeisel, A.; Hochgerner, H.; Rinwa, P.; Jakobsson, J.E.T.; Lönnerberg, P.; La Manno, G.; Sharma, N.; Borgius, L.; Kiehn, O.; et al. Neuronal atlas of the dorsal horn defines its architecture and links sensory input to transcriptional cell types. *Nat. Neurosci.* **2018**, *21*, 869–880. [CrossRef]

200. Sathyamurthy, A.; Johnson, K.R.; Matson, K.J.E.; Dobrott, C.I.; Li, L.; Ryba, A.R.; Bergman, T.B.; Kelly, M.C.; Kelley, M.W.; Levine, A.J. Massively Parallel Single Nucleus Transcriptional Profiling Defines Spinal Cord Neurons and Their Activity during Behavior. *Cell Rep.* **2018**, *22*, 2216–2225. [CrossRef]

201. Wu, L.J.; Duan, B.; Mei, Y.D.; Gao, J.; Chen, J.G.; Zhuo, M.; Xu, L.; Wu, M.; Xu, T.L. Characterization of acid-sensing ion channels in dorsal horn neurons of rat spinal cord. *J. Biol. Chem.* **2004**, *279*, 43716–43724. [CrossRef]

202. Wang, Y.; Fu, X.; Huang, L.; Wang, X.; Lu, Z.; Zhu, F.; Xiao, Z. Increased Asics Expression via the Camkii-CREB Pathway in a Novel Mouse Model of Trigeminal Pain. *Cell. Physiol. Biochem.* **2018**, *46*, 568–578. [CrossRef]

203. Li, X.; Xu, L.S.; Xu, Y.F.; Yang, Q.; Fang, Z.X.; Yao, M.; Chen, W.Y. The gene regulatory network in different brain regions of neuropathic pain mouse models. *Eur. Rev. Med. Pharm. Sci.* **2020**, *24*, 5053–5061. [CrossRef]

204. Huang, C.; Hu, Z.-l.; Wu, W.-N.; Yu, D.-F.; Xiong, Q.-J.; Song, J.-R.; Shu, Q.; Fu, H.; Wang, F.; Chen, J.-G. Existence and distinction of acid-evoked currents in rat astrocytes. *Glia* **2010**, *58*, 1415–1424. [CrossRef] [PubMed]

205. Yu, X.-W.; Hu, Z.-L.; Ni, M.; Fang, P.; Zhang, P.-W.; Shu, Q.; Fan, H.; Zhou, H.-Y.; Ni, L.; Zhu, L.-Q.; et al. Acid-sensing ion channels promote the inflammation and migration of cultured rat microglia. *Glia* **2015**, *63*, 483–496. [CrossRef] [PubMed]

206. Cegielski, V.; Chakrabarty, R.; Ding, S.; Wacker, M.J.; Monaghan-Nichols, P.; Chu, X.-P. Acid-Sensing Ion Channels in Glial Cells. *Membranes* **2022**, *12*, 119. [CrossRef]

207. Yang, F.; Sun, X.; Ding, Y.; Ma, H.; Yang, T.O.; Ma, Y.; Wei, D.; Li, W.; Xu, T.; Jiang, W. Astrocytic Acid-Sensing Ion Channel 1a Contributes to the Development of Chronic Epileptogenesis. *Sci. Rep.* **2016**, *6*, 31581. [CrossRef]

208. Feldman, D.H.; Horiuchi, M.; Keachie, K.; McCauley, E.; Bannerman, P.; Itoh, A.; Itoh, T.; Pleasure, D. Characterization of acid-sensing ion channel expression in oligodendrocyte-lineage cells. *Glia* **2008**, *56*, 1238–1249. [CrossRef]

209. Baumann, T.K.; Burchiel, K.J.; Ingram, S.L.; Martenson, M.E. Responses of adult human dorsal root ganglion neurons in culture to capsaicin and low pH. *Pain* **1996**, *65*, 31–38. [CrossRef]

210. Baumann, T.K.; Chaudhary, P.; Martenson, M.E. Background potassium channel block and TRPV1 activation contribute to proton depolarization of sensory neurons from humans with neuropathic pain. *Eur. J. Neurosci.* **2004**, *19*, 1343–1351. [CrossRef]

211. North, R.Y.; Li, Y.; Ray, P.; Rhines, L.D.; Tatsui, C.E.; Rao, G.; Johansson, C.A.; Zhang, H.; Kim, Y.H.; Zhang, B.; et al. Electrophysiological and transcriptomic correlates of neuropathic pain in human dorsal root ganglion neurons. *Brain* **2019**, *142*, 1215–1226. [CrossRef]

212. Li, M.; Inoue, K.; Branigan, D.; Kratzer, E.; Hansen, J.C.; Chen, J.W.; Simon, R.P.; Xiong, Z.-G. Acid-Sensing Ion Channels in Acidosis-Induced Injury of Human Brain Neurons. *J. Cereb. Blood Flow Metab.* **2010**, *30*, 1247–1260. [CrossRef]

213. Xu, Y.; Jiang, Y.Q.; Li, C.; He, M.; Rusyniak, W.G.; Annamdevula, N.; Ochoa, J.; Leavesley, S.J.; Xu, J.; Rich, T.C.; et al. Human ASIC1a mediates stronger acid-induced responses as compared with mouse ASICIa. *FASEB J.* **2018**, *32*, 3832–3843. [CrossRef]

214. Neuhof, A.; Tian, Y.; Reska, A.; Falkenburger, B.H.; Gründer, S. Large Acid-Evoked Currents, Mediated by ASIC1a, Accompany Differentiation in Human Dopaminergic Neurons. *Front. Cell. Neurosci.* **2021**, *15*, 668008. [CrossRef] [PubMed]

215. Kalinovskii, A.P.; Osmakov, D.I.; Koshelev, S.G.; Lubova, K.I.; Korolkova, Y.V.; Kozlov, S.A.; Andreev, Y.A. Retinoic Acid-Differentiated Neuroblastoma SH-SY5Y Is an Accessible In Vitro Model to Study Native Human Acid-Sensing Ion Channels 1a (ASIC1a). *Biology* **2022**, *11*, 167. [CrossRef] [PubMed]

216. Steen, K.H.; Steen, A.E.; Reeh, P.W. A dominant role of acid pH in inflammatory excitation and sensitization of nociceptors in rat skin, in vitro. *J. Neurosci.* **1995**, *15*, 3982–3989. [CrossRef]

217. McMahon, S.B.; Jones, N.G. Plasticity of pain signaling: Role of neurotrophic factors exemplified by acid-induced pain. *J. Neurobiol.* **2004**, *61*, 72–87. [CrossRef]

218. Heber, S.; Ciotu, C.I.; Hartner, G.; Gold-Binder, M.; Ninidze, N.; Gleiss, A.; Kress, H.-G.; Fischer, M.J.M. TRPV1 antagonist BCTC inhibits pH 6.0-induced pain in human skin. *Pain* **2020**, *161*, 1532–1541. [CrossRef]

219. Morgan, D.L.; Borys, D.J.; Stanford, R.; Kjar, D.; Tobleman, W. Texas Coral Snake (Micrurus tener) Bites. *South. Med. J.* **2007**, *100*, 152–156. [CrossRef]

220. Lopes, D.M.; Denk, F.; McMahon, S.B. The Molecular Fingerprint of Dorsal Root and Trigeminal Ganglion Neurons. *Front. Mol. Neurosci.* **2017**, *10*, 304. [CrossRef]

221. Yan, J.; Edelmayer, R.M.; Wei, X.; De Felice, M.; Porreca, F.; Dussor, G. Dural afferents express acid-sensing ion channels: A role for decreased meningeal pH in migraine headache. *Pain* **2011**, *152*, 106–113. [CrossRef]

222. Holton, C.M.; Strother, L.C.; Dripps, I.; Pradhan, A.A.; Goadsby, P.J.; Holland, P.R. Acid-sensing ion channel 3 blockade inhibits durovascular and nitric oxide-mediated trigeminal pain. *Br. J. Pharmacol.* **2020**, *177*, 2478–2486. [CrossRef]

223. Dussor, G. ASICs as therapeutic targets for migraine. *Neuropharmacology* **2015**, *94*, 64–71. [CrossRef] [PubMed]

224. Holland, P.R.; Akerman, S.; Andreou, A.P.; Karsan, N.; Wemmie, J.A.; Goadsby, P.J. Acid-sensing ion channel 1: A novel therapeutic target for migraine with aura. *Ann. Neurol.* **2012**, *72*, 559–563. [CrossRef] [PubMed]

225. Zhu, H.; Ding, J.; Wu, J.; Liu, T.; Liang, J.; Tang, Q.; Jiao, M. Resveratrol attenuates bone cancer pain through regulating the expression levels of ASIC3 and activating cell autophagy. *Acta Biochim. Et Biophys. Sin.* **2017**, *49*, 1008–1014. [CrossRef]

226. Morgan, M.; Thai, J.; Trinh, P.; Habib, M.; Effendi, K.N.; Ivanusic, J.J. ASIC3 inhibition modulates inflammation-induced changes in the activity and sensitivity of Aδ and C fiber sensory neurons that innervate bone. *Mol. Pain* **2020**, *16*, 1744806920975950. [CrossRef]

227. Ikeuchi, M.; Kolker, S.J.; Sluka, K.A. Acid-Sensing Ion Channel 3 Expression in Mouse Knee Joint Afferents and Effects of Carrageenan-Induced Arthritis. *J. Pain* **2009**, *10*, 336–342. [CrossRef] [PubMed]

228. Jurczak, A.; Delay, L.; Barbier, J.; Simon, N.; Krock, E.; Sandor, K.; Agalave, N.M.; Rudjito, R.; Wigerblad, G.; Rogóż, K.; et al. Antibody-induced pain-like behavior and bone erosion: Links to subclinical inflammation, osteoclast activity, and acid-sensing ion channel 3–dependent sensitization. *Pain* **2022**, *163*, 1542–1559. [CrossRef]

229. Pidoux, L.; Delanoe, K.; Barbier, J.; Marchand, F.; Lingueglia, E.; Deval, E. Single Subcutaneous Injection of Lysophosphatidyl-Choline Evokes ASIC3-Dependent Increases of Spinal Dorsal Horn Neuron Activity. *Front. Mol. Neurosci.* **2022**, *15*, 880651. [CrossRef]

230. Holzer, P. Acid-sensing ion channels in gastrointestinal function. *Neuropharmacology* **2015**, *94*, 72–79. [CrossRef]

231. Yiangou, Y.; Facer, P.; Smith, J.A.; Sangameswaran, L.; Eglen, R.; Birch, R.; Knowles, C.; Williams, N.; Anand, P. Increased acid-sensing ion channel ASIC-3 in inflamed human intestine. *Eur. J. Gastroenterol. Hepatol.* **2001**, *13*, 891–896. [CrossRef]

232. Han, X.; Zhang, Y.; Lee, A.; Li, Z.; Gao, J.; Wu, X.; Zhao, J.; Wang, H.; Chen, D.; Zou, D.; et al. Upregulation of acid sensing ion channels is associated with esophageal hypersensitivity in GERD. *FASEB J.* **2021**, *36*, e22083. [CrossRef]

233. Olesen, A.E.; Nielsen, L.M.; Larsen, I.M.; Drewes, A.M. Randomized clinical trial: Efficacy and safety of PPC-5650 on experimental esophageal pain and hyperalgesia in healthy volunteers. *Scand. J. Gastroenterol.* **2014**, *50*, 138–144. [CrossRef] [PubMed]

234. Sluka, K.A.; Gregory, N.S. The dichotomized role for acid sensing ion channels in musculoskeletal pain and inflammation. *Neuropharmacology* **2015**, *94*, 58–63. [CrossRef] [PubMed]

235. Walder, R.Y.; Rasmussen, L.A.; Rainier, J.D.; Light, A.R.; Wemmie, J.A.; Sluka, K.A. ASIC1 and ASIC3 Play Different Roles in the Development of Hyperalgesia After Inflammatory Muscle Injury. *J. Pain* **2010**, *11*, 210–218. [CrossRef]

236. Hung, C.-H.; Lee, C.-H.; Tsai, M.-H.; Chen, C.-H.; Lin, H.-F.; Hsu, C.-Y.; Lai, C.-L.; Chen, C.-C. Activation of acid-sensing ion channel 3 by lysophosphatidylcholine 16:0 mediates psychological stress-induced fibromyalgia-like pain. *Ann. Rheum. Dis.* **2020**, *79*, 1644–1656. [CrossRef]

237. Taguchi, T.; Katanosaka, K.; Yasui, M.; Hayashi, K.; Yamashita, M.; Wakatsuki, K.; Kiyama, H.; Yamanaka, A.; Mizumura, K. Peripheral and spinal mechanisms of nociception in a rat reserpine-induced pain model. *Pain* **2015**, *156*, 415–427. [CrossRef]

238. Yen, L.-T.; Hsieh, C.-L.; Hsu, H.-C.; Lin, Y.-W. Preventing the induction of acid saline-induced fibromyalgia pain in mice by electroacupuncture or APETx2 injection. *Acupunct. Med.* **2020**, *38*, 188–193. [CrossRef]

239. Gao, M.; Yan, X.; Lu, Y.; Ren, L.; Zhang, S.; Zhang, X.; Kuang, Q.; Liu, L.; Zhou, J.; Wang, Y.; et al. Retrograde nerve growth factor signaling modulates tooth mechanical hyperalgesia induced by orthodontic tooth movement via acid-sensing ion channel 3. *Int. J. Oral Sci.* **2021**, *13*, 18. [CrossRef]

240. Long, H.; Wang, Y.; Jian, F.; Liao, L.-N.; Yang, X.; Lai, W.-L. Current advances in orthodontic pain. *Int. J. Oral Sci.* **2016**, *8*, 67–75. [CrossRef]

241. Yang, H.; Shan, D.; Jin, Y.; Liang, H.; Liu, L.; Guan, Y.; Chen, C.; Luo, Q.; Yang, Y.; Lai, W.; et al. The Role of Acid-sensing Ion Channel 3 in the Modulation of Tooth Mechanical Hyperalgesia Induced by Orthodontic Tooth Movement. *Neuroscience* **2020**, *442*, 274–285. [CrossRef]

242. Osada, A.; Hitomi, S.; Nakajima, A.; Hayashi, Y.; Shibuta, I.; Tsuboi, Y.; Motoyoshi, M.; Iwata, K.; Shinoda, M. Periodontal acidification contributes to tooth pain hypersensitivity during orthodontic tooth movement. *Neurosci. Res.* **2022**, *177*, 103–110. [CrossRef]

243. Zhang, S.; Divaris, K.; Moss, K.; Yu, N.; Barros, S.; Marchesan, J.; Morelli, T.; Agler, C.; Kim, S.J.; Wu, D.; et al. The Novel ASIC2 Locus Is Associated with Severe Gingival Inflammation. *JDR Clin. Transl. Res.* **2016**, *1*, 163–170. [CrossRef] [PubMed]

244. Matriconl, J.; Gelotl, A.; Etiennel, M.; Lazdunskil, M.; Mullerl, E.; Ardidl, D. Spinal cord plasticity and acid-sensing ion channels involvement in a rodent model of irritable bowel syndrome. *Eur. J. Pain* **2011**, *15*, 335–343. [CrossRef] [PubMed]

245. Duan, B.; Wang, Y.Z.; Yang, T.; Chu, X.P.; Yu, Y.; Huang, Y.; Cao, H.; Hansen, J.; Simon, R.P.; Zhu, M.X.; et al. Extracellular Spermine Exacerbates Ischemic Neuronal Injury through Sensitization of ASIC1a Channels to Extracellular Acidosis. *J. Neurosci.* **2011**, *31*, 2101–2112. [CrossRef] [PubMed]

246. Xiong, Z.-G.; Zhu, X.-M.; Chu, X.-P.; Minami, M.; Hey, J.; Wei, W.-L.; MacDonald, J.F.; Wemmie, J.A.; Price, M.P.; Welsh, M.J.; et al. Neuroprotection in Ischemia. *Cell* **2004**, *118*, 687–698. [CrossRef]

247. Johnson, M.B.; Jin, K.; Minami, M.; Chen, D.; Simon, R.P. Global Ischemia Induces Expression of Acid-Sensing Ion Channel 2a in Rat Brain. *J. Cereb. Blood Flow Metab.* **2016**, *21*, 734–740. [CrossRef]

248. Gao, J.; Duan, B.; Wang, D.-G.; Deng, X.-H.; Zhang, G.-Y.; Xu, L.; Xu, T.-L. Coupling between NMDA Receptor and Acid-Sensing Ion Channel Contributes to Ischemic Neuronal Death. *Neuron* **2005**, *48*, 635–646. [CrossRef]

249. Pignataro, G.; Simon, R.P.; Xiong, Z.G. Prolonged activation of ASIC1a and the time window for neuroprotection in cerebral ischaemia. *Brain* **2006**, *130*, 151–158. [CrossRef]

250. Ren, Y.; Li, C.; Chang, J.; Wang, R.; Wang, Y.; Chu, X.-P. Hi1a as a Novel Neuroprotective Agent for Ischemic Stroke by Inhibition of Acid-Sensing Ion Channel 1a. *Transl. Stroke Res.* **2017**, *9*, 96–98. [CrossRef]

251. Hu, R.; Duan, B.; Wang, D.; Yu, Y.; Li, W.; Luo, H.; Lu, P.; Lin, J.; Zhu, G.; Wan, Q.; et al. Role of Acid-Sensing Ion Channel 1a in the Secondary Damage of Traumatic Spinal Cord Injury. *Ann. Surg.* **2011**, *254*, 353–362. [CrossRef]

252. Vergo, S.; Craner, M.J.; Etzensperger, R.; Attfield, K.; Friese, M.A.; Newcombe, J.; Esiri, M.; Fugger, L. Acid-sensing ion channel 1 is involved in both axonal injury and demyelination in multiple sclerosis and its animal model. *Brain* **2011**, *134*, 571–584. [CrossRef]

253. Friese, M.A.; Craner, M.J.; Etzensperger, R.; Vergo, S.; Wemmie, J.A.; Welsh, M.J.; Vincent, A.; Fugger, L. Acid-sensing ion channel-1 contributes to axonal degeneration in autoimmune inflammation of the central nervous system. *Nat. Med.* **2007**, *13*, 1483–1489. [CrossRef] [PubMed]

254. Arun, T.; Tomassini, V.; Sbardella, E.; de Ruiter, M.B.; Matthews, L.; Leite, M.I.; Gelineau-Morel, R.; Cavey, A.; Vergo, S.; Craner, M.; et al. Targeting ASIC1 in primary progressive multiple sclerosis: Evidence of neuroprotection with amiloride. *Brain* **2013**, *136*, 106–115. [CrossRef] [PubMed]

255. Mango, D.; Nisticò, R. Neurodegenerative Disease: What Potential Therapeutic Role of Acid-Sensing Ion Channels? *Front. Cell. Neurosci.* **2021**, *15*, 730641. [CrossRef] [PubMed]

256. Ortega-Ramírez, A.; Vega, R.; Soto, E. Acid-Sensing Ion Channels as Potential Therapeutic Targets in Neurodegeneration and Neuroinflammation. *Mediat. Inflamm.* **2017**, *2017*, 3728096. [CrossRef]

257. Arias, R.L.; Sung, M.-L.A.; Vasylyev, D.; Zhang, M.-Y.; Albinson, K.; Kubek, K.; Kagan, N.; Beyer, C.; Lin, Q.; Dwyer, J.M.; et al. Amiloride is neuroprotective in an MPTP model of Parkinson's disease. *Neurobiol. Dis.* **2008**, *31*, 334–341. [CrossRef] [PubMed]

258. Sun, X.; Cao, Y.-B.; Hu, L.-F.; Yang, Y.-P.; Li, J.; Wang, F.; Liu, C.-F. ASICs mediate the modulatory effect by paeoniflorin on alpha-synuclein autophagic degradation. *Brain Res.* **2011**, *1396*, 77–87. [CrossRef]

259. Gu, X.-S.; Wang, F.; Zhang, C.-Y.; Mao, C.-J.; Yang, J.; Yang, Y.-P.; Liu, S.; Hu, L.-F.; Liu, C.-F. Neuroprotective Effects of Paeoniflorin on 6-OHDA-Lesioned Rat Model of Parkinson's Disease. *Neurochem. Res.* **2016**, *41*, 2923–2936. [CrossRef]

260. Lai, K.; Song, X.-L.; Shi, H.-S.; Qi, X.; Li, C.-Y.; Fang, J.; Wang, F.; Maximyuk, O.; Krishtal, O.; Xu, T.-L.; et al. Bilirubin enhances the activity of ASIC channels to exacerbate neurotoxicity in neonatal hyperbilirubinemia in mice. *Sci. Transl. Med.* **2020**, *12*, eaax1337. [CrossRef]

261. Cuesta, A.; del Valle, M.E.; García-Suárez, O.; Viña, E.; Cabo, R.; Vázquez, G.; Cobo, J.L.; Murcia, A.; Alvarez-Vega, M.; García-Cosamalón, J.; et al. Acid-sensing ion channels in healthy and degenerated human intervertebral disc. *Connect. Tissue Res.* **2014**, *55*, 197–204. [CrossRef]

262. Ding, J.; Zhang, R.; Li, H.; Ji, Q.; Cheng, X.; Thorne, R.F.; Hondermarck, H.; Liu, X.; Shen, C. ASIC1 and ASIC3 mediate cellular senescence of human nucleus pulposus mesenchymal stem cells during intervertebral disc degeneration. *Aging* **2021**, *13*, 10703–10723. [CrossRef]

263. Li, X.; Wu, F.R.; Xu, R.S.; Hu, W.; Jiang, D.L.; Ji, C.; Chen, F.H.; Yuan, F.L. Acid-sensing ion channel 1a-mediated calcium influx regulates apoptosis of endplate chondrocytes in intervertebral discs. *Expert Opin. Ther. Targets* **2013**, *18*, 1–14. [CrossRef] [PubMed]

264. Yang, Y.; Ding, J.; Chen, Y.; Ma, G.; Wei, X.; Zhou, R.; Hu, W. Blockade of ASIC1a inhibits acid-induced rat articular chondrocyte senescence through regulation of autophagy. *Hum. Cell* **2022**, *35*, 665–677. [CrossRef] [PubMed]

265. Li, X.; Xu, R.-S.; Jiang, D.-L.; He, X.-L.; Jin, C.; Lu, W.-G.; Su, Q.; Yuan, F.-L. Acid-sensing ion channel 1a is involved in acid-induced osteoclastogenesis by regulating activation of the transcription factor NFATc1. *FEBS Lett.* **2013**, *587*, 3236–3242. [CrossRef]

266. Wang, Z.; Han, L.; Chen, H.; Zhang, S.; Zhang, S.; Zhang, H.; Li, Y.; Tao, H.; Li, J. Sa12b Improves Biological Activity of Human Degenerative Nucleus Pulposus Mesenchymal Stem Cells in a Severe Acid Environment by Inhibiting Acid-Sensitive Ion Channels. *Front. Bioeng. Biotechnol.* **2022**, *10*, 816362. [CrossRef]

267. Han, L.; Wang, Z.; Chen, H.; Li, J.; Zhang, S.; Zhang, S.; Shao, S.; Zhang, Y.; Shen, C.; Tao, H. Sa12b-Modified Functional Self-Assembling Peptide Hydrogel Enhances the Biological Activity of Nucleus Pulposus Mesenchymal Stem Cells by Inhibiting Acid-Sensing Ion Channels. *Front. Cell Dev. Biol.* **2022**, *10*, 822501. [CrossRef]

268. Abboud, F.M.; Benson, C.J. ASICs and cardiovascular homeostasis. *Neuropharmacology* **2015**, *94*, 87–98. [CrossRef]

269. Fromy, B.; Lingueglia, E.; Sigaudo-Roussel, D.; Saumet, J.L.; Lazdunski, M. Asic3 is a neuronal mechanosensor for pressure-induced vasodilation that protects against pressure ulcers. *Nat. Med.* **2012**, *18*, 1205–1207. [CrossRef]

270. Weavil, J.C.; Kwon, O.S.; Hughen, R.W.; Zhang, J.; Light, A.R.; Amann, M. Gene and protein expression of dorsal root ganglion sensory receptors in normotensive and hypertensive male rats. *Am. J. Physiol.-Regul. Integr. Comp. Physiol.* **2022**, *323*, R221–R226. [CrossRef] [PubMed]

271. Faraci, F.M.; Taugher, R.J.; Lynch, C.; Fan, R.; Gupta, S.; Wemmie, J.A. Acid-Sensing Ion Channels. *Circ. Res.* **2019**, *125*, 907–920. [CrossRef]

272. Gupta, S.C.; Singh, R.; Asters, M.; Liu, J.; Zhang, X.; Pabbidi, M.R.; Watabe, K.; Mo, Y.Y. Regulation of breast tumorigenesis through acid sensors. *Oncogene* **2015**, *35*, 4102–4111. [CrossRef]

273. Yang, C.; Zhu, Z.; Ouyang, X.; Yu, R.; Wang, J.; Ding, G.; Jiang, F. Overexpression of acid-sensing ion channel 1a (ASIC1a) promotes breast cancer cell proliferation, migration and invasion. *Transl. Cancer Res.* **2020**, *9*, 7519–7530. [CrossRef] [PubMed]

274. Bychkov, M.; Shulepko, M.; Osmakov, D.; Andreev, Y.; Sudarikova, A.; Vasileva, V.; Pavlyukov, M.S.; Latyshev, Y.A.; Potapov, A.A.; Kirpichnikov, M.; et al. Mambalgin-2 Induces Cell Cycle Arrest and Apoptosis in Glioma Cells via Interaction with ASIC1a. *Cancers* **2020**, *12*, 1837. [CrossRef]

275. Bychkov, M.L.; Kirichenko, A.V.; Shulepko, M.A.; Mikhaylova, I.N.; Kirpichnikov, M.P.; Lyukmanova, E.N. Mambalgin-2 Inhibits Growth, Migration, and Invasion of Metastatic Melanoma Cells by Targeting the Channels Containing an ASIC1a Subunit Whose Up-Regulation Correlates with Poor Survival Prognosis. *Biomedicines* **2021**, *9*, 1324. [CrossRef] [PubMed]

276. Bychkov, M.L.; Shulepko, M.A.; Vasileva, V.Y.; Sudarikova, A.V.; Kirpichnikov, M.P.; Lyukmanova, E.N. ASIC1a Inhibitor mambalgin-2 Suppresses the Growth of Leukemia Cells by Cell Cycle Arrest. *Acta Nat.* **2020**, *12*, 101–116. [CrossRef]

277. Smoller, J.W.; Gallagher, P.J.; Duncan, L.E.; McGrath, L.M.; Haddad, S.A.; Holmes, A.J.; Wolf, A.B.; Hilker, S.; Block, S.R.; Weill, S.; et al. The Human Ortholog of Acid-Sensing Ion Channel Gene ASIC1a Is Associated With Panic Disorder and Amygdala Structure and Function. *Biol. Psychiatry* **2014**, *76*, 902–910. [CrossRef] [PubMed]

278. Gugliandolo, A.; Gangemi, C.; Caccamo, D.; Currò, M.; Pandolfo, G.; Quattrone, D.; Crucitti, M.; Zoccali, R.A.; Bruno, A.; Muscatello, M.R.A. The RS685012 Polymorphism of ACCN2, the Human Ortholog of Murine Acid-Sensing Ion Channel (ASIC1) Gene, is Highly Represented in Patients with Panic Disorder. *NeuroMolecular Med.* **2015**, *18*, 91–98. [CrossRef]

279. Coryell, M.W.; Wunsch, A.M.; Haenfler, J.M.; Allen, J.E.; Schnizler, M.; Ziemann, A.E.; Cook, M.N.; Dunning, J.P.; Price, M.P.; Rainier, J.D.; et al. Acid-Sensing Ion Channel-1a in the Amygdala, a Novel Therapeutic Target in Depression-Related Behavior. *J. Neurosci.* **2009**, *29*, 5381–5388. [CrossRef]

280. Taugher, R.J.; Lu, Y.; Fan, R.; Ghobbeh, A.; Kreple, C.J.; Faraci, F.M.; Wemmie, J.A. ASIC1A in neurons is critical for fear-related behaviors. *Genes Brain Behav.* **2017**, *16*, 745–755. [CrossRef]

281. Pidoplichko, V.I.; Aroniadou-Anderjaska, V.; Prager, E.M.; Figueiredo, T.H.; Almeida-Suhett, C.P.; Miller, S.L.; Braga, M.F.M. ASIC1a Activation Enhances Inhibition in the Basolateral Amygdala and Reduces Anxiety. *J. Neurosci.* **2014**, *34*, 3130–3141. [CrossRef]

282. Ziemann, A.E.; Schnizler, M.K.; Albert, G.W.; Severson, M.A.; Howard, M.A., 3rd; Welsh, M.J.; Wemmie, J.A. Seizure termination by acidosis depends on ASIC1a. *Nat. Neurosci.* **2008**, *11*, 816–822. [CrossRef]

283. Lv, R.-j.; He, J.-s.; Fu, Y.-h.; Zhang, Y.-q.; Shao, X.-q.; Wu, L.-w.; Lu, Q.; Jin, L.-r.; Liu, H. ASIC1a polymorphism is associated with temporal lobe epilepsy. *Epilepsy Res.* **2011**, *96*, 74–80. [CrossRef] [PubMed]

284. Guo, W.; Chen, X.; He, J.-J.; Wei, Y.-J.; Zang, Z.-L.; Liu, S.-Y.; Yang, H.; Zhang, C.-Q. Down-Regulated Expression of Acid-Sensing Ion Channel 1a in Cortical Lesions of Patients with Focal Cortical Dysplasia. *J. Mol. Neurosci.* **2014**, *53*, 176–182. [CrossRef] [PubMed]

285. Cao, Q.; Wang, W.; Gu, J.; Jiang, G.; Wang, K.; Xu, Z.; Li, J.; Chen, G.; Wang, X. Elevated Expression of Acid-Sensing Ion Channel 3 Inhibits Epilepsy via Activation of Interneurons. *Mol. Neurobiol.* **2014**, *53*, 485–498. [CrossRef] [PubMed]

286. Stein, C.; Kopf, A. Pain therapy—Are there new options on the horizon? *Best Pract. Res. Clin. Rheumatol.* **2019**, *33*, 101420. [CrossRef]

Article

Chemical Synthesis of a Functional Fluorescent-Tagged α-Bungarotoxin

Oliver Brun [1,2,3], Claude Zoukimian [4], Barbara Oliveira-Mendes [1], Jérôme Montnach [1], Benjamin Lauzier [1], Michel Ronjat [1], Rémy Béroud [4], Frédéric Lesage [3,5], Didier Boturyn [6] and Michel De Waard [1,4,7,*]

[1] L'institut Du Thorax, Nantes Université, INSERM, CNRS, INSERM UMR 1087/CNRS UMR 6291, 8 Quai Moncousu, F-44007 Nantes, France; oliver.brun@univ-nantes.fr (O.B.); Barbara.Ribeire@univ-nantes.fr (B.O.-M.); jerome.montnach@univ-nantes.fr (J.M.); benjamin.lauzier@univ-nantes.fr (B.L.); michel.ronjat@univ-nantes.fr (M.R.)
[2] Faculty of Medicine, Université de Montréal, Montréal, QC H3T 1J4, Canada
[3] Research Center, Montréal Heart Institute, Montréal, QC H1T 1C8, Canada; frederic.lesage@polymtl.ca
[4] Smartox Biotechnology, 6 Rue des Platanes, F-38120 Saint-Egrève, France; claude.zoukimian@smartox-biotech.com (C.Z.); remy.beroud@smartox-biotech.com (R.B.)
[5] École Polytechnique Montréal, Montreal, QC H3C 3A7, Canada
[6] UMR 5250, Département de Chimie Moléculaire, CNRS, Université Grenoble Alpes, CEDEX 09, F-38058 Grenoble, France; didier.boturyn@univ-grenoble-alpes.fr
[7] LabEx "Ion Channels, Science & Therapeutics", F-06560 Valbonne, France
[*] Correspondence: michel.dewaard@univ-nantes.fr; Tel.: +33-228-080-076

Abstract: α-bungarotoxin is a large, 74 amino acid toxin containing five disulphide bridges, initially identified in the venom of *Bungarus multicinctus* snake. Like most large toxins, chemical synthesis of α-bungarotoxin is challenging, explaining why all previous reports use purified or recombinant α-bungarotoxin. However, only chemical synthesis allows easy insertion of non-natural amino acids or new chemical functionalities. Herein, we describe a procedure for the chemical synthesis of a fluorescent-tagged α-bungarotoxin. The full-length peptide was designed to include an alkyne function at the amino-terminus through the addition of a pentynoic acid linker. Chemical synthesis of α-bungarotoxin requires hydrazide-based coupling of three peptide fragments in successive steps. After completion of the oxidative folding, an azide-modified Cy5 fluorophore was coupled by click chemistry onto the toxin. Next, we determined the efficacy of the fluorescent-tagged α-bungarotoxin to block acetylcholine (ACh)-mediated currents in response to muscle nicotinic receptor activation in TE671 cells. Using automated patch-clamp recordings, we demonstrate that fluorescent synthetic α-bungarotoxin has the expected nanomolar affinity for the nicotinic receptor. The blocking effect of fluorescent α-bungarotoxin could be displaced by incubation with a 20-mer peptide mimicking the α-bungarotoxin binding site. In addition, TE671 cells could be labelled with fluorescent toxin, as witnessed by confocal microscopy, and this labelling was partially displaced by the 20-mer competitive peptide. We thus demonstrate that synthetic fluorescent-tagged α-bungarotoxin preserves excellent properties for binding onto muscle nicotinic receptors.

Keywords: toxins; peptide chemistry; native chemical ligation; α-bungarotoxin; click chemistry; automated patch-clamp; fluorescent peptide; TE671 cells; nicotinic acetylcholine receptor

Key Contribution: The first report of the chemical synthesis of a fluorescent α-bungarotoxin that leads to a bioactive peptide.

Citation: Brun, O.; Zoukimian, C.; Oliveira-Mendes, B.; Montnach, J.; Lauzier, B.; Ronjat, M.; Béroud, R.; Lesage, F.; Boturyn, D.; De Waard, M. Chemical Synthesis of a Functional Fluorescent-Tagged α-Bungarotoxin. *Toxins* **2022**, *14*, 79. https://doi.org/10.3390/toxins14020079

Received: 5 January 2022
Accepted: 17 January 2022
Published: 21 January 2022

Publisher's Note: MDPI stays neutral with regard to jurisdictional claims in published maps and institutional affiliations.

1. Introduction

α-bungarotoxin is a 74-mer peptide containing five disulphide bridges. It was originally identified and purified from the venom of the snake *Bungarus multicinctus*, an elapid snake from Taiwan [1]. Its unique properties were soon recognized through its ability to act as an antagonist of acetylcholine (ACh) for the nicotinic acetylcholine receptor (AChR) on which it binds, mainly by interacting with the α subunits of AChR with very high affinity and irreversibly [2]. As member of the type II α-neurotoxins (five disulphide bridges as opposed to type I with four disulphide bridges), it belongs to the family of three-finger toxins. It binds both the muscle type AChR (mAChR) at the motor endplate and the neuronal α7 AChR. While two and five binding sites are available, respectively, for these two types of AChR, respectively, the binding of a single α-bungarotoxin is sufficient to block channel opening of the ligand-gated channels, indicating a non-competitive mechanism of antagonism [3]. Structural data reveal that glycosylation of AChR is important for the high-affinity binding of the toxin [4]. Among all the interesting applications developed with this toxin are the improved long-term imaging of zebrafish embryo development using genetically encoded α-bungarotoxin, allowing for a complete immobilization of the embryo [5]. While label-free α-bungarotoxin-binding assays have been developed, namely using a BIAcore sensor chip technology [6], most applications require labelling of the toxin. Very early on, the purified peptide could be iodinated to characterize the properties of the toxin receptor at the level of the skeletal muscle [7]. This allowed for a precise quantification of mAChR numbers at the motor end plate of rat diaphragm [8]. As usual for toxins employed as radiotracers, the iodinated version of α-bungarotoxin allowed tracking the purification of AChR from *Torpedo californica* electroplax [9]. The scope of investigation on AChR being very dynamic, it did not take long before an FITC-, TRITC- or rhodamine-derived fluorescent version of the purified α-bungarotoxin was identified [10] and used to visualize the distribution of AChR onto vertebrate skeletal muscle fibres [11,12]. Most of this labelling involves N-hydroxysuccinimide (NHS) ester-mediated derivatisation of an amine-reactive group of α-bungarotoxin. Interestingly, in vivo injection of fluorescent α-bungarotoxin was shown to improve the efficiency of motor endplate labelling [13]. Yet, it was clearly recognized that, depending on the ratio of dye/toxin, this labelling decreased the blocking potency of α-bungarotoxin for AChR [11], since fluorescence labelling occurs randomly along the sequence onto free amines, some possibly required for pharmacophore integrity. Similarly, a method was presented for the efficient conjugation of horseradish peroxidase to α-bungarotoxin, which is useful for the histochemical staining of mAChR on muscle fibres for light and electron microscopy [14] and for the coupling to gold particles for electron microscopy applications [15]. Yet, in spite of a claimed 1:1 molar conjugation, labelled α-bungarotoxin seems to lose part of its activity. The importance of tagged α-bungarotoxin is nevertheless substantiated by the fact that the α-bungarotoxin binding site (BS), an optimized small peptide of about 13 amino acids derived from a combinatorial phage-display peptide library, fully preserves the unique properties of high affinity binding of the toxin [16,17]. Hence, in addition to the fact that it may serve as a lead compound for the development of antidotes [18], this BBS peptide can also be used as a tag onto poorly immunogenic proteins to visualize their cell behaviour with tagged α-bungarotoxin [19–28].

Most of the applications reporting α-bungarotoxin usage have been performed with the toxin purified directly from snake venom. There are two reasons for this: (i) first, high quantities of this venom can be obtained by milking this snake species, and (ii) α-bungarotoxin is abundantly represented in this venom. As a consequence, it took quite some time before anyone published a method for the production of recombinant α-bungarotoxin [28]. However, recombinant α-bungarotoxin comes with additional amino acids at the N-terminus. Efforts to produce synthetic α-bungarotoxin have also been delayed considerably because of the complexity of the chemical synthesis; the size of the peptide is substantial, as is the number of disulphide bridges. Yet, this effort is crucial if one desires to produce analogues containing non-natural amino acids or if site-defined tagging onto the toxin is the goal. A recent study reported the first chemical synthesis of α-bungarotoxin using peptide-hydrazide-based native chemical ligation (NCL) between two α-bungarotoxin fragments [29]. While this is an important achievement, there was no conclusive evidence in this report that the synthetic α-bungarotoxin was functionally active on mAChR or that this chemical synthesis could be used to produce fluorescent analogues. Herein, our aim was to produce synthetic α-bungarotoxin with an alkyne function at the N-terminus of the chain in order to selectively conjugate any fluorophore or molecule of interest like biotin via click chemistry. With regard to the large size of this toxin, we decided to synthesize the linear chain thanks to two successive ligations of three α-bungarotoxin fragments. In addition, we kept the hydrazide-based ligation strategy to synthesize the linear α-bungarotoxin, as it can be considered the most convenient native chemical ligation method and currently is one of the most used [30]. The oxidative folding leads to folded α-bungarotoxin functionalized with an alkyne group. We performed a wealth of functional tests to illustrate that tagged synthetic α-bungarotoxin maintains potent activity compared to the native purified peptide. Our data illustrate that we capably produced for the first time a synthetic fluorescent α-bungarotoxin that preserves all the functionalities of purified α-bungarotoxin, opening an avenue for new applications that require specific sequence peptide toxin modifications, which are not possible on the natural peptide or only partially on the recombinant peptide.

2. Results and Discussion

2.1. Strategy of Synthesis and Chemical Production of Intermediate Fragments of α-Bungarotoxin

The amino acid length of α-bungarotoxin is too long to allow a full one-shot peptide synthesis by solid-phase peptide synthesis (SPPS). For this reason, we first selected three α-bungarotoxin peptide fragments with sizes most favourable for performing peptide-hydrazide-based native chemical ligation [31] by considering the different ligation sites (Xaa-Cys, Xaa being any kind of amino acid) available (Figure 1). Three fragments turned out to be favourable for this synthesis approach. PA-α-BgTx$_{1-28}$ (**1**; IVCHTTATSPISAVTCPP-GENLCYRKMW) with a pentynoic acid (PA) at the amino terminus appears suitable because it offers a Trp-Cys point of ligation. The alkyne moiety instead of its azide counterpart was chosen for later click chemistry in order to avoid the reduction of the azide function that may potentially occur during oxidative folding of full-length α-BgTx. A shorter α-BgTx$_{1-22}$ fragment was also possible to synthesize (**2**; IVCHTTATSPISAVTCPP-GENL) but the ligation point Leu-Cys is less favourable [32]. The intermediate fragment α-BgTx$_{29-43}$ (**3**; CDAFCSSRGKVVELG) was preferred over the lengthier fragment α-BgTx$_{29-47}$ (**4**; CDAFCSSRGKVVELGCAAT) because it provides an ideal Gly-Cys ligation point despite a shorter length than **4**. The third fragment was α-BgTx$_{44-74}$ (**5**; CAATCP-SKKPYEEVTCCSTDKCNPHPKQRPG). The first two fragments were synthesized with a hydrazide function at the C-terminus in order to achieve two successive ligations. The use of hydrazide fragments, known as crypto-thioesters (i.e., masked peptide thioesters, which, unless activated, are inert in NCL), allows the ligation to occur in the N-to-C direction without requiring temporary orthogonal protection of the Cys residues, as in the case of the Dawson method [33]. While o-aminoanilide fragments [34] could also have been used as masked thioesters, this would have implied coupling a Gly residue as the first amino acid onto a Dawson resin with the α-BgTx$_{29-43}$ fragment. This reaction can easily acylate the two amine functions of this resin, even if a second generation of the Dawson resin has addressed this issue [35].

The syntheses of these three fragments (**1**, **3** and **5**) were performed by automated SPPS, yielding crude material with clear major analytical reversed-phase high-performance liquid chromatography (RP-HPLC) peaks and yields after purification of 26% (**1**), 57% (**3**) and 44% (**5**) (Figure 2A–C). The products had the appropriate masses (insets in Figure 2).

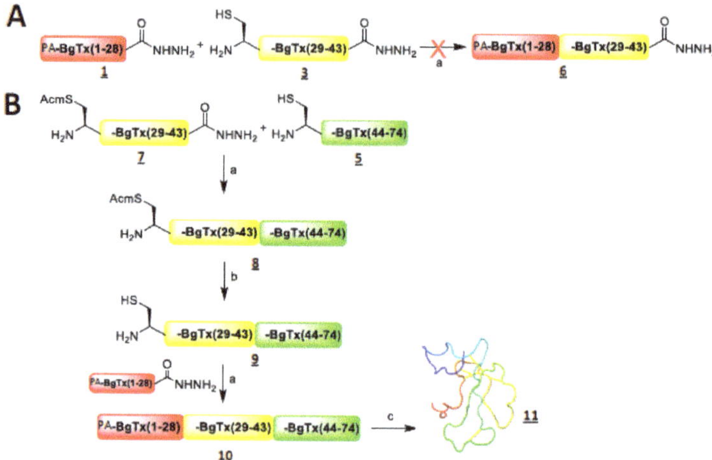

Figure 1. Syntheses strategies of PA-α-BgTx by NCL. (**A**) Synthesis strategy from the N-terminus towards the C-terminus. Only the first step is shown, since it was a dead-end procedure. a = 0.2 M NaH$_2$PO$_4$, H$_2$O, 6 M Gn.HCl, pH 3, −15 °C, 10 eq. NaNO$_2$, 20 min, then 100 eq. 4-mercatophenylacetic acid (MPAA), pH 6.5, room temperature (RT), overnight. (**B**) Synthesis strategy from the C-terminus towards the N-terminus. b = Acm deprotection: 0.2 M NaH$_2$PO$_4$, H$_2$O, 6 M Gn.HCl, pH 7.0, 10 eq. PdCl$_2$, 10 min, then 100 eq. dithiothreitol (DTT). c = oxidative folding: 0.1 M Tris pH 8.2, 5 mM glutathione (GSH), 0.5 mM glutathione disulphide (GSSG), 48 h.

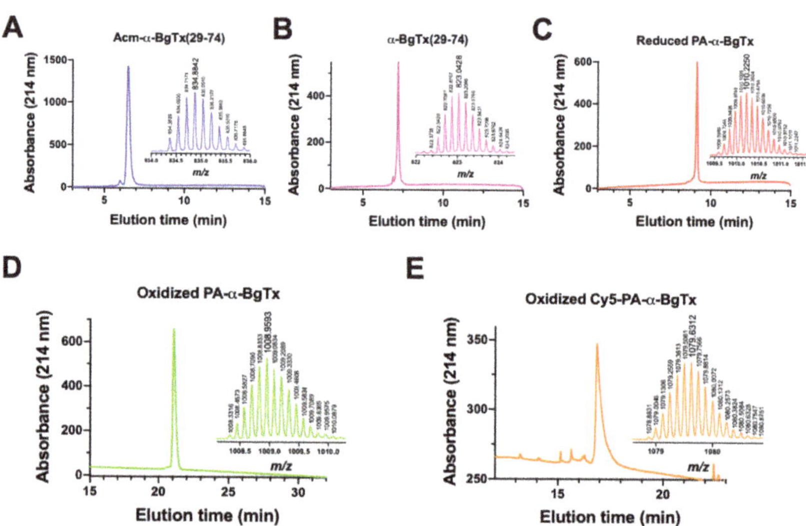

Figure 2. RP-HLPC and MS profiles of synthetic peptides. RP-HPLC A to C were performed on an AdvanceBio Peptide C18 column (10 cm, 214 nm, 5–65% B solvent in 12 min), RP-HPLC D on an AdvanceBio Peptide C18 column (25 cm, 214 nm, 15–65% B solvent in 25 min) and RP-HPLC E on a XSelect Peptide CSH C18 column (15 cm, 214 nm, 5–60% B solvent in 20 min). Retention times were t$_R$ = 6.5 min (**A**), 7.2 min (**B**), 9.1 min (**C**), 21.1 min (**D**) and 16.9 min (**E**).

2.2. Chemical Synthesis of Full-Length α-Bungarotoxin

Our initial aim was to assemble the linear α-BgTx peptide starting from the N- to the C-terminus by ligating PA-α-BgTx$_{1-28}$ (**1**) with α-BgTx$_{29-43}$ (**3**) to obtain PA-α-BgTx$_{1-43}$ (**6**), followed by PA-α-BgTx$_{1-43}$ ligation with α-BgTx$_{44-74}$ (Figure 1A). This was attempted by first selectively activating the hydrazide function of **1** with NaNO$_2$ (15 min, $-20\ ^\circ$C), which was later neutralized by an excess of MPAA. The subsequent addition of fragment **3** leads to the expected intermediate fragment **6**, PA-α-BgTx$_{1-43}$, by NCL after one night (Figure 1A). Unfortunately, the first ligation leads to a mixture of the desired product, with truncated peptides corresponding to the hydrolysis of one or two amino acids before and after the ligation site, as determined by mass spectroscopy. High-resolution mass spectrometry (HRMS) analyses allowed the identification of the amide bonds that were cleaved within the peptide; they were all located at one or two residue distance from the ligation site, either towards the N-terminus (amide bond Met^{27}-Trp^{28}) or towards the C-terminus (amide bonds Asp^{30}-Ala^{31} and Ala^{31}-Phe^{32}). Fragments **1** and **3** were not sensitive to the NCL buffer and, interestingly, when the NCL reaction was performed between fragments **1** and **5** (instead of **3**), the resulting peptide PA-α-BgTx$_{1-28,44-74}$ was stable (data not shown). These controls indicate that the degradation of **6** is an intrinsic property of its sequence.

Based on the observation that such a degradation is not observed with the full length α-BgTx, we then decided to shift towards a second strategy by assembling the fragments from the C-terminus to the N-terminus (Figure 1B). This avoids the formation of the intermediate peptide **6**. Therefore, α-BgTx$_{29-43}$ was synthesized again but with an Acm protection on the N-terminal cysteine of the fragment to yield Acm-α-BgTx$_{29-43}$ (**7**) in order to selectively link this fragment to **5**, α-BgTx$_{44-74}$ (Figure 1B). This protection is fully compatible with hydrazide-based NCL conditions [36] and prevents the cyclisation or polymerisation of the fragment in the NCL conditions once the hydrazide function is converted into a thioester. Following activation of the hydrazide function by NaNO$_2$ (15 min, $-20\ ^\circ$C), the NCL was conducted with fragments **7** and **5** at RT overnight and in the presence of MPAA. After purification, the ligation product **8** (Acm-α-BgTx$_{29-74}$) was obtained with a yield of 60% and with the expected mass (Figure 2A). No degradation product was observed. Next, the Acm protecting group was removed from the N-terminal Cys residue of **8**. For that purpose, we used a published method based on the use of palladium and DTT [37]. Fragment **8** was solubilized in NCL buffer at neutral pH and incubated with 10 equivalents of PdCl$_2$ for 10 min at 37 $^\circ$C. Next, 100 equivalents of DTT were added to neutralize the excess palladium and eliminated by centrifugation before purification by RP-HPLC. Fragment **9** (α-BgTx$_{29-74}$) was hence obtained pure, with a yield of 53% (Figure 2B). This last fragment could therefore be used for a second NCL with the PA-α-BgTx$_{1-28}$ fragment (**1**); **1** was therefore first activated by oxidation with NaNO$_2$ and later put in presence of MPAA and **9**. Following purification and lyophilisation, unfolded and reduced PA-α-BgTx (**10**) was obtained with a yield of 42% (Figure 2C). In order to oxidize all the ten cysteine residues into the five natural disulphide bonds, several oxidizing buffers were used. The alkaline buffer containing 0.1 M tris(hydroxymethyl)aminomethane (Tris), pH 8.2 and the redox couple reduced glutathione (GSH, 5 mM)/oxidized glutathione (GSSG, 0.5 mM) yielded the best results. After 48 h, and following RP-HPLC purification, the folded/oxidized PA-α-BgTx (**11**) was obtained with a yield of 23% (Figures 1B and 2D). The total yield after 4 steps was 3%.

2.3. Coupling of a Fluorochrome on PA-α-BgTx by Click Chemistry

To check the feasibility of using click chemistry on **11**, we decided to selectively couple Cy5-N3 (**12**) onto **11** by copper-catalysed azide-alkyne cycloaddition (CuAAC) [38,39]. Following solubilization of **11**, 10 equivalents of **12**, 10 equivalents of CuBr(CH$_3$)$_2$S and 5 equivalents of tris(3-hydroxypropyltriazolylmethyl)amine (THPTA) were added and incubated at RT overnight and under inert atmosphere. Cy5-α-BgTx (**13**) was purified by RP-HPLC, with a yield of 21% based on the dosing at 646 nm of Cy5 emission (Figure 2E). The excess of dye **12** was also collected to be used again in further click reactions. To test whether **13** was bioactive, we launched a series of functional tests to validate the synthetic product.

2.4. High-Throughput Evaluation of ACh and Cy5-α-Bungarotoxin Effects on the TE671 Cell Line

Following chemical synthesis of Cy5-α-BgTx, the next step was to assess whether this compound retained the inhibitory effect and affinity for mAChR. This was performed by recording mAChR currents in TE671 cells elicited by ACh application in the presence of either native purified α-BgTx or synthetic Cy5-α-BgTx using a high-throughput automated patch-clamp system (Syncropatch 384, Nanion, Munich, Germany). This system allows the simultaneous recording of 384 cells, permitting a quantitative evaluation of the effects of ACh and peptides. The resting voltage was set to -60 mV to record inward currents through mAChR. The use of TE671 human cell line, derived from rhabdomyosarcoma, for this study is appropriate since it endogenously expresses mAChR [40] and binds α-BgTx [41]. Each cell was maintained at a resting membrane potential of -60 mV to record inward currents through mAChR. The effects of a first pulse of 3.33 µM ACh, in the absence of toxins, and of a second 3.33 µM ACh pulse following a 15 min incubation with one of the α-BgTx analogues were monitored. Only cells considered responsive, with a minimal current amplitude of -100 pA after the first ACh application, were selected for analyses in order to clearly discriminate the effects of the peptides. As expected from a channel that tends to desensitize the second application of ACh, it provided a lower current amplitude (Figure 3A). Indeed, the maximal current amplitude recorded at the second dose averaged -230.5 ± 22.9 pA versus -424.2 ± 45.0 pA for the first ACh pulse ($n = 82$; $p < 0.005$). Since this level of desensitization is strictly correlated with ACh concentration and time of application [42–44], we decided to normalise the current elicited by the second dose of ACh against that obtained for the first pulse using individual cells as their own intrinsic control, thereby accounting for intercellular variability. ACh concentration and time of application could not be further optimised, as they correspond roughly to the EC_{50} ligand's value on mAChR [45] and a short 5 s respectively. Therefore, to assess the effects of native α-BgTx and synthetic Cy5-α-BgTx, the pulse 2/pulse 1 (P2/P1) ratio for each cell was normalised against the mean P2/P1 obtained for control cells in the absence of toxins (Figure 3B). Clear evidence was thus provided that both peptides, at concentrations above 0.03 nM, were able to block the response of TE671 cells to a second application of ACh (dashed arrows illustrating the reduced current levels from dotted horizonal lines). From the current amplitudes recorded using this protocol, concentration-response curves for purified α-BgTx and Cy5-α-BgTx were generated (Figure 3C). Cy5-α-BgTx potently blocks mAChR currents with an IC_{50} value of 1.64 ± 0.19 nM ($n = 19–28$ cells/concentration). In comparison, the natural purified and non-modified α-BgTx peptide inhibits mAChR with a slightly better potency, with an IC_{50} value of 0.25 ± 0.02 nM ($n = 8–41$ cells/concentration). The Hill coefficient remained unaltered by the addition of the Cy5 tag, with an estimation of 1.45 ± 0.17 for native α-BgTx and 1.38 ± 0.20 for Cy5-α-BgTx. Both affinity and Hill coefficient estimations are consistent with previous records of α-BgTx binding on TE671 cells [46]. The slight difference in IC_{50} values observed between native and synthetic modified α-BgTx can be accounted for by the difficulties in reliably assessing the concentrations of purified α-BgTx and the potential presence of the fluorescent tag at the N-terminus of the peptide. This is thus the first report illustrating the activity on its receptor of a synthetic α-BgTx modified at its N-terminus with a fluorescent tag.

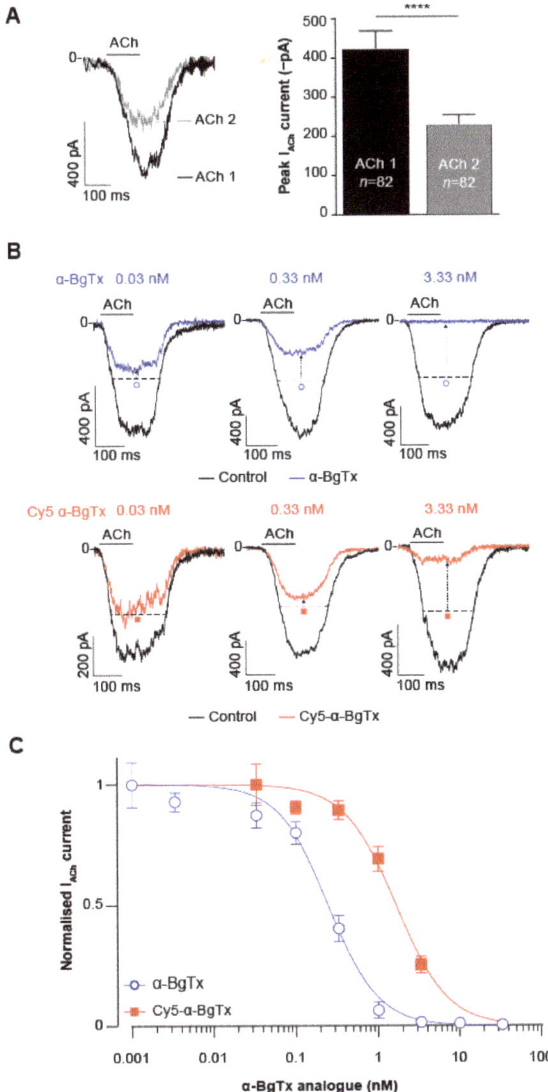

Figure 3. Effect of native α-BgTx and Cy5-α-BgTx on ACh-mediated current in TE671 cells. (**A**) Representative ACh-mediated currents following two consecutive 3.33 µM ACh doses. First ACh pulse illustrated in black, second dose in grey. Mean ± SEM (*n* = 82 cells, **** *p* < 0.0005). The reduction in current amplitude with the second ACh application is due to desensitisation of mAChR. (**B**) Representative current traces in response to a first application of 3.33 ACh (black trace), or a second identical application in presence of increasing concentrations of native α-BgTx (blue) or Cy5-α-BgTx (red). The dotted line represents the expected current after a second ACh application in the absence of any toxin. (**C**) Concentration-response curves for the native α-BgTx and Cy5-α-BgTx-mediated block of ACh response (*n* = 8–41 cells/concentration, total *N* = 339 cells), **** *p* < 0.0005.

2.5. Neutralisation of α-BgTx-Mediated Inhibition of ACh Response by a Binding Site-Mimicking Peptide

As discussed previously, peptides representing the mAChR binding site for α-BgTx are known, and a mimicking peptide can be synthesized. We added, to the optimum binding sequence (WRYYESSLEPYPD-OH), 3 amino acids (GSG) at both the N- and C-terminus, and an additional biotin tag onto an extra C-terminal K residue. The resulting peptide used for our experiments is GSGWRYYESSLEPYPDGSGK(biotin)-OH (termed BSpep). Due to the pre-established irreversibility of α-BgTx binding to mAChR, it was hypothesised than pre-incubating the toxin with the binding site analogue should neutralise it, thereby preventing its inhibition of ACh activity on the muscle-type mAChR. We first assessed the binding of variable concentrations of BSpep to 5 nM Cy5-α-BgTx by microscale thermophoresis (MST) (Figure 4). Fit of the data yields an apparent K_d of 1.04 μM.

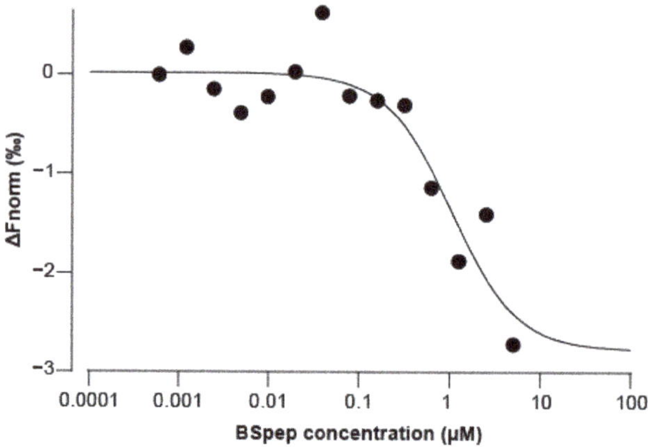

Figure 4. BgTx as measured by MST. A total of 5 nM Cy5-α-BgTx was incubated for 15 min at room temperature with a range of BSpep concentrations. The change in thermophoretic signal yields a K_D = 1.04 μM in these experimental conditions.

Next, the neutralising effect of BSpep was assessed in vitro on TE671 cells using the aforementioned patch-clamp assay. 1 μM BSpep was preincubated with 3.3 nM native α-BgTx or 10 nM Cy5-α-BgTx at 4 °C for 2 h, prior to incubation with the TE671 cells for 15 min. Incubation with 1 μM BSpep completely restored normal ACh response in the presence of maximally blocking concentrations of either natural purified α-BgTx or synthetic Cy5-α-BgTx (Figure 5A). Inhibition with 3.33 nM α-BgTx resulted in a complete blockade of the ACh-mediated current, reaching only 0.1 ± 2% of the control current elicited by ACh alone (n = 13) (Figure 5B). However, preincubation of the natural α-BgTx peptide with 1 μM BSpep rescued currents up to 100 ± 6% of the normal current (n = 27). Similarly, blocking ACh-mediated currents with 10 nM Cy5-α-BgTx only yielded 2 ± 1% (n = 41) of the control current, which was restored when this synthetic peptide was preincubated with 1 μM BSpep to 101 ± 2% (n = 44) (Figure 5B). The possibility of an α-BgTx-independent effect of 1 μM BSpep on the ACh-mediated response was also investigated by incubating BSpep with the cells. This resulted in no significant increase from the control current (125 ± 11%, n = 14). Overall, these results validate the expected effect of BSpep, occurring through the capture of α-BgTx, preventing its inhibitory action on the muscle-type nAChR. The data also indicate that the inhibition of mAChR by the synthetic Cy5-α-BgTx occurs through the binding of the peptide on the same mAChR site as native α-BgTx.

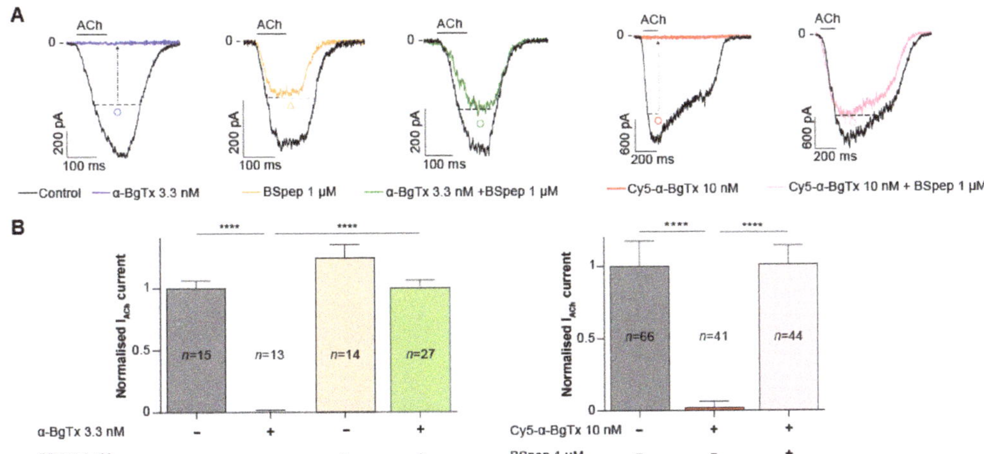

Figure 5. Effect of the BSpep on native α-BgTx and Cy5-α-BgTx inhibition of ACh-mediated current in TE671 cells. (**A**) Representative current traces in response to a first application of 3.33 μM ACh (black) or a second identical application in the presence of α-BgTx (blue), BSpep (yellow), α-BgTx + BSpep (green), Cy5-α-BgTx (red) or Cy5-α-BgTx + BSpep (pink). The dotted line represents the expected current after a second ACh control dose in the absence of toxins. (**B**) Mean normalised current after the second 3.33 μM ACh dose in the presence of α-BgTx (blue), BSpep (yellow), α-BgTx + BSpep (green), Cy5-α-BgTx (red) or Cy5-α-BgTx + BSpep (pink). Mean ± SEM. Kruskal-Wallis test with Dunn's multiple comparison post-test, **** $p < 0.0005$.

2.6. Characterisation of the Cellular Distribution of Muscle-Type mAChR with Fluorescent Cy5-α-BgTx

Using fluorescent Cy5-α-BgTx, we aimed at visualising the distribution of mAChR on fixed TE671 cells by confocal microscopy. At a Cy5-α-BgTx concentration of 200 nM, a fluorescent signal was observed at the expected excitation/emission wavelengths (638/700 nm, respectively), although fluorescence was mostly observed intracellularly (Figure 6A). It has previously been reported that TE671 express mAChR as an immature foetal form, known to localise preferentially in intracellular pools, accounting for up to 80% of the total count [47–49]. To validate the specificity of the observed fluorescence, Cy5-α-BgTx labelling was also performed after neutralising the toxin with 100 μM BSpep through an overnight incubation at 4 °C. This significantly lowered the recorded cell fluorescence, reaching only $41 \pm 6\%$ ($n = 3$ regions of interest; total cell number = 973) of the signal obtained with free Cy5-α-BgTx ($n = 2$ regions of interest; total cell number = 707; $p < 0.05$) (Figure 6B). This result suggests that the major part of the signal observed with the toxin alone is mAChR-specific, whereas the remaining part may also be due to peptide complex internalization or simply an unwanted dissociation of Cy5-α-BgTx from BSpep, occurring as a result of competition with mAChR localized at the plasma membrane. The affinity of BSpep for Cy5-α-BgTx being significantly lower than the one of mAChR dissociation from the peptide is likely a contributing factor. Altogether with its previously discussed electrophysiological characterisation, this contributes to validate the use of Cy5-α-BgTx as a highly potent and selective fluorescent probe for the visualisation of muscle-type mAChR.

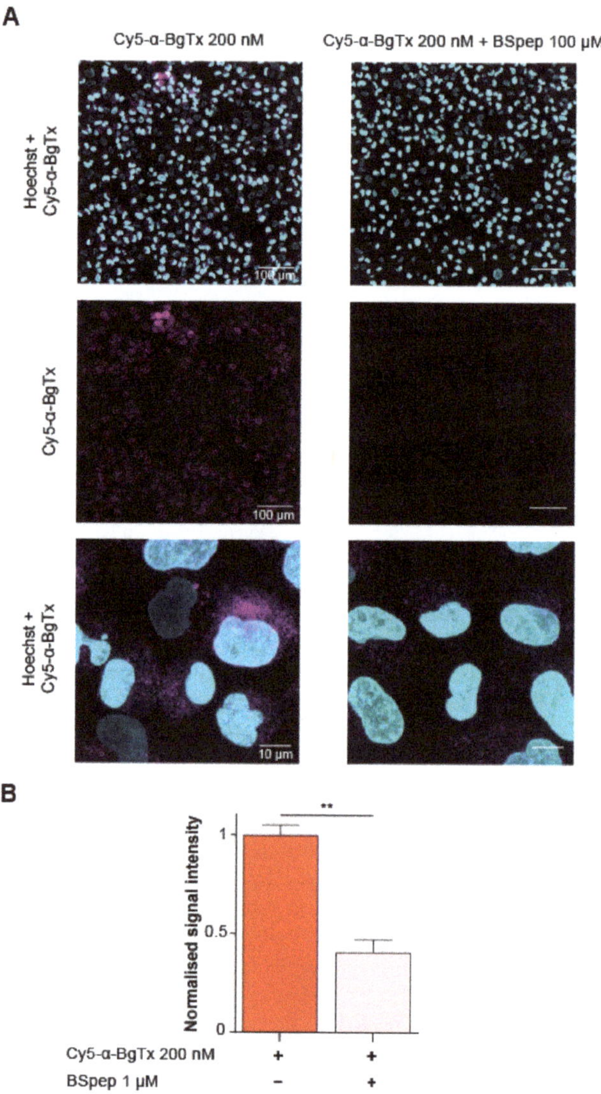

Figure 6. Distribution of nAChR in TE671 cells as labelled with Cy5-α-BgTx. (**A**) Confocal microscopic imaging of TE671 cells labelled with 5 μg/mL Hoechst and either 200 nM Cy5-α-BgTx (left) or 200 nM Cy5-α-BgTx pre-incubated with 100 μM BSpep (right). Top: X60 magnification, Hoechst (cyan) and Cy5-α-BgTx (magenta) displayed; Middle: X60 magnification, only Cy5-α-BgTx displayed; Bottom: X240 magnification, Hoechst and Cy5-α-BgTx displayed. (**B**) Quantification of Cy5-α-BgTx fluorescence after labelling with either 200 nM Cy5-α-BgTx (red) or 200 nM Cy5-α-BgTx pre-incubated with 100 μM BSpep (pink). Mean ± SEM. Two-tailed *t*-test, ** $p < 0.05$.

3. Conclusions

With 74 amino acids and no less than five disulphide bridges, the chemical synthesis of α-BgTx represents a real technical challenge for two reasons. First, the length of the peptide is considerable, and therefore its chemical synthesis requires a strategy of peptide fragment assembly. Second, the presence of 10 cysteine residues theoretically allows for no less than 45 combinations of disulphide bridge arrangements. Our data indicate that we were able to successfully assemble α-BgTx sequence using three carefully chosen peptide fragments and using two successive hydrazide-based ligations of toxin peptide fragments. Thanks to the addition of an alkyne function at the N-terminus of the toxin, a regioselective addition of an azide-modified fluorochrome was performed. In addition, the linear peptide underwent a proper oxidative folding; a single RP-HPLC peak arose from the process, demonstrating that a single combination of disulphide bridge pairing was largely favoured, without further experimental control. We assume that we naturally reached the proper pattern of disulphide bridging, although we did not check this point specifically. Obviously, it appears that the additional presence of pentynoic acid at the N-terminus did not interfere with the folding process. To ensure that α-BgTx was properly synthesised and that the addition of a fluorescent tag at the N-terminus by click chemistry did not prevent bioactivity, we performed a number of functional tests with Cy5-α-BgTx. We used an automated patch-clamp system along with an optimized protocol to assess the inhibition of ACh-mediated currents on TE671 cells. The results illustrate a slight increase in the IC_{50} value for the synthetic fluorescent α-BgTx compared to the IC_{50} value produced by purified α-BgTx, but this difference can easily be attributed to the difficulties in reliably evaluating the concentration of purified peptide and/or the impact of the fluorescent probe on the activity of the synthetic compound. The difference in the IC_{50} values observed was, however, quite reasonable and indicates complete bioactivity of the synthetic product. Of course, it may be of interest in the future to perform a full comparison between native, non-labelled synthetic and tagged-synthetic α-bungarotoxin to check how the fluorescent tag affects activity directly or indirectly. In a study examining the effect of the fluorescent tag on the pharmacology of protoxin II, it was shown that ATTO488 fluorescent tag was more conservative in the native bioactivity of the peptide than the Cy5 tag, indicating that the nature of the tag matters [50]. Here, the use of Cy5 may yield the same result as protoxin II; however, this can be tested later, as the use of click chemistry allows for an easy change in the tag identity. Remarkably, Cy5-α-BgTx inhibition of ACh-mediated currents could be blocked by preincubation with the BSpep, which mimics the mAChR binding site of the peptide. These results further highlight that the synthetic peptide, similarly to the purified one, blocks mAChR by binding onto a specific and well-defined site on the receptor. The applicability of Cy5-α-BgTx for labelling mAChR was further demonstrated by labelling a receptor on TE671 cells, labelling that could be partially prevented by incubation with the BSpep inhibitory sequence. The synthesis described herein will mainly be of interest for those that need to modify natural amino acids within α-bungarotoxin at the expense of an increased production cost of about five to seven-fold compared to labelled-purified α-bungarotoxin.

4. Materials and Methods

4.1. Reagents and Materials

All chemical reagents and solvents were purchased from Sigma-Aldrich (Saint-Quentin-Fallavier, France) except the 2-chlorotrityl chloride resin, which was purchased from Iris Biotech GmbH (Marktredwitz, Germany).

4.2. Resin Loading for Solid-Phase Peptide Syntheses

Two types of resin loading were performed for the synthesis of α-bungarotoxin. First, we loaded 2-chlorotrityl chloride resin for carboxylated peptides. The first amino acid was loaded manually by treating 1 g of 2-chlorotrityl chloride resin (1.6 mmol/g of Cl) swollen in DCM with a solution of 0.6 mmol amino acid and 1 mL of iPr_2Net in 5 mL of DCM for 1 h, followed by adding 5 mL of MeOH and agitating for an additional 1 h. The resin was washed twice with DCM and twice with DMF. Next, it was treated three times with 10 mL of a 20% piperidine solution in DMF. The resulting liquid phases were poured together, and the loading was determined by measuring the absorbance of dibenzofulvene-piperidine adducts at 299 nm ($\varepsilon = 7800$ L·mol^{-1}·cm^{-1}). Second, we loaded 2-chlorotrityl chloride resin for hydrazide peptides. A total of 1 g of 2-chlorotrityl chloride resin (1.6 mmol/g of Cl) was washed twice with DCM and twice with NMP and treated with 10 mL of a 5% hydrazine hydrate solution in NMP for 1 h. The resin was filtered and treated with 10 mL of MeOH/NH$_2$NH$_2$/NMP 10:5:85 for 30 min, filtered again, and washed twice with NMP, twice with DCM and twice with NMP. The first amino acid was loaded manually by treating the resin with a solution of 0.6 mmol amino acid, 0.6 mmol HCTU and 1.2 mmol of iPr2Net in 5 mL of NMP for 1 h. The resin was filtered and treated with 10 mL of Ac$_2$O/iPr$_2$Net/NMP 10:5:85 for 15 min, and then filtered and washed twice with DCM and twice with DMF. The resin was treated three times with 10 mL of a 20% piperidine solution in DMF, the resulting liquid phases were poured together, and the loading was determined by measuring the absorbance of dibenzofulvene-piperidine adduct at 299 nm.

4.3. Solid-Phase Peptide Syntheses

Solid-phase peptide syntheses (SPPS) of α-bungarotoxin fragments and BSpep (GSGW RYYESSLEPYPDGSGK(biotin)-OH) were run on an automated Symphony synthesizer from Gyros Protein Technologies using Fmoc/tBu chemistry at a 0.1 mmol scale, with HCTU as coupling reagent and 2-chlorotrityl chloride resin as solid support. The elongation was carried out using a 10-fold excess of protected amino acids and HCTU and 20-fold excess of iPr$_2$NEt in NMP. The side chain protecting groups used were Asn(Trt), Arg(Pbf), Cys(Trt), Gln(Trt), Glu(O*t*Bu), Lys(Boc), Ser(*t*Bu), Thr(*t*Bu) and Tyr(*t*Bu). Fmoc deprotection was performed using a 20% piperidine solution in DMF. Each coupling lasted 15 min and was repeated three times, followed by capping with a 10% acetic anhydride in NMP. The crude peptides were released from the resin with TFA/iPr3SiH/1,3-dimethoxybenzene/2,2′-(ethylenedioxy)diethanethiol/H$_2$O (85:2.5:3.75:3.75:5) for 3 h, and the peptides were precipitated with ice-cold diethyl ether, recovered by centrifugation and washed twice with diethyl ether. HPLC and MS analyses were performed on the crude peptides before purification.

4.4. HPLC and Mass Spectrometry Analyses

The peptides were analysed by HPLC and high-resolution ESI-MS mass spectrometry. HPLC analyses were carried out on an Agilent system equipped with a Chromolith$^{®}$ High Resolution RP-18e column (150 Å, 100 × 4.6 mm) at a flow rate of 3 mL/min, with a XSelect Peptide CSH C18 column (130 Å, 2.5 µm, 150 × 4.6 mm) or with an Agilent AdvanceBio Peptide column (2.7 µm, 100 × 2.1 mm or 250 × 2.1 mm) at a flow rate of 1 mL/min and 214 nm. Solvents A and B were 0.1% TFA in H$_2$O and 0.1% TFA in MeCN. The acquisition of the LC-ESI-MS data was carried out on a Waters Q-TOF Xevo G2S mass spectrometer with an Acquity UHPLC system and Lockspray source equipped with a Waters Acquity UPLC BEH300 C18 column (1.7 µm, 2.1 × 150 mm). Peptide elution was performed at a flow rate of 0.4 mL/min with a 10–70% gradient of buffer B over 10 min. Solvents A and B were 0.1% formic acid (FA) in H$_2$O and 0.1% FA in MeCN. All calculated and found masses were monoisotopic. Purification of the peptides was performed by HPLC on a preparative Agilent 1260 system equipped with a Phenomenex Jupiter column (4 µm Proteo 90 Å, C12, 250 × 4.6mm) at a flow rate of 20 mL/min or a semi-preparative Agilent 1260 system equipped with an Agilent Poroshell EC-C18 column (120 Å, 4 µm, 250 × 9.4 mm) at a flow rate of 4.5 mL/min.

4.5. Reduced PA-α-BgTx Synthesis by Native Chemical Ligation

PA-α-BgTx$_{1-43}$ (**6**)—10 mg of PA-α-BgTx$_{1-28}$ (**1**, 1 equiv.) was dissolved in 500 µL of NCL buffer consisting of 200 mM NaH$_2$PO$_4$ and 6 M guanidinium hydrochloride (Gn.HCl) at pH 3.0 and cooled in an ice/salt bath at −15 °C for 5 min. Then, 56 µL of 0.5 M aqueous NaNO$_2$ solution (10 equiv.) was added, and the mixture was agitated for 15 min at −15 °C. A total of 47 mg of MPAA (100 equiv.) was dissolved in 300 µL of NCL buffer at pH 7.0 and added to the reaction mixture, followed by 6.6 mg of α-BgTx$_{29-43}$ (**3**, 1.3 equiv.) dissolved in 500 µL of NCL buffer at pH 3.0. The pH was adjusted to 6.8, and the reaction was agitated overnight under inert atmosphere at room temperature. Then, 80 mg of TCEP (100 equiv.)

was dissolved in 8 mL of 6 M Gn.HCl and the pH was adjusted to 5.0; this solution was added to the reaction mixture and mixed for 30 min. The pH was lowered with 500 µL of TFA, and the mixture was washed 3 times with 20 mL of Et$_2$O and purified by preparative RP-HPLC using a linear 20–70% buffer B gradient over 40 min; **6** was not pure enough to determine the ligation yield.

Acm-α-BgTx$_{29-74}$ (**8**)—18.2 mg of Acm-α-BgTx$_{29-43}$ (**7**, 1 equiv.) was dissolved in 1 mL of NCL buffer at pH 3.0 and cooled in an ice/salt bath at −15 °C for 5 min. Then, 193 µL of 0.5 M aqueous NaNO$_2$ solution (10 equiv.) was added, and the mixture was agitated for 15 min at −15 °C. A total of 162 mg of MPAA (100 equiv.) was dissolved in 500 µL of NCL buffer at pH 7.0 and added to the reaction mixture, followed by 43 mg of α-BgTx$_{44-74}$ (**5**, 1.1 equiv.) dissolved in 1 mL of NCL buffer at pH 3.0. The pH was adjusted to 6.8, and the reaction was agitated overnight under inert atmosphere at room temperature. Then, 277 mg of TCEP (100 equiv.) was dissolved in 10 mL of 6 M Gn.HCl and the pH was adjusted to 5.0; this solution was added to the reaction mixture and mixed for 30 min. The pH was lowered with 500 µL of TFA, and the mixture was washed 3 times with 20 mL of Et$_2$O and purified by preparative RP-HPLC using a linear 20–70% buffer B gradient over 40 min. After lyophilization, 33.2 mg of **8** was obtained (yield 60%). RP-HPLC retention time (RT) was 6.5 min (AdvanceBio Peptide C18 10 cm, 5–65% B in 12 min); ESI-HRMS C$_{207}$H$_{334}$N$_{62}$O$_{68}$S$_7$ calculated 5000.26; found 5000.29.

α-BgTx$_{29-74}$ (**9**)—27.5 mg of **8** was dissolved in 1 mL of NCL buffer at pH 7.0 and 10 mg of PdCl$_2$ (10 equiv.) in 200 µL of NCL buffer at pH 7.0 was added. The mixture was agitated 10 min at 37 °C, and 85 mg of DTT (100 equiv.) in 200 µL of water was added. The reaction mixture was centrifugated, the pellet was washed with 200 µL of NCL buffer, and the liquid phases were poured together, acidified with formic acid, and purified by preparative RP-HPLC using a linear 10–60% buffer B gradient for 40 min. After lyophilization, 14.5 mg of **9** was obtained (yield 53%). RP-HPLC RT 7.2 min (AdvanceBio Peptide C18 10 cm, 5–65% B in 12 min); ESI-HRMS C$_{204}$H$_{329}$N$_{61}$O$_{67}$S$_7$ calculated 4929.23; found 4929.16.

Reduced PA-α-BgTx (**10**)—16.8 mg of PA-α-BgTx$_{1-28}$ (**1**, 1.5 equiv.) was dissolved in 600 µL of NCL buffer at pH 3.0 and cooled in an ice/salt bath at −15 °C for 5 min. Then, 95 µL of 0.5 M aqueous NaNO$_2$ solution (10 equiv.) was added, and the mixture was agitated for 15 min at −15 °C. A total of 79 mg of MPAA (100 equiv.) was dissolved in 200 µL of NCL buffer at pH 7.0 and added to the reaction mixture, followed by 18.0 mg of α-BgTx$_{29-74}$ (**9**, 1 equiv.) dissolved in 600 µL of NCL buffer at pH 3.0. The pH was adjusted to 6.8, and the reaction was agitated overnight under inert atmosphere at room temperature. Then, 135 mg of TCEP (100 equiv.) was dissolved in 6 mL of 6 M Gn.HCl and the pH was adjusted to 5.0; this solution was added to the reaction mixture and mixed for 30 min. The pH was lowered with 500 µL of TFA, and the mixture was washed 3 times with 20 mL of Et$_2$O and purified by preparative RP-HPLC using a linear 20–70% buffer B gradient for 40 min. After lyophilization, 12.3 mg of **10** was obtained (yield 42%). RP-HPLC RT 9.1 min (AdvanceBio Peptide C18 10 cm, 5–65% B in 12 min); ESI-HRMS C$_{343}$H$_{543}$N$_{97}$O$_{106}$S$_{11}$ calculated 8068.70; found 8068.68.

4.6. Folding of Reduced PA-α-BgTx

A total of 3.0 mg of reduced PA-α-BgTx (**10**) was dissolved in 3 mL of water and added dropwise to 30 mL of a 0.1 M Tris.HCl buffer at pH 8.2 containing 5 mM GSH and 0.5 mM GSSG while stirring. After 48 h at room temperature, the reaction mixture was acidified with 1 mL of formic acid, filtered and purified by semi-preparative RP-HPLC using a linear 20–40% buffer B gradient over 30 min at 40 °C. After lyophilization, 680 µg of oxidized PA-α-BgTx (**11**, quantified by UV measurement, ε = 9105 L·mol^{-1}·cm^{-1}) was obtained (yield 23%). RP-HPLC RT 21.1 min (AdvanceBio Peptide C18 25 cm, 5–65% B in 12 min); ESI-HRMS C$_{343}$H$_{533}$N$_{97}$O$_{106}$S$_{11}$ calculated 8058.62; found 8058.60.

4.7. Coupling of Cy5 on PA-α-BgTx

A total of 110 µg of oxidized PA-α-BgTx (**11**, 1 equiv.) was dissolved in 12 µL of (NMP/*t*BuOH 9:1)/H$_2$O/(0.4 M HEPES, 50 mM aminoguanidine hydrochloride) 1:1:1, and 0.1 mg of Cy5-N$_3$ (**12**, 10 equiv.) dissolved in 10 µL of NMP/*t*BuOH 9:1 was added. Then, 2.8 mg of CuBr.(CH$_3$)$_2$S and 3 mg of THPTA were dissolved in 1 mL of NMP/*t*BuOH; 10 µL of this solution was added to the reaction mixture (10 equiv. of CuBr.(CH$_3$)$_2$S and 5 equiv. of THPTA) and agitated overnight under inert atmosphere at room temperature. The reaction mixture was acidified with 1 drop of formic acid and purified by semi-preparative RP-HPLC using a linear 20–70% buffer B gradient for 30 min at 40 °C. After lyophilization, 25 µg of Cy5-α-BgTx (**13**, quantified by UV measurement, ε = 250,000 L·mol^{-1}·cm^{-1} at 646 nm) was obtained (yield 21%). RP-HPLC RT 16.9 min (XSelect Peptide CSH C18, 5–60% B in 20 min); ESI-HRMS C$_{378}$H$_{578}$N$_{103}$O$_{107}$S$_{11}$ calculated 8623.99; found 8623.98.

4.8. Cell Culture

TE671 cells were maintained in Dulbecco's Modified Eagle's Medium (DMEM)-high glucose (4.5 g/L) supplemented with 10% foetal calf serum, 1 mM pyruvic acid, 4 mM glutamine, 10 IU/mL penicillin and 10 µg/mL streptomycin (Gibco, Grand Island, NY, USA), and incubated at 37 °C in a 5% CO$_2$ atmosphere. For electrophysiological recordings, cells were detached with trypsin, and floating single cells were diluted (~300 k cells/mL) in extracellular medium containing (in mM): 4 KCl, 140 NaCl, 5 Glucose, 10 HEPES, 2 CaCl$_2$, 1 MgCl$_2$ (pH 7.4, osmolarity 298 mOsm).

4.9. Ligand Binding Studies

Binding of BSpep to Cy5-α-BgTx was investigated by microscale thermophoresis (MST) using the Monolith NT 115 (NanoTemper Technologies GmbH, Germany). All dilutions were performed in proprietary MST NT.115 buffer, supplemented with 0.05% Tween 20. Briefly, 5 nM of Cy5-α-bungarotoxin was incubated with a range of BSpep concentrations for 15 min at room temperature, before being loaded in Monolith Premium capillaries (NanoTemper Technologies GmbH, Germany) and mounted in the Monolith NT 115 thermophoresis chamber. MST was monitored using the dedicated MO Control software. Readings were performed using the built-in picoRED illumination source (600 to 650 nm) at 20% illumination power, and MST power was set as medium. MST fluorescence (F1) was averaged from 1.5 to 2.5 min through MST and normalised against initial fluorescence (F0), which was recorded 1 min before the start of MST. The resulting normalised fluorescence values were expressed as the difference from initial fluorescence (ΔFnorm), which was plotted against BSpep concentration. Dose-ΔFnorm relationships were fit to a sigmoidal equation: Y = Bottom + (Top − Bottom)/(1 + 10^((LogIC$_{50}$ − X) × HillSlope)).

4.10. Automated Patch-Clamp Recordings

Whole-cell recordings were used to investigate the effects of purified α-bungarotoxin and synthetic Cy5-α-bungarotoxin on ACh-induced currents in TE671 cells, by adapting a protocol previously used by Taiwe et al. [45]. Automated patch-clamp recordings were performed using a SyncroPatch 384PE (Nanion, München, Germany). Chips with single-hole high resistance (~4–5 MΩ) were used for TE671 cell attachment and recordings. Pulse generation and data collection were performed with PatchControl384 v1.6.6_B18 (Nanion) and Biomek v4.1_B31 (Beckman Coulter, Brea, CA, USA). Whole-cell recordings were conducted according to the Nanion procedure. Cells were stored in a cell hotel reservoir at 10 °C, with shaking speed at 60 RPM. After initiating the experiment, cell catching, sealing, whole-cell formation, liquid application, recording, and data acquisition were performed sequentially. The intracellular solution contained (in mM): 10 KCl, 110 KF, 10 NaCl, 10 EGTA and 10 HEPES (pH 7.2, osmolarity 280 mOsm). The extracellular filling solution contained (in mM): 140 NaCl, 4 KCl, 5 glucose and 10 HEPES (pH 7.4, osmolarity 290 mOsm). The extracellular solution contained (in mM): 140 NaCl, 4 KCl, 2 CaCl$_2$, 1 MgCl$_2$, 5 glucose and 10 HEPES (pH 7.4, osmolarity 298 mOsm). Whole-cell patch experiments were performed at a holding potential of -60 mV at room temperature (20–22 °C). Currents were sampled at 10 kHz. Each molecule, acetylcholine (100 µM, Sigma, Saint-Quentin Fallavier, France), purified α-bungarotoxin and Cy5-α-bungarotoxin and BSpep 1, was prepared in extracellular solution supplemented with 0.3% bovine serum albumin and 0.86 mM Ca^{2+} in 384-well compound plates. Working compound solution was diluted 3 times in recording wells by adding 30 µL compound to 60 µL external solution to reach final concentration. Where required, BSpep was preincubated with native α-bungarotoxin or Cy5-α-bungarotoxin for 2 h at 4 °C. ACh was applied as a 5 s pulse and rapidly washed using stacked addition to avoid excess mAChR desensitisation. In order to obtain single cell-specific controls, an initial pulse of ACh was performed, and the elicited current was recorded, followed by 6 washes with extracellular buffer. Washes were performed as sequential addition and removal of 30 µL extracellular medium. Following these washes, native α-bungarotoxin or Cy5-α-bungarotoxin $\pm$ preincubation with BSpep was added and allowed to incubate for 15 min in working solution. A second stacked addition of ACh was then performed, and the current response was recorded. Data were analysed with DataController384 V1.8.0_B24 and GraphPad Prism. Dose-response relationships were fit to a sigmoidal equation: $Y =$ Bottom $+$ (Top $-$ Bottom)$/(1 + 10^{\wedge}((\text{LogIC}_{50} - X) \times \text{HillSlope}))$.

4.11. Cell Imaging by Confocal Microscopy

TE671 were incubated overnight at 37 °C in a 5% CO_2 atmosphere in DMEM (Gibco, Grand Island, NY, USA) on Ibidi µ-slide 18-well plates (Ibidi GmbH, Germany) in order to obtain a cell confluency of approximately 70% on the day of fixation. Cells were washed with non-supplemented DMEM, followed by fixation with 50 µL 4% paraformaldehyde (PFA) over 5 min. Exposure time to the fixative was closely monitored to minimise cell permeabilization. Fixed cells were then washed with phosphate buffer solution (PBS), prior to labelling with 50 µL staining solution (PBS with 1% Bovine Serum Albumin, 5 µg/mL Hoechst and 200 nM Cy5-α-bungarotoxin). Then, 200 nM Cy5-α-bungarotoxin was preincubated overnight at 4 °C with 100 µM BSpep for competition experiments. Incubation was performed over 30 min at RT. Labelling was followed by 3 PBS washes of 5 min each. Wells were ultimately bathed in PBS $+$ 0.1% PFA. Image acquisition was performed with a NIKON A1 RSi confocal microscope (Nikon, Champigny sur Marne, France), using a $\times 60$ magnification. Hoechst fluorescence was excited at 403 nm and recorded at 450 nm. Excitation/recording wavelengths for Cy5 were 638/700 nm, respectively. Images were analysed, and fluorescence was quantified using ImageJ.

4.12. Statistical Analyses

Data analyses were performed using GraphPad Prism. All data were presented as mean $\pm$ s.e.m. Comparison of peak current amplitude was performed by a Kruskal-Wallis test with Dunn's multiple comparison correction. Comparison of fluorescence intensity was performed by a two-tailed t-test, following normality confirmation by a Shapiro-Wilk test. p values were considered significant below 0.05.

Author Contributions: Conceptualization, D.B. and M.D.W.; methodology, C.Z., B.O.-M. and B.L.; software, J.M.; validation, R.B., F.L., D.B. and M.D.W.; formal analysis, O.B., C.Z., J.M. and M.R.; investigation, O.B., C.Z., B.O.-M. and J.M.; resources, R.B. and M.D.W.; data curation, M.D.W.; writing—original draft preparation, O.B. and M.D.W.; writing—review and editing, M.D.W.; visualization, J.M. and M.D.W.; supervision, B.O.-M., J.M., M.R., F.L., D.B. and M.D.W.; project administration, R.B., D.B. and M.D.W.; funding acquisition, R.B., F.L. and M.D.W. All authors have read and agreed to the published version of the manuscript.

Funding: This research was funded by the French Association Nationale de la Recherche et de la Technologie (ANRT, 525/2016) and by Smartox Biotechnology for a PhD CIFRE fellowship to Claude Zoukimian. Support from the Agence Nationale de la Recherche (ANR-11-LABX-0015) the Fondation Leducq in the frame of its program of *"Equipement de recherche et plateformes technologiques"*, the Région Pays de la Loire (nouvelle équipe, 2016-11092/11093), and the European

FEDER (2017/FEDER/PL0014592) is also acknowledged. Jérôme Montnach is supported by the Agence Nationale de la Recherche (ANR-18-CE19-0024-02).

Institutional Review Board Statement: Not applicable.

Informed Consent Statement: Not applicable.

Data Availability Statement: Raw data are available on https://github.com/jerome-montnach/Cy5-BgTx.git repository accessed on 2 January 2022.

Conflicts of Interest: The authors Claude Zoukimian, Rémy Béroud and Michel De Waard declare the following competing interest: employee, CEO and founder of Smartox Biotechnology, respectively. The funders had no role in the design of the study; in the collection, analyses, or interpretation of data; in the writing of the manuscript or in the decision to publish the results.

References

1. Mebs, D.; Narita, K.; Iwanaga, S.; Samejima, Y.; Lee, C.Y. Purification, properties and amino acid sequence of alpha-bungarotoxin from the venom of Bungarus multicinctus. *Hoppe-Seyler's Z. Für Physiol. Chem.* **1972**, *353*, 243–262. [CrossRef] [PubMed]
2. Changeux, J.P.; Kasai, M.; Lee, C.Y. Use of a snake venom toxin to characterize the cholinergic receptor protein. *Proc. Natl. Acad. Sci. USA* **1970**, *67*, 1241–1247. [CrossRef]
3. daCosta, C.J.; Free, C.R.; Sine, S.M. Stoichiometry for alpha-bungarotoxin block of alpha7 acetylcholine receptors. *Nat. Commun.* **2015**, *6*, 8057. [CrossRef] [PubMed]
4. Dellisanti, C.D.; Yao, Y.; Stroud, J.C.; Wang, Z.Z.; Chen, L. Crystal structure of the extracellular domain of nAChR alpha1 bound to alpha-bungarotoxin at 1.94 A resolution. *Nat. Neurosci.* **2007**, *10*, 953–962. [CrossRef] [PubMed]
5. Swinburne, I.A.; Mosaliganti, K.R.; Green, A.A.; Megason, S.G. Improved Long-Term Imaging of Embryos with Genetically Encoded alpha-Bungarotoxin. *PLoS ONE* **2015**, *10*, e0134005. [CrossRef]
6. Paulo, J.A.; Hawrot, E. A radioisotope label-free alpha-bungarotoxin-binding assay using BIAcore sensor chip technology for real-time analysis. *Anal. Biochem.* **2009**, *389*, 86–88. [CrossRef]
7. Berg, D.K.; Kelly, R.B.; Sargent, P.B.; Williamson, P.; Hall, Z.W. Binding of -bungarotoxin to acetylcholine receptors in mammalian muscle (snake venom-denervated muscle-neonatal muscle-rat diaphragm-SDS-polyacrylamide gel electrophoresis). *Proc. Natl. Acad. Sci. USA* **1972**, *69*, 147–151. [CrossRef]
8. Fambrough, D.M.; Hartzell, H.C. Acetylcholine receptors: Number and distribution at neuromuscular junctions in rat diaphragm. *Science* **1972**, *176*, 189–191. [CrossRef]
9. Raftery, M.A. Isolation of acetylcholine receptor—α-Bungarotoxin complexes from Torpedo californica electroplax. *Arch. Biochem. Biophys.* **1973**, *154*, 270–276. [CrossRef]
10. Chazotte, B. Labeling Acetylcholine Receptors in Live Cells Using Rhodamine {alpha}-Bungarotoxin for Imaging. *CSH Protoc.* **2008**, *2008*, pdb-prot4928. [CrossRef]
11. Anderson, M.J.; Cohen, M.W. Fluorescent staining of acetylcholine receptors in vertebrate skeletal muscle. *J. Physiol.* **1974**, *237*, 385–400. [CrossRef] [PubMed]
12. Fertuck, H.C.; Salpeter, M.M. Localization of acetylcholine receptor by 125I-labeled alpha-bungarotoxin binding at mouse motor endplates. *Proc. Natl. Acad. Sci. USA* **1974**, *71*, 1376–1378. [CrossRef] [PubMed]
13. Chen, W.; Yu, T.; Chen, B.; Qi, Y.; Zhang, P.; Zhu, D.; Yin, X.; Jiang, B. In vivo injection of alpha-bungarotoxin to improve the efficiency of motor endplate labeling. *Brain Behav.* **2016**, *6*, e00468. [CrossRef]
14. Vogel, Z.; Towbin, M.; Daniels, M.P. Alpha-bungarotoxin-horseradish peroxidase conjugate: Preparation, properties and utilization for the histochemical detection of acetylcholine receptors. *J. Histochem. Cytochem.* **1979**, *27*, 846–851. [CrossRef] [PubMed]
15. Jones, I.W.; Barik, J.; O'Neill, M.J.; Wonnacott, S. Alpha bungarotoxin-1.4 nm gold: A novel conjugate for visualising the precise subcellular distribution of alpha 7* nicotinic acetylcholine receptors. *J. Neurosci. Methods* **2004**, *134*, 65–74. [CrossRef]
16. Harel, M.; Kasher, R.; Nicolas, A.; Guss, J.M.; Balass, M.; Fridkin, M.; Smit, A.B.; Brejc, K.; Sixma, T.K.; Katchalski-Katzir, E.; et al. The binding site of acetylcholine receptor as visualized in the X-Ray structure of a complex between alpha-bungarotoxin and a mimotope peptide. *Neuron* **2001**, *32*, 265–275. [CrossRef]
17. Bernini, A.; Ciutti, A.; Spiga, O.; Scarselli, M.; Klein, S.; Vannetti, S.; Bracci, L.; Lozzi, L.; Lelli, B.; Falciani, C.; et al. NMR and MD studies on the interaction between ligand peptides and alpha-bungarotoxin. *J. Mol. Biol.* **2004**, *339*, 1169–1177. [CrossRef]
18. Bracci, L.; Lozzi, L.; Pini, A.; Lelli, B.; Falciani, C.; Niccolai, N.; Bernini, A.; Spreafico, A.; Soldani, P.; Neri, P. A branched peptide mimotope of the nicotinic receptor binding site is a potent synthetic antidote against the snake neurotoxin alpha-bungarotoxin. *Biochemistry* **2002**, *41*, 10194–10199. [CrossRef] [PubMed]
19. Tabor, G.T.; Park, J.M.; Murphy, J.G.; Hu, J.H.; Hoffman, D.A. A novel bungarotoxin binding site-tagged construct reveals MAPK-dependent Kv4.2 trafficking. *Mol. Cell Neurosci.* **2019**, *98*, 121–130. [CrossRef] [PubMed]
20. Watschinger, K.; Horak, S.B.; Schulze, K.; Obermair, G.J.; Wild, C.; Koschak, A.; Sinnegger-Brauns, M.J.; Tampe, R.; Striessnig, J. Functional properties and modulation of extracellular epitope-tagged Ca(V)2.1 voltage-gated calcium channels. *Channels* **2008**, *2*, 461–473. [CrossRef] [PubMed]

21. McCann, C.M.; Bareyre, F.M.; Lichtman, J.W.; Sanes, J.R. Peptide tags for labeling membrane proteins in live cells with multiple fluorophores. *Biotechniques* **2005**, *38*, 945–952. [CrossRef] [PubMed]
22. Hannan, S.; Wilkins, M.E.; Thomas, P.; Smart, T.G. Tracking cell surface mobility of GPCRs using alpha-bungarotoxin-linked fluorophores. *Methods Enzymol.* **2013**, *521*, 109–129. [CrossRef] [PubMed]
23. Morton, R.A.; Luo, G.; Davis, M.I.; Hales, T.G.; Lovinger, D.M. Fluorophore assisted light inactivation (FALI) of recombinant 5-HT(3)A receptor constitutive internalization and function. *Mol. Cell Neurosci.* **2011**, *47*, 79–92. [CrossRef]
24. Guo, J.; Chen, H.; Puhl, H.L., 3rd; Ikeda, S.R. Fluorophore-assisted light inactivation produces both targeted and collateral effects on N-type calcium channel modulation in rat sympathetic neurons. *J. Physiol.* **2006**, *576*, 477–492. [CrossRef]
25. Sekine-Aizawa, Y.; Huganir, R.L. Imaging of receptor trafficking by using alpha-bungarotoxin-binding-site-tagged receptors. *Proc. Natl. Acad. Sci. USA* **2004**, *101*, 17114–17119. [CrossRef] [PubMed]
26. Caffery, P.M.; Krishnaswamy, A.; Sanders, T.; Liu, J.; Hartlaub, H.; Klysik, J.; Cooper, E.; Hawrot, E. Engineering neuronal nicotinic acetylcholine receptors with functional sensitivity to alpha-bungarotoxin: A novel alpha3-knock-in mouse. *Eur. J. Neurosci.* **2009**, *30*, 2064–2076. [CrossRef] [PubMed]
27. Sanders, T.; Hawrot, E. A novel pharmatope tag inserted into the beta4 subunit confers allosteric modulation to neuronal nicotinic receptors. *J. Biol. Chem.* **2004**, *279*, 51460–51465. [CrossRef]
28. Xu, J.; Li, J.; Wu, X.; Song, C.; Lin, Y.; Shen, Y.; Ye, W.; Sun, C.; Wang, X.; Li, Z.; et al. Expression and refolding of bioactive alpha-bungarotoxin V31 in E. coli. *Protein Expr. Purif.* **2015**, *110*, 30–36. [CrossRef] [PubMed]
29. Xiao-Qi, G.; Jun, L.; Ying, L.; Yong, Z.; Dongliang, H.; Changlin, T. Total synthesis of snake toxin alpha-bungarotoxin and its analogues by hydrazide-based native chemical ligation. *Chin. Chem. Lett.* **2018**, *29*, 1139–1142. [CrossRef]
30. Agouridas, V.; El Mahdi, O.; Cargoet, M.; Melnyk, O. A statistical view of protein chemical synthesis using NCL and extended methodologies. *Bioorg. Med. Chem.* **2017**, *25*, 4938–4945. [CrossRef]
31. Fang, G.M.; Li, Y.M.; Shen, F.; Huang, Y.C.; Li, J.B.; Lin, Y.; Cui, H.K.; Liu, L. Protein chemical synthesis by ligation of peptide hydrazides. *Angew. Chem. Int. Ed. Engl.* **2011**, *50*, 7645–7649. [CrossRef] [PubMed]
32. Hackeng, T.M.; Griffin, J.H.; Dawson, P.E. Protein synthesis by native chemical ligation: Expanded scope by using straightforward methodology. *Proc. Natl. Acad. Sci. USA* **1999**, *96*, 10068–10073. [CrossRef] [PubMed]
33. Blanco-Canosa, J.B.; Dawson, P.E. An efficient Fmoc-SPPS approach for the generation of thioester peptide precursors for use in native chemical ligation. *Angew. Chem. Int. Ed. Engl.* **2008**, *47*, 6851–6855. [CrossRef] [PubMed]
34. Wang, J.X.; Fang, G.M.; He, Y.; Qu, D.L.; Yu, M.; Hong, Z.Y.; Liu, L. Peptide o-aminoanilides as crypto-thioesters for protein chemical synthesis. *Angew. Chem. Int. Ed. Engl.* **2015**, *54*, 2194–2198. [CrossRef]
35. Blanco-Canosa, J.B.; Nardone, B.; Albericio, F.; Dawson, P.E. Chemical Protein Synthesis Using a Second-Generation N-Acylurea Linker for the Preparation of Peptide-Thioester Precursors. *J. Am. Chem. Soc.* **2015**, *137*, 7197–7209. [CrossRef]
36. Agouridas, V.; El Mahdi, O.; Diemer, V.; Cargoet, M.; Monbaliu, J.M.; Melnyk, O. Native Chemical Ligation and Extended Methods: Mechanisms, Catalysis, Scope, and Limitations. *Chem. Rev.* **2019**, *119*, 7328–7443. [CrossRef] [PubMed]
37. Laps, S.; Sun, H.; Kamnesky, G.; Brik, A. Palladium-Mediated Direct Disulfide Bond Formation in Proteins Containing S-Acetamidomethyl-cysteine under Aqueous Conditions. *Angew. Chem. Int. Ed. Engl.* **2019**, *58*, 5729–5733. [CrossRef] [PubMed]
38. Tornoe, C.W.; Christensen, C.; Meldal, M. Peptidotriazoles on solid phase: [1,2,3]-triazoles by regiospecific copper(i)-catalyzed 1,3-dipolar cycloadditions of terminal alkynes to azides. *J. Org. Chem.* **2002**, *67*, 3057–3064. [CrossRef] [PubMed]
39. Hong, V.; Presolski, S.I.; Ma, C.; Finn, M.G. Analysis and optimization of copper-catalyzed azide-alkyne cycloaddition for bioconjugation. *Angew. Chem. Int. Ed. Engl.* **2009**, *48*, 9879–9883. [CrossRef]
40. Stratton, M.R.; Darling, J.; Pilkington, G.J.; Lantos, P.L.; Reeves, B.R.; Cooper, C.S. Characterization of the human cell line TE671. *Carcinogenesis* **1989**, *10*, 899–905. [CrossRef]
41. Schoepfer, R.; Luther, M.; Lindstrom, J. The human medulloblastoma cell line TE671 expresses a muscle-like acetylcholine receptor. Cloning of the alpha-subunit cDNA. *FEBS Lett.* **1988**, *226*, 235–240. [CrossRef]
42. Shao, Z.; Mellor, I.R.; Brierley, M.J.; Harris, J.; Usherwood, P.N. Potentiation and inhibition of nicotinic acetylcholine receptors by spermine in the TE671 human muscle cell line. *J. Pharmacol. Exp. Ther.* **1998**, *286*, 1269–1276. [PubMed]
43. Dudel, J.; Franke, C.; Hatt, H. Rapid activation and desensitization of transmitter-liganded receptor channels by pulses of agonists. *Ion Channels* **1992**, *3*, 207–260. [CrossRef] [PubMed]
44. Franke, C.; Koltgen, D.; Hatt, H.; Dudel, J. Activation and desensitization of embryonic-like receptor channels in mouse muscle by acetylcholine concentration steps. *J. Physiol.* **1992**, *451*, 145–158. [CrossRef] [PubMed]
45. Taiwe, G.S.; Montnach, J.; Nicolas, S.; De Waard, S.; Fiore, E.; Peyrin, E.; El-Aziz, T.M.A.; Amar, M.; Molgo, J.; Ronjat, M.; et al. Aptamer Efficacies for In Vitro and In Vivo Modulation of alphaC-Conotoxin PrXA Pharmacology. *Molecules* **2019**, *24*, 229. [CrossRef] [PubMed]
46. Syapin, P.J.; Salvaterra, P.M.; Engelhardt, J.K. Neuronal-like features of TE671 cells: Presence of a functional nicotinic cholinergic receptor. *Brain Res.* **1982**, *231*, 365–377. [CrossRef]
47. Cetin, H.; Liu, W.; Cheung, J.; Cossins, J.; Vanhaesebrouck, A.; Maxwell, S.; Vincent, A.; Beeson, D.; Webster, R. Rapsyn facilitates recovery from desensitization in fetal and adult acetylcholine receptors expressed in a muscle cell line. *J. Physiol.* **2019**, *597*, 3713–3725. [CrossRef]
48. Conroy, W.G.; Saedi, M.S.; Lindstrom, J. TE671 cells express an abundance of a partially mature acetylcholine receptor alpha subunit which has characteristics of an assembly intermediate. *J. Biol. Chem.* **1990**, *265*, 21642–21651. [CrossRef]

49. Albuquerque, E.X.; Pereira, E.F.; Alkondon, M.; Rogers, S.W. Mammalian nicotinic acetylcholine receptors: From structure to function. *Physiol. Rev.* **2009**, *89*, 73–120. [CrossRef]
50. Montnach, J.; De Waard, S.; Nicolas, S.; Burel, S.; Osorio, N.; Zoukimian, C.; Mantegazza, M.; Boukaiba, R.; Beroud, R.; Partiseti, M.; et al. Fluorescent- and tagged-protoxin II peptides: Potent markers of the Nav 1.7 channel pain target. *Br. J. Pharmacol.* **2021**, *178*, 2632–2650. [CrossRef]

MDPI

St. Alban-Anlage 66

4052 Basel

Switzerland

Tel. +41 61 683 77 34

Fax +41 61 302 89 18

www.mdpi.com

Toxins Editorial Office

E-mail: toxins@mdpi.com

www.mdpi.com/journal/toxins

www.ingramcontent.com/pod-product-compliance
Lightning Source LLC
LaVergne TN
LVHW071257170726
843515LV00006BA/1423